Decentralized Spatial Computing

Matt Duckham

Decentralized Spatial Computing

Foundations of Geosensor Networks

 Springer

Matt Duckham
Department of Infrastructure Engineering
The University of Melbourne
Melbourne, Australia

ISBN 978-3-642-30852-9   ISBN 978-3-642-30853-6 (eBook)
DOI 10.1007/978-3-642-30853-6
Springer Heidelberg New York Dordrecht London

Library of Congress Control Number: 2012945010

ACM Computing Classification (1998): H.2, J.2, E.1, F.2

Cover illustration: The artwork on the book cover derives from Walking Drawing No. 3, a digital drawing by John Stell.

Printed on acid-free paper

Springer is part of Springer Science+Business Media (www.springer.com)

To Mike

Foreword

The fact that a computation occurs in some place is self-evident, but has not been obvious or relevant to researchers for a long time, for as long as the "place" has not itself been involved in the computation as an input. So, almost all tools and techniques (algorithmic, analytical, design) developed for solving problems and performing computations correctly and efficiently in serial and (to a lesser extent) parallel systems do not take the place of the computation into account.

Things are different in distributed and network computing: the "network" (the interacting multiplicity of autonomous computational entities) is the computing device, the "place"; for tasks such as routing and broadcasting, which are functions of the network itself, the network is also the input. The first type of solution employed was to hide this fact by collecting all the needed information at one site (so that the "place" is no longer the device, and traditional techniques can be used); this rather conservative trivial approach—the *centralized* approach—was quickly dismissed because it was highly inefficient, poorly scalable, and dangerously insecure. New computational tools and techniques and strategies had to be devised, following a *decentralized* approach, leading to highly efficient *localized* solutions, not only for network functions, but for all kind of computations and tasks. Indeed, although a centralized approach is still technically a possible solution to problem solving in distributed computing systems, in the core distributed computing field in which I operate, "distributed" is taken to mean exclusively "decentralized." Outside the core, the distinction must still be made, as done in this book.

The advent of mobile and wireless communication has shown to researchers that the universe of distributed computing is indeed much wider than previously thought. In particular, *spatial* information is central in many new distributed computational environments such as mobile sensor networks, delay-tolerant networks, mobile ad hoc networks, and swarms of autonomous mobile robots. Indeed, in the unexpectedly large number of these new applications dealing with spatial computing, decentralized approaches are the natural tools for efficient and reliable problem solving. To date, in spite of the very large body of literature on these new systems, very little is available to researchers in terms of foundations, models, and design and analytical techniques following this approach. This book fills this gap, and does so with clarity, rigor, and elegance.

This book is a comprehensive, systematic, and rigorous introduction to decentralized spatial computing systems, showing how to apply the existing decentralized ideas to the domain of spatial computing. Easy to read, easy to follow, with clear and detailed explanations and examples, it provides a handbook of tools and techniques for designing decentralized spatial algorithms; starting from examples of key decentralized spatial algorithms, it shows how they are developed, and how to employ them to build more complex and sophisticated algorithms.

This book is a timely contribution, as more and more people from domains like spatial systems will need to understand and apply fundamental principles from distributed computing. At the same time, it is an important contribution also for distributed computing researchers, possibly for advancing fundamental research, and definitely for opening to investigation the specific issues presented by spatial algorithms.

I am sure this book will provide general readers with a proper field of vision, and researchers with a solid starting point for further discoveries in decentralized spatial computing. There are many concepts that need to be added to distributed computing to deal specifically with spatial computing problems; there are new tools that will need to be discovered and designed; there are new insights that will be developed that I am sure will enrich the entire distributed computing field. As Matt wrote to me: *spatial is special*; after reading the book, I know he is right.

Nicola Santoro

School of Computer Science
Carleton University, Ottawa, Canada

22 November 2011

Contents

List of Protocols

Acknowledgments

I owe the opportunity to write this book to the Australian Research Council (ARC), which awarded me a Future Fellowship (FT0990531) for research into decentralized spatial computing and ambient spatial intelligence. This exceptional ARC scheme has provided the one thing that many other funding schemes cannot: ample time to pursue basic research. I am very fortunate to enjoy the freedom afforded to me by this funding, and sincerely hope the scheme will be copied by other funding agencies around the world.

There are also many individuals to acknowledge for their contributions, big and small, in shaping this book. I hope the reader will forgive my acknowledging these people in full. My research only progresses within the context of an extensive network of collaborators and supporters.

The University of Melbourne Geomatics students of Integrated Spatial Systems 2 2009 (Ben Anderson, Alexander Masendycz, Christopher Moore, Long Zhang, Eric McCowan, Bradley Davies, Yongene Leaw, Katie Potts, Kai Dai, Nicholas Lawson, Rothana Kim, Ashlea Davy, David Measki, Shumaila Pottachira, Jie Guo, Hui Teo) and 2010 (Nhial Majak, Imran Bin Abdullah, Hongyu Chen, Beau Mikus, Brodie Richards, Peteris Lazdins, Ezra Trotter, Cameron Holley, Eric Chang, Noriko Hirashima, Marley Mumford, Samantha Hayes, Alvin Tian, Jessica Gommers, Paul Munoz, Peter Barstow, Andrew Novak, Peter Carnovale, Adrian Wong, Shura Shukor, Zhong Goh, Zhixu Li, Simon Bambridge, Nathan Hull, John Carmichael, Myles Sewell, Lachlan McCleary, Andrew Gunn, Lucas Polhe, Michael Chionna, Ashleigh Sier, William Woodgate, Tasman Schofield, William Lai, Richard South, James Carracher, Kate Aulich, Richard Arndt, David Vaughan, Alexander Gibb, Jarryd Poyner, Damian Yeung, Ahmad Tengku) were willing guinea pigs for early manuscripts. I would particularly like to thank Andrew Caldwell, Amelia French, Robert Pearson, and Michael Stephens for their interest and assistance beyond the interaction in class.

I have also benefited greatly from the support of talented young researchers in my group, in particular Jafar Sadeq, Lin-Jie Guan, Myeonghun Jeong, Jeremy Yeoman, and Mingshen Shi, and more senior researchers in my department, including Allison Kealy, Alexander Klippel (since moved to PSU), Patrick Laube (since moved to University of Zürich), Kai-Florian Richter, and

Stephan Winter. I am also very grateful to Sarah Jarrod for her hard work and inexhaustible patience in helping with the administrative side of life in a large organization.

Other collaborators who through their comments and collaboration have had important impacts on the development of the material in this book include Susanne Bleisch, Tony Cohn, Christian Freksa, Antony Galton, Sanjiang Li, Harvey Miller, Doron Nussbaum, David O'Sullivan, Jochen Renz, Jörg-Rüdiger Sack, Nicola Santoro, John Stell, and Tim Wark. In the development of environmental monitoring applications of ambient spatial intelligence, I gratefully acknowledge the support and encouragement of Peter Bardsley, Jarod Lyon, Stuart Kininmonth, Graeme Newell, and Gary Stoneham.

I would particularly like to acknowledge Silvia Nittel, whose inspirational and visionary talks in the early days of geosensor networks first set me on a path to exploring the fascinating problems of decentralized spatial computing.

I am also extremely grateful to Alan Both and to Hairuo Xie, who together developed most of the NetLogo code that complements this book, but went much further in uncovering, pursuing, and helping to correct errors in the protocol specifications. Any remaining errors are mine alone!

Of course, my professional support network would be meaningless without my extended family in Melbourne: Ingrid, Laurie, Allison, Tim, Nola, and Geoff.

Finally, this book is dedicated to Mike Worboys. I owe far more to Mike than words can express or space can permit. Suffice it to say, Mike is a true scholar, a true gentleman, and a true friend.

Thank you all.

Preface

This book was born out of the simple conviction: that there is a right way and a wrong way to design decentralized spatial algorithms. This conviction grew from many frustrating months designing decentralized spatial algorithms in, what I now believe to be, the *wrong* way. Conventional centralized approaches to algorithm design cannot be reliably applied to decentralized systems because they do not account adequately for the critical features of these systems: the interaction between nodes, and the limited, local knowledge of individual system components. The issues of decentralization have long been a topic of study in the domain of distributed computing. However, advances in technology, such as geosensor networks and smart phones, are bringing these issues to the fore of spatial computing too. Like so many other books in the field of geographic information science, the key question is once again: "What's special about spatial?" More specifically, why should we study the problems of designing decentralized *spatial* algorithms separately from the more general problems of decentralized algorithms? The answers to this question begin in Chapter 1 (on the second page of Chapter 1), but continue throughout the book, to the final chapter.

How to read this book

This book has a simple story to tell, and so is best read in chapter order. The chapters form a logical progression, each chapter building on the previous: Chapter 1 setting the scene; Chapter 2 specifying an abstract formal model of decentralized spatial information systems; Chapter 3 introducing our decentralized spatial algorithm design technique; Chapter 4 exploring the design of increasingly sophisticated decentralized algorithms, with minimal spatial information; Chapter 5 introducing more sophisticated spatial capabilities; Chapter 6 adding time into the mix; Chapter 7 verifying empirically the efficiency; Chapter 8 verifying the veracity and robustness of those algorithms; and Chapter 9 finally outlining some of the important topics for further work in this area.

The book is written for two (intersecting) groups of students and researchers: computer scientists and geographic information scientists. GIScience is a big tent that delights in welcoming others from a plethora of disciplines, from geography to geostatistics, economics to ecology, philosophy to physics, and cognitive science to cartography as well as from traditional strongholds of computing—computer science, computational geometry, and mathematics. (These examples are not chosen at random: I have enjoyed close research collaborations with individuals from each and every one of these disciplines.) Some familiarity with basic discrete mathematics (sets, relations, functions, graphs) is certainly an advantage in reading this book. To help those who have forgotten, or even never encountered discrete mathematics before, the book contains a concise appendix to the key discrete mathematics topics required for this book, as well as recommended references to further readings in this area (Appendix A). There is also a brief introduction to SQL, the standard language for manipulating and querying databases, in Appendix B. For those new to GIScience, the core expertise in geographic information that unites GIScientists—what makes spatial special—can be found in a number of books on the topic, including [76] and a book I coauthored with Mike Worboys[119]. Those who wish to delve deeper into the more theoretical aspects of distributed systems and decentralized computing, recommended texts include [77, 82] and [99], the latter forming part of the foundations of this current book.

Throughout this book I have tried to write to my "student"—someone who wants to learn about the *process* of decentralized spatial algorithm design, what works and what doesn't, what the basic principles are, and what the "take-home messages" are. This "big picture" can easily be obscured by the technical details. Where these two—technical rigor and procedural understanding—are in conflict, I have always sacrificed the focus, depth, and detail found in an original research article in favor of the broader message. A more comprehensive coverage of these technical details can be found in the many scholarly articles this book draws upon.

Understanding distributed and decentralized spatial algorithms requires one to be exposed not only to the theory, but also to the practice. To assist in the transition from theory to practice, from reading to internalizing, simulation models and associated source code for every algorithm presented in this book can be found (freely available) on the Web site that complements the book (http://book.ambientspatial.net).

The simulation models are written in NetLogo, a popular and easy-to-use agent-based simulation system that rewards handsomely the small investment in time required to learn how to use it. Despite the simplicity of NetLogo, developing a complete code corpus was a major effort. We have attempted to verify carefully all the code developed. However, given the scale and complexity of the task of developing the protocols and simulation models, it is entirely possible that some mistakes or problems may come to light. My research group and I will continue improving and extending the code base at least until 2014, and hopefully even further into the future. Consequently, if you think you have found an error in the code, we would be keen to hear from you (contact details available through the Web site) and we will do our best to fix any errors.

At the end of each chapter you will also find a few review questions. These questions are designed to help you test your own understanding of the material in the chapter, as well as to provide some direction for further work or research. At the time of writing, I have not compiled a list of worked answers to these questions, but as and when these become available, they will also be placed on the book's Web site.

Finally, I welcome any comments or feedback in connection with this book and the Web site (for example, alternative or improved protocols or NetLogo code; examples of worked answers to review questions; and other comments on the book). I will endeavor to respond as quickly as my other commitments allow, and will post any errata on the book Web site.

Matt Duckham

Department of Infrastructure Engineering
University of Melbourne, Parkville, Australia

25 April 2012

Artwork

Walking Drawing No. 3, digital drawing. John Stell, 2011.

The image derives from a digital drawing process in which marks are accumulated while walking. It is analogous to drawing by moving the paper, rather than the drawing instrument. The subsequent combination of the fragments is essentially a collage of parts of many maps, each from a different viewpoint. The means of gathering the components of the image suggests a kind of decentralized recording in which the various places on the walk generate their own descriptions without there being any privileged overall view as in conventional map-making.

John Stell

www.johnstell.com

Part I

Foundations of Decentralized Spatial Computing

Part I of this book introduces the foundations of decentralized spatial computing—the basic concepts, structures, and design techniques for decentralized computing with spatial information. Traditionally, the study of spatial computing typically assumes the existence of a logically centralized spatial information repository, capable of collating all the relevant spatial information and controlling the computational procedure. But what happens when we relax this assumption? How can we compute in spatial information systems where there is no single component with global knowledge, no system element with responsibility for controlling the computation?

In three chapters, Part I of this book sets out to answer these questions, providing the motivation and background for decentralized spatial computing (Chapter 1); a precise formal model for representing decentralized spatial information systems (Chapter 2); and a design technique for developing decentralized algorithms capable of generating useful spatial information without any centralized control (Chapter 3).

When Computing Happens Somewhere

1

Summary: The procedures by which we capture and compute with spatial information are changing. This chapter sets the scene for these changes, focusing on three questions: what? why? and how? What is changing is the location where computing with spatial information occurs. Spatial information has traditionally been stored and processed in physical locations that are unrelated to the actual locations referred to in that information. However, a new paradigm is emerging, where spatial computing capabilities are embedded in geographic space. The development of new spatial computing technologies is a key reason why these changes are occurring. These new technologies are capable of capturing as well as computing with and communicating spatial information. The question of how to deal with these changes is the central question addressed in this book. What models, tools, and techniques are needed for the new paradigm?

COMPUTING increasingly happens *somewhere*, with that geographic location being relevant to the computational process itself. This book is concerned with computing with information *about* dynamic geographic phenomena at the same time as computing *in* geographic space.

In this chapter we introduce the "computing somewhere" paradigm, and examine what makes this paradigm different from traditional models of computing with spatial information (§1.1). New technologies (in particular, *geosensor networks*, introduced later in §1.2) are motivating the investigation of "computing somewhere" and its application to a wide range of practical problems (§1.4). The primary focus of this book is on the topic of decentralized spatial computing (§1.3) as a mechanism for "computing somewhere" within these new technologies and applications.

M. Duckham, *Decentralized Spatial Computing*, DOI 10.1007/978-3-642-30853-6_1,
© Springer-Verlag Berlin Heidelberg 2013

1.1 Computing Somewhere

Computing simultaneously *in* and *about* geographic space is a challenge for traditional models of spatial computation. Conventional approaches to spatial computing are founded on the assumption that spatial information is stored, collated, and processed in large information repositories, such as GIS or spatial databases. The location of an information repository is irrelevant to the information stored in that repository, or to the processing tasks performed by it. Even in cases where a repository may be composed of multiple components connected by a digital communications network (e.g., in a distributed GIS), the procedures used for processing spatial information normally treat the distributed repository as a discrete logical unit, and are not concerned with the locations of that repository's components.

By contrast, the idea behind "*computing somewhere*" is to examine the effects of retracting this assumption. When information *about* geographic space is distributed *throughout* geographic space (as opposed to a virtual, communications network space), building and maintaining centralized spatial information repositories quickly become impractical or undesirable.

As a result, there are increasingly scenarios that demand models of spatial computation which can take into account *where* information is located, and the geographic constraints to moving information to new locations (Fig. 1.1). While traditional computing models store and process spatial information in a centralized repository (Fig. 1.1a), the "computing somewhere" paradigm assumes information is related to its location in geographic space. Further, spatial constraints to the movement of information give rise to "geospatial information neighborhoods," where information exchange is easier between stores that are spatially closer together.

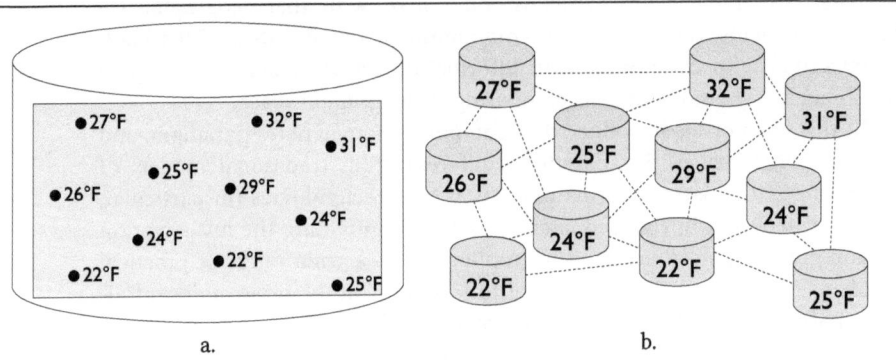

Fig. 1.1. Computing happens somewhere. a. Traditional models of computing with spatial information assume information is stored and processed in centralized information repositories. b. Increasingly, information storage and processing is dependent on location in geographic space, with spatial constraints to movement of information leading to geospatial information neighborhoods

1.1.1 Example: Humans Computing Somewhere

As an analogy for the "computing somewhere" paradigm, we can imagine the question of whether people at a soccer match constitute a "crowd" (which might be defined as, let us say, more than 1,000 people in close proximity). One way to investigate this would be to collect some centralized data about the soccer fans. For example, an observer with a camera could take a picture, and then count the people in the picture. However, a different approach would be to involve the people themselves. An individual fan could start by placing a tally mark on a piece of paper. She can then pass this paper to a randomly selected neighbor, and ask her to add another tally. The paper then continues to be passed to neighbors subject to three rules: 1. please add a tally to the paper only if you have not already done so; 2. check the tally to see if it contains 1,000 tally marks; 3. if it does, shout out "Crowd!"; if not, just pass the paper to another randomly selected neighbor. Assuming the individuals in the crowd do as they are instructed, and if the crowd is large enough, sooner or later someone will shout "Crowd!" (Fig. 1.2).

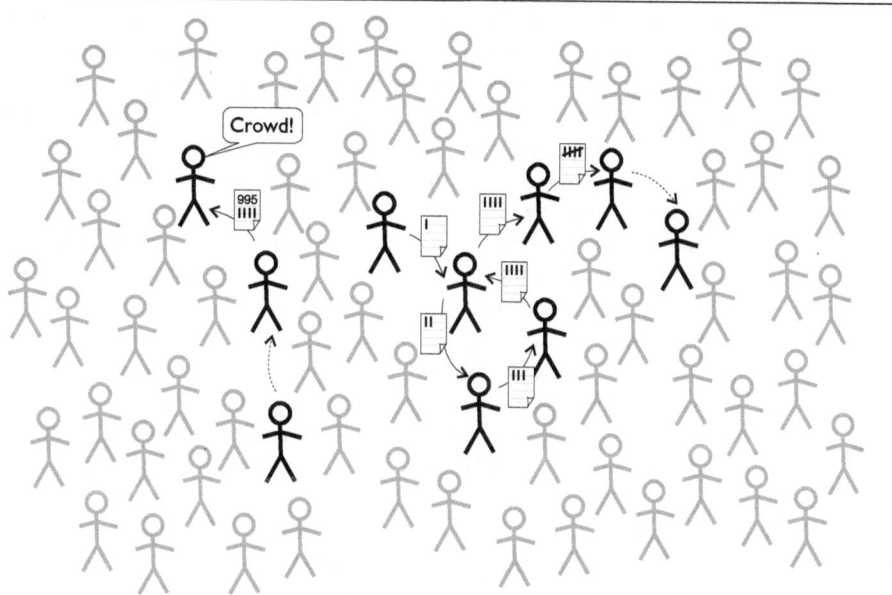

Fig. 1.2. "Crowd!" computing: individuals can cooperate to compute whether a "crowd" of more than 1,000 people exists

Although deliberately simplified, this analogy does provide an intuitive example of "computing somewhere." At root, the analogy describes an *algo-*

rithm: a specification of a computational procedure for solving a problem. This particular algorithm is *decentralized*: there is no single person you can point to who is *controlling* the procedure (although the individual who starts the process does play a special role). Further, although this algorithm does not involve coordinate positions, it does rely on the spatial constraints on movement of information—in this case the physical limitations of handing a neighbor a piece of paper. These spatial constraints are fundamental to the generation of a sensible answer; if we break these constraints, the computation no longer operates correctly. We might imagine a non-geographic version of the algorithm, where people instead SMS the message and rules to arbitrarily remote "neighbors" in their mobile phone address book. Such a version would not tell you much about geographic space, since it presents no geographic constraints to the movement of information (although arguably it might still tell you something about *social* network space).

The analogy highlights a number of further issues important for "computing somewhere" that are some of the topics of later chapters.

- *What if the people move?* Our algorithm assumes the stadium is closed, with no people allowed to leave or enter the crowd for the duration of the computation. Dealing with *dynamism*, where people can move at the same time the computation is occurring would require a more sophisticated algorithm and crowd definition. (See Chapter 6.)
- *Will the computation end?* In cases where the soccer match is watched by less than 1,000 people, our simple algorithm will never terminate; people will continue to pass the message around to one another indefinitely. Properties such as *termination* and *deadlock* (where it is possible for the system to be in a state where no action can be performed) are important characteristics of algorithms for "computing somewhere." (See Chapter 8.)
- *How long will the algorithm take?* Relying on randomness to move the message has the advantage of simplicity. But the resulting algorithm may not make particularly efficient use of each individual's time or effort. The algorithm may lead to long delays in generating an answer (termed *latency*) as well require a large number of actions, where individuals have to receive, process, and forward the message many times. (See Chapters 3 and 7.)
- *What is a "crowd"?* In our analogy, a crowd is defined arbitrarily to be 1,000 or more people. Many geographic phenomena, such as crowds or flocks, mountains or urban areas, woodlands or cyclones, are similarly difficult to delineate with certainty. Dealing with *uncertainty* both in data and in definitions is an important feature of any spatial computation, and "computing somewhere" is no exception. (See Chapter 8.)

1.2 Geosensor Networks

The previous section frames the "computing somewhere" paradigm, and some of its associated problems, independently of any specific technology. However, the motivation to investigate "computing somewhere" is increasing as a result of the development of new technologies, in particular *geosensor networks*. A *geosensor network* (GSN) is:

> "a wireless sensor network that monitors phenomena in geographic space." [86]

In turn, a *wireless sensor network* (WSN) is:

> "a wireless network of miniaturized sensor-enabled computers, called *sensor nodes* (or *sensor motes*)." [121]

This book is concerned with information about geographic space, so we are primarily concerned with *geosensor* networks (the special case of wireless sensor networks that monitor phenomena in geographic space) rather than wireless sensor networks.

Figure 1.3 shows an example of a sensor node. Every sensor node must have at least:

- a microcontroller (a simple CPU) for *computing with* data;
- a wireless radio (transceiver) for *communicating* data; and
- one or more sensors for *capturing* data about its environment.

Fig. 1.3. Example sensor node (SunSPOT)

A sensor is any device for measuring the physical qualities of the environment, such as humidity, CO_2 concentration, or even location. Sensors effectively convert environmental stimuli, such as heat, light, or electromagnetic radiation, into digital signals that can be processed by a computer. Table 1.1 provides a summary of the different sensor types that exist, classified according to the environmental stimulus they respond to.

Sensor stimulus	Examples of sensors
Acoustic	Acoustic sensors detect sounds in the environment. *Passive* acoustic sensors, such as microphones, ultrasound sensors, and some seismic sensors, detect "naturally" occurring ambient sounds. *Active* acoustic sensors, such as SONAR (sound navigation and ranging), emit sound energy and detect returning signals from the environment.
Chemical	Today most sensors for *detecting* chemicals are not themselves chemical sensors (i.e. they respond to one of the five other stimuli in this table). However, an important emerging class of sensors are stimulated directly by ambient chemicals. For example, some sensors capable of detecting chemicals such as ethanol and acetone, directly detect specific chemical reactions using carefully constructed metallized membranes over miniaturized hotplates.
Electromagnetic	Electromagnetic sensors include (passive) magnetic sensors, electrical current sensors, as well as (active) RADAR (radio detection and ranging) sensors. For example, magnetic sensors can detect changes or disturbances in magnetic fields.
Optical	Optical sensors include a wide range of familiar sensors, including visible light sensors, digital cameras (arrays of light sensors), and infrared sensors. Many less familiar sensors, such as some carbon dioxide sensors which rely on optical stimuli, such as near infrared (NIR) optical sensors that measure the absorption of CO_2.
Thermal	Thermal sensors measure temperature or temperature gradients. Many different types of digital thermal sensor exist, for example, *thermistors*, which are a type of electrical resistor with variable resistance proportional to environmental temperature.
Mechanical	Mechanical sensors respond to direct mechanical pressure, flows, or movements. Mechanical sensors include accelerometers, altimeters, anemometers, barometers, a range of different types of flow meters, as well as many humidity sensors.

Table 1.1. Six categories of sensors, based on stimuli

It is important to make a clear distinction between the *sensor* (the device that measures environmental qualities) and the *sensor node* (the computing device that includes one or more on-board sensors). In this book we will use "sensor node" or just "node" to refer to the computing device; "sensor" always refers specifically to the measuring device.

Together, scientific and technological advances in the design and manufacture of mobile computing devices, in wireless digital communication, and in microelectromechanical system (MEMS) sensors have provided the technology for geosensor networks. These advances are continuing rapidly, with new hardware and software for sensor nodes, new communication protocols and transceivers, and new types and designs of digital sensors appearing almost daily. These advances are helping to make wireless sensor network technology cheaper, more powerful, simpler, and cleaner.

Irrespective of inevitable refinements to these technologies, in the future we can expect to encounter more and more devices and systems that are capable of the "three geographic Cs": capturing, communicating, and computing with information about geographic space. In the following sections we look more closely at what, exactly, makes geosensor networks special: their uniqueness as a new source of spatial information, and the unique resource constraints imposed by this technology.

1.2.1 Unique Information Source

Geosensor networks provide a unique source of information about our natural and built environments. The key features of geosensor networks that make them unique information sources are:

1. the high spatial and temporal detail of information they can generate;
2. their automated operation; and
3. their flexibility to be deployed across a wide range of different scenarios and environments.

While other technologies may offer some of these features, only geosensor networks are capable of combining all these features.

High detail

Geosensor networks can generate information about an environment at high levels of both spatial and temporal detail, in other words, with fine spatial and temporal *granularity*. Individual nodes in a geosensor network may be meters, centimeters, or even millimeters apart, potentially capturing very fine-grained spatial detail about an environment. Some other spatial data capture technologies can offer comparable levels of spatial detail, such as some remote sensing systems. However, these technologies typically only offer limited temporal detail about changes over time. Nodes in a geosensor network can be programmed to sense data at frequencies of months, days, hours, minutes, seconds, or less, providing fine-grained temporal detail about environmental changes in combination with fine spatial detail. The network may even adapt its temporal or spatial granularity to changes occurring in the environment, effectively "zooming in" on interesting phenomena.

Automated

Sensor nodes use on-board power sources and wireless digital communication. As a result, geosensor networks can be deployed to monitor sensitive, hazardous, or remote environments automatically. Many applications are too costly, dangerous, or delicate to be effectively monitored using other, more invasive or permanent data capture technologies that require human intervention or operation.

Fig. 1.4. A Leach's storm-petrel, *Oceanodroma leucorhoa*, the rare seabird whose nesting habits were studied by [109] using a geosensor network. (Source: US Fish and Wildlife Service, C. Schlawe)

For example, a landmark application of a geosensor network involves monitoring the nesting habitats of a rare seabird (Leach's storm-petrel, *Oceanodroma leucorhoa*) on Great Duck Island, a remote island off the east coast of the US (see Fig. 1.4). Previous studies had indicated that such birds can be especially sensitive to disturbance—even relatively short periods of human activity around the birds at nesting time were shown to lead to marked reductions in breeding success. Further, the remoteness of the island made direct human observations expensive and somewhat hazardous. In 2003, scientists deployed a network of more than 100 nodes on the island for monitoring a range of environmental parameters, including temperature, humidity, and light. The network was deployed in and around storm-petrel nesting sites before the breeding season, ready to monitor *in situ* once nesting commenced [109]. The information gathered by the network provided invaluable insights into the process of nest selection in the sea birds, ultimately informing efforts to protect the birds and their habitat. With other technologies it would simply not be possible to gather such information without disturbing the birds, and so damaging the very ecosystem that scientists were trying to understand.

Flexible

Geosensor networks are adaptable to a wide range of application scenarios and requirements, in three ways. First, sensor node technology is designed to be low-cost, and is expected to reduce greatly in cost in the future. Second, sensor nodes are configurable to specific sensing requirements of an application. An ever-widening range of types of MEMS sensors are being developed, allowing geosensor networks to be used to sense any combination of environmental parameters, from light to sound, chemicals to temperature, pressure to magnetism (see Table 1.1 and question 1.3). Third, geosensor networks can be targeted at specific spatial regions, or deployed or activated in response to specific occurrences. In emergency applications, for example, a geosensor network might be deployed in a building following a fire or explosion.

Summary: "Close sensing"

While geosensor networks offer unique sources of information about our geographic environment, they *complement* rather than *replace* existing spatial data capture technologies. For example, remote sensing techniques such as aerial photogrammetry, satellite remote sensing, and LIDAR (light detection and ranging) can match, or in some cases even exceed, the spatial detail of geosensor networks. Remote sensing can also be used to monitor larger spatial extents than geosensor networks (at least in the short- to medium-term future). Satellite remote sensing already provides near-global coverage. The technology to support geosensor networks large enough to cover entire regions or countries is still many years away.

However, remote sensing today cannot offer the level of temporal detail of a geosensor network; it is typically technically or economically infeasible to near-continuously monitor changes in an environment using remote sensing. Similarly, manual ground survey methods are typically too expensive and time-consuming to capture fine temporal granularity changes. Remote sensing also involves the collection of information over a (large) distance. Many optical and electromagnetic and some thermal stimuli are propagated well over large distances. However, acoustic, chemical, and mechanical stimuli typically require sensors in close physical proximity to be detected. In comparison to remote sensing, geosensor networks can be characterized as a technology for "*close sensing*" (Fig. 1.5).

1.2.2 Unique Resource Constraints

Geosensor networks are a key technology for "computing somewhere." The information generated by an individual sensor node is directly related to its location. Further, geosensor networks present spatial constraints on the movement of information. These constraints arise through the constraints on resources inherent in wireless communication: communication between more proximal nodes generally requires less resources; communication between more distal nodes places higher demands upon network resources. *Resource constrained computing* is therefore a fundamental feature of geosensor networks.

Fig. 1.5. "Close sensing": Geosensor networks complement but do not replace traditional spatial data capture technologies such as satellite remote sensing

Nodes in a geosensor network will be, in the majority, *untethered*, in the sense that they have no wired connections (for example, to a mains power supply or to wired digital communications networks). Untethered nodes are essential in applications where nodes may be mobile, for example, attached to humans, animals, or robots, or carried by water or air currents. Even when nodes are immobile, their capability of being deployed without requiring complex or expensive wiring is essential to achieving high detail, automated, and flexible networks (cf. p. 10).

Being untethered presents several significant resource constraints on sensor nodes. The most important constraint is on node energy resources. Nodes are primarily reliant on battery power for all system operations, including sensing, processing, and communicating information. Harvesting energy from the environment (sometimes also referred to as *energy scavenging*), for example, using solar panels or thermal converters, can supplement battery power. However, in the majority of applications, the levels of energy that can be harvested from the environment are orders of magnitude below the energy that can be generated by conventional batteries. Figure 1.6 compares typical power densities (milliwatts per cm^2, mW/cm^2) for common environmental energy sources (solar, vibration and acoustic noise) with conventional battery energy densities (milliwatts hours per cm^2, mW/cm^2)[1].

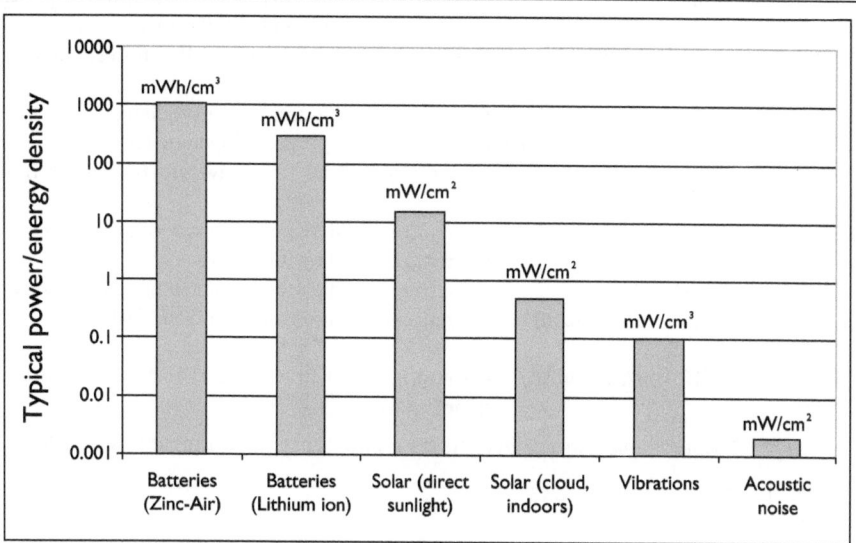

Fig. 1.6. Comparison of typical power density (mW/cm^2) for common environmental energy sources and energy density for batteries (mWh/cm^2). Source: [90]

Although developments in existing (e.g., battery) or new (e.g., fuel cell or nuclear) energy technologies may revolutionize future energy sources, recent history suggests that such quantum leaps in energy technology are extremely rare events. As a result, it seems highly probable that energy resources will remain a constraint on geosensor networks in at least the short to medium term.

Given that constraints on energy are a feature of sensor nodes, conservation of energy resources is a major consideration in the design of a geosensor network. For example, *duty cycling* (powering down nodes to an inactive "sleep mode" for a—potentially large—proportion of the time they are operating) is a common technique for reducing the energy budget of sensor nodes, and so extending their battery life. Providing exact numbers for the energy required for different node operations is difficult, as these operations are highly dependent on the rapidly developing hardware and specific operation details. However, some approximate figures are given in Table 1.2.

---

[1] Energy harvesting *generates* power, in watts (energy per unit time); batteries *store* energy, in watt hours (or joules).

Operation	Energy budget	Notes
Computing	2.77×10^{-10} mWh (1nJ)	Energy required by a microcontroller to execute a single instruction.
	1.2mW	Power required by microcontroller in fully operational mode (Texas Instruments MSP 430).
	3×10^{-4} mW (0.3μW)	Power required by microcontroller in sleep mode (Texas Instruments MSP 430).
Communicating	2.77×10^{-7} mWh (1μJ)	Energy required to transmit a single bit.
	1.38×10^{-7} mWh (0.5μJ)	Energy required to receive a single bit.
	200mW	Power required by transceiver to receive or transmit data.
Sensing	1.2mA	Current required by photoresistor (light sensor).
	0.7mA	Current required by humidity sensor.
	0.1mA	Current required by thermistor (temperature sensor).
Storing	1×10^{-6} mAh (1nAh)	Current required to read data from flash memory (Mica node).
	8×10^{-5} mAh (80nAh)	Current required to write data to flash memory (Mica node).

Table 1.2. Approximate energy budgets for typical sensor node operations [55, 60]

Communication constraints

An important message of Table 1.2 is that wireless communication is the most energy intensive operation of all performed by a sensor node. The power or energy required for wirelessly communicating data is typically orders of magnitude more than the energy required for processing data. For example, the energy required to execute a single microcontroller instruction (2.77×10^{-10} mWh) is approximately 1,000 times less than the energy required to transmit a single bit of information (2.77×10^{-7} mWh). As a consequence, communication places important constraints upon any geosensor network.

The constraints imposed by communication also have explicitly spatial characteristics. The physics of electromagnetic waves (such as the radio frequency waves used for wireless communication) causes their energy to dissipate as the wave propagates through space. This phenomenon, known as *path loss*, means that even in idealized conditions the energy lost by a wireless signal is proportional to the square of the distance between the transmitter and a receiver. However, wireless signals are also attenuated by the physical media through which they must travel, such as air, water, vegetation, and buildings. As a result, in practice the energy lost by a wireless signal on earth is typically of the order of d^4, where d is the distance between transmitter and receiver.

Path loss and signal attenuation combined with the resource constraints in a geosensor network mean that in practice wireless sensor nodes can only transmit over relatively short distances in geographic space (with today's technology, typically of the order of tens of meters). Thus, to monitor a larger geographic area, it is not possible for all nodes to communicate directly with one another: only spatially nearby nodes have the energy resources to communicate directly. Communications between more distant nodes must be relayed via intermediate neighbors, termed *multi-hop communication* (see Fig. 1.7).

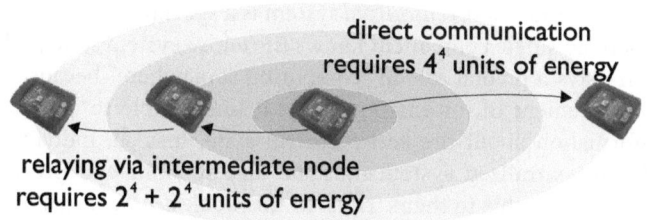

direct communication
requires 4^4 units of energy

relaying via intermediate node
requires $2^4 + 2^4$ units of energy

Fig. 1.7. The total energy required to transmit a signal can be reduced by using multi-hop wireless communication rather than direct wireless communication

These constraints on communication are one of the root causes of the constraints on the movement of information in geosensor networks. As we have already seen, these spatial constraints on the movement of information are fundamental to "computing somewhere."

1.3 Decentralized Spatial Computing

The previous sections in this chapter have presented the paradigm of "computing somewhere" (computing simultaneously *in* and *about* geographic space, §1.1); and argued that geosensor networks are a key technology for "computing somewhere." This section introduces the concept of *decentralized spatial computing*, the central topic of this book.

In this book, the term "spatial computing" means "computing with information about (geographic) space."[2] Over the past three or four decades, the field of geographic information science has developed an impressive arsenal of techniques and technologies for capturing, storing, managing, processing, and using spatial information.

These tools are now commonly being used within *distributed* computing architectures. In a *distributed system*, multiple information systems coop-

---

[2] Although the term "spatial computing" has long been used to mean "computing with information about space" (e.g., see [21, 64]), recently the term has begun to be used synonymously with "computing somewhere" (i.e., "computing with information *in* and *about* space", cf. [9, 10]).

erate synchronously in order to complete some computing task [119]. For example, most of today's applications of spatial information systems, such as Web mapping, location-based services, spatial decision support, spatial data mining, and so forth, already rely on distributed computing. In these applications, individual distributed system elements, such as GIS, spatial databases, map browsers, Web servers, mobile phones, and PDAs, are connected by a digital communications network, and must interact seamlessly as part of the application, for example, by delivering real-time road condition information to drivers.

In contrast, a *decentralized* system is a special case of a distributed system where no single component knows the entire system state [77]. Decentralization plays a pivotal role in "computing somewhere" because the constraints on movement of information make it undesirable to collate and centralize information about the entire system state. Instead, individual components of a decentralized system must be able to filter and process information *locally* accessible to them. Thus, in this book *decentralized spatial computing* is defined as:

> the study of the decentralized algorithms, data structures, and techniques for computing with spatial information.

Decentralization implies a high degree of independent operation and autonomous control. Nodes in a decentralized system are expected in important ways to "do their own thing" within the system. However, independence and autonomy are not always easy to define precisely. Nodes in a decentralized system will usually all be issued with a single program at initialization or deployment. From time to time, all the elements in the system may also be reprogrammed to perform some new task, termed *retasking*. One might argue that this programming and retasking is a form of *centralized* control, even though the tasks issued involve nodes acting autonomously. Thus, while independence and autonomy are important features of a decentralized system, the defining characteristic for decentralization remains the absence of any single component with global knowledge of the system state.

Although our definition of decentralization provides a crisp delineation between centralized and decentralized systems, in most cases we are interested in *degrees* of decentralization rather than the extremes. In a *fully* decentralized system, no information is shared between individual computing components, which severely limits the range and usefulness of computing tasks that can satisfied by such architectures. Conversely, in a fully *centralized* system, *all* information must be contained in a single computer: today few applications adopt such an architecture. As a result, this book is primarily concerned with highly (but not entirely) decentralized systems, where individual components have knowledge of a relatively small proportion of the entire system state (typically their own state and the state of some of their immediate spatial neighbors). The objective of decentralized spatial computing, then, is to generate useful knowledge about the whole picture, even when individual system components possess only small pieces of the puzzle.

1.3.1 Why Use Decentralized Spatial Computing? Part I

Given that today we already have many centralized tools and techniques for sophisticated manipulation and analysis of spatial information, a fundamental question facing decentralized spatial computing is therefore: why attempt *decentralized* spatial computing? Why not simply communicate all spatial information back to a centralized server, and use all our existing tools to handle this information?

In the specific example of geosensor networks, one part of the answer to this question has already been explored in the context of resource constrained computing (§1.2.2). Since executing a single microcontroller instruction requires several orders of magnitude less energy than transmitting one bit of information, computing with data instead of communicating data can lead to dramatic improvements in the energy budget of sensor nodes, and so increase the longevity of a geosensor network. In short, in a geosensor network it is normally advantageous to *trade communication for computation*: to process information in the network in order to reduce the amount of data that must be communicated. Decentralized computing of spatial information is a way to trade communication for computation, and so to make more efficient use of limited network energy resources than communicating information back to a centralized information system.

In addition to energy constraints, there are five further reasons for our interest in using decentralized spatial computing in place of centralized spatial computing: information overload, scalability, sensor/actuator networks, latency, and information privacy, discussed below, and summarized in Fig. 1.9.

Information overload

First, decentralized spatial computing can help in managing *information overload*. A geosensor network may be composed of thousands or millions of nodes. Each node may have dozens of on-board sensors (some of which, as in digital cameras, may themselves be arrays of many individual sensors). Each sensor may be generating data at a frequency of hours, minutes, seconds, or even many times a second for an extended period of months or even years. In the future, data from many thousands or millions of such networks may be combined into a suite of applications. Further, the highly dynamic nature of data from a geosensor network means that data items rapidly become out-of-date, with newer data constantly being generated. In such scenarios, managing large volumes of data about dynamic geographic phenomena presents substantial risks of information overload.

Even if we imagine a future where data storage is practically unlimited, there are still important information overload issues connected with technologies such as geosensor networks. Individual data items generated by nodes in a geosensor network are often almost meaningless. Individual data items are akin to single pixels from one frame of a movie; the data from across the entire network is analogous to a single frame from a movie; data from all nodes over a period of time is required to see the complete moving picture (see Fig. 1.8). Looking at individual pixels from a movie tells you very little about the objects and occurrences in that movie. Similarly, only when individual data items are combined into salient spatial and spatiotemporal patterns, such as boundaries and the evolution of regions, do data items collectively acquire meaning.

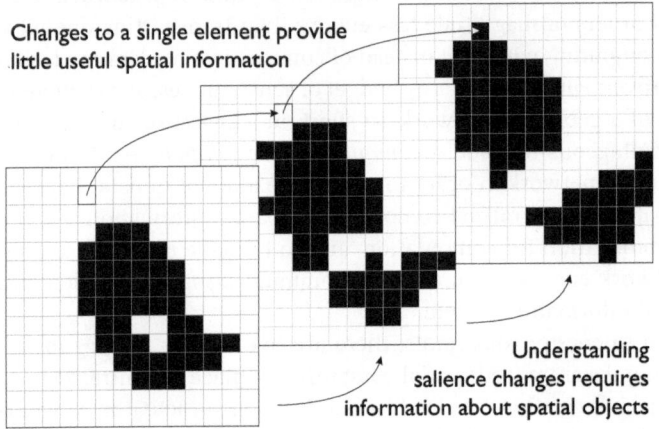

Fig. 1.8. Geosensor networks as a movie camera—meaningful occurrences can only be detected while watching the entire picture

Our information society is increasingly faced with problems of too much information. In many applications today, avoiding being overloaded with information is a more challenging task than actually generating or gathering information. Decentralized spatial computing can assist in managing information overload by filtering and processing data in the network, helping us to identify salient spatial patterns in the data. For example, a decentralized spatial algorithm might monitor changes to the area of a region of high temperature in a geosensor network. Embedding the intelligence required to identify these changes in the geosensor network itself potentially alleviates the need to centrally collate and store large volumes of spatiotemporal temperature data.

Scalability

Today's networks are typically composed of relatively small numbers of nodes, of the order of tens or hundreds. However, as the technology matures, the objective is to construct networks with thousands, millions, or more nodes as a matter of routine. The issue of network size is especially important in spatial computing, because the number of nodes is directly related to the level of spatial detail that the network can provide. As networks scale to larger and larger numbers of nodes, centralized coordination of the network behavior becomes increasingly difficult.

The problem of scalability is exacerbated by the inherent unreliability of individual sensor nodes. In order for the cost to be low enough to allow networks of thousands or millions of nodes to be constructed, sensor nodes must be amenable to mass production, and not rely on expensive, high-precision components. As a result, individual sensor nodes are highly prone to failure. It is therefore important for geosensor networks to adapt automatically to node failure, and to replacement nodes being added to the network. Centralized control makes it harder to adapt and reconfigure networks where sensor nodes are continually dying and being replaced.

The solution to these problems is decentralization. Decentralization enables the system behavior to be defined through interactions between individual nodes. Such an approach is inherently more scalable than attempting to define the behavior of the entire system. In a decentralized system, as more nodes are added to the system, the individual rules each node follows remain unchanged. In a centralized system, new nodes added to the system must be recorded and coordinated centrally.

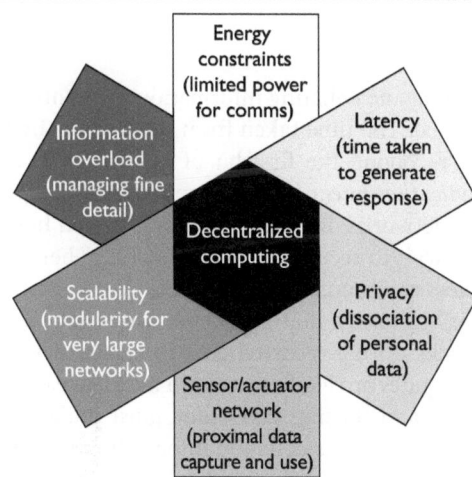

Fig. 1.9. Six motivations for decentralized approaches to spatial computing

Sensor/actuator networks

In many cases, information generated by nodes in a geosensor network may be required by other nodes or systems in the immediate spatial vicinity of that information. A farmer requesting information about the soil salinity in a field in which she is standing and a police officer requiring information about crowd movements at an event he is attending are examples of situations where individuals need information about the environment in which they are physically located. In other cases, the information generated by a geosensor network may be integrated with fully automated systems, as in irrigation systems that supply water only to those locations that are sensed as being driest (e.g., [19]).

A *sensor/actuator network* not only *senses* information about the environment, but is also capable of initiating actions to *change* the state of the environment. For example, a sensor/actuator network might be used to reroute vehicles away from traffic congestion, control a precision agricultural irrigation system, or automatically reconfigure water courses or sewers to reduce the risk of flooding or overflows. In all these examples it is potentially less efficient and less robust to relay all sensed information from sensor nodes in the network to a centralized server for collation and processing, and then subsequently *return* processed information to actuators in the network. Given the spatial constraints to the movement of information inherent in "computing somewhere" it is preferable to avoid, where possible, communicating that information to a remote centralized computing system only later to redisseminate processed information back into the network. Again, decentralization can help satisfy this goal, by retaining information close to where it is generated and used.

Latency

Communicating and collating information in a centralized information repository takes time. The time taken from sending the first bit of a message from a source to receiving the first bit of a message at a destination is termed the *communication latency*. Multi-hop communication, such as that used in a geosensor networks, increases latency. At each hop, an intermediate node may have to wait to receive the entire message before relaying it to the next node. Wireless communication protocols may also need first to communicate "handshakes" to verify that neighbors are ready to receive data, and then perform checks on the received data to ensure it is not corrupted. Further, intermediate nodes are likely to need to interleave communication with other computational and sensing tasks they must undertake. When a centralized information repository is collating information from multiple sources, it will be the last message received (i.e., the *slowest* communication) that dominates the time required by the system to act on the received information, termed the *system* or *operational latency*. In a sensor/actuator network, such delays are only magnified by the need to then communicate processed data back to nodes within the network.

As a result, in a geosensor network the level of communication required by a task is an important factor in the overall time taken to complete that task. In real-time applications, operational latency can be of critical importance. To a driver approaching a road intersection, for example, even a second's additional delay in receiving up-to-date information about road or weather conditions further along the route could be critical to his or her decision to follow one route as opposed to another. Potentially, decentralized computing can assist in reducing the amount of data communicated, and the distances over which data is communicated. In turn, this can help decrease the operational latency of a system, ultimately improving its efficacy and usability.

Information privacy

Information privacy concerns the right of individuals to control access to their personal information. Personal information can include information collected by other individuals or organizations, such as medical records or bank statements. Personal information may also be automatically sensed, such as the changes to an elderly person's heart rate and blood sugar or a person's current location. *Location privacy* is a special case of information privacy that concerns the right to control information about one's geographic location and movements. Some applications of geosensor networks, such as environmental monitoring, may not involve monitoring any personal information. However, applications such as location-based services, mobile facilities management, security services, elder care, and emergency response, may generate personal information about individuals using or monitored by the system.

There are many different approaches to privacy protection, and a discussion of these is beyond the scope of this book (but see [34] for a review of location privacy protection techniques). However, one important mechanism for protecting information privacy is simply to avoid communicating personal information when possible. As a result, decentralized computing can sometimes help to protect information privacy in geosensor networks that monitor or generate personal information. Decentralized computing can ensure personal information is only shared with a small number of frequently changing nearby neighbors, or avoided altogether. In turn, this can make finding, collating, and linking personal information about an individual required for invasion of privacy much harder than when information is stored in a few centralized information repositories (e.g., [6]).

1.3.2 Why Use Decentralized Spatial Computing? Part II

The discussion in the previous section has answered the question "Why use *decentralized* spatial computing?" from the perspective of the field of geographic information science. GIScience has a long history of developing *centralized* tools for computing with spatial information. However, in other established fields, such as distributed computing, many of the reasons for using decentralization discussed in the previous section are well rehearsed. In these fields, decentralized computing is a well-established topic. Thus, there is another side to our question "Why use decentralized *spatial* computing?" What makes decentralized *spatial* computing as distinct from decentralized computing worth studying?

Part of the answer to this question has, again, already been discussed. A decentralized system is not, in general, subject to the constraints of "computing somewhere": the constraints to movement of information in a decentralized system may not be spatial. Similarly, the information stored by each node may not be related to the location of that node. Table 1.3 classifies different types of decentralized computing according to these spatial characteristics: whether the movement of information is spatially constrained, and whether the information held by a node at a particular location is related to that location.

Information storage	Constraints to movement of information	
	Spatial	Non-spatial
Related to location	"Computing somewhere" e.g., geosensor networks	"Computing everywhere" e.g., location-based services (such as Google latitude)
Unrelated to location	"Computing anywhere" e.g., Intranet, MANETs	"Computing nowhere" e.g., Internet, Web mapping, distributed GIS

Table 1.3. Examples of decentralized computing, classified according to spatial characteristics

For example, the Internet effectively provides no spatial constraints to the movement of information. At my desk in Melbourne I can equally easily access Web sites in Australia or Austria. Nor is the information stored at a particular Internet node constrained by its location: there is nothing to stop information about Austria being physically located on a server in Australia. Similarly, many distributed spatial information systems, such as Web mapping systems and Web-based GIS, place no spatial constraints on the movement or location of storage of information. In contrast to "computing somewhere" this approach to information systems can be thought of as "computing nowhere": the location of the computation is immaterial to the computational process itself.

Intranets and MANETs (mobile ad hoc networks) also provide no constraints on the information stored at a particular location; I can store information about Austria in my private Australian intranet. However, such distributed systems do provide examples of systems that present spatial constraints to the movement of information. For example, security restrictions are designed to make it difficult to access information on an intranet from outside the locations serviced by that intranet; the short range wireless communications used by MANETS also means that physical location determines whether a node can be part of a particular MANET. In our classification, this class of systems is referred to as "computing anywhere": there may be spatial constraints to the movement of information, but information may be unrelated to the locations where it is processed.

By contrast, applications such as location-based services involve computing with information that is related to its location but with no spatial constraints on the storage and movement of that information. For example, location-based social networking applications, such as Google Latitude and Foursquare, can allow individuals to make their current location generally available on the Web to users anywhere in the world (subject to certain privacy restrictions). This class of systems we call "computing everywhere": the location is important to the computation, but these systems assume ubiquitous access to and unconstrained movement of information.

Thus, what makes "computing somewhere" different is the combined spatial constraints on both the movement of information and what information is held at a particular geographic location. These differences distinguish decentralized *spatial* computing applications, such as geosensor networks, from the other applications of decentralized computing discussed above. Together, these constraints on the location and movement of information present simultaneously a challenge to and an opportunity for designing decentralized spatial algorithms.

Decentralized spatial computing: A challenge and an opportunity

Many basic spatial queries concern information purely about point locations (a *point query*). Decentralized spatial algorithms are typically relatively easy to design for point queries, because the information required to satisfy the query is likely to be stored or generated at or near the location referred to by the query. However, the challenge facing decentralized spatial computing is that queries often concern entities that have spatial *extents* (such as lines, regions, clusters, or surfaces). Such spatial entities cannot efficiently be represented as sets of points: instead, they have complex spatial structures and relationships (e.g., polygons, representing objects such as regions of high temperature, or raster tessellations, representing fields such as temperature surfaces). Computing information about entities with spatial extents using decentralized approaches is difficult because this will usually require the combination of information from geographically dispersed locations.

On the other hand, decentralized spatial computing also presents an important opportunity. The inherent structure of spatial information means that "nearby things are more related" (Tobler's so-called first law of geography). Because each node's state is partly dependent on location, more proximal information is more likely to be more relevant in answering spatial queries. For example, when constructing the boundary of a region (e.g., as a first step in responding to a query about the area or connectivity of that region), a component that locally detects that it is part of the boundary can expect that its neighbors will also include other parts of the boundary. The opportunity, then, is to design decentralized algorithms that take advantage of spatial structure and contiguity.

Related topics in GIScience

Finally in this section, it is worth highlighting, in the context of this discussion of the background of decentralized spatial computing, its relationship to some other topics within GIScience and more generally within computer science. *Parallel computing*, for example, is concerned with using multiple processors simultaneously to provide efficient solutions to computationally intensive problems. Parallel computing has remained an important topic in GIScience for many years because many spatial algorithms are complex enough to require substantial computing resources. However, parallel computing does not usually present spatial constraints to the movement of information (information may be exchanged at little or no cost between processors) nor to where information is processed. Similarly, related work on *grid computing*, *cloud computing*, and *distributed GIS* typically assume no spatial constraints on the movement of information.

Multi-agent simulation is concerned with developing goal-oriented and autonomous programs that can sense and react to their environment. Often the simulated environment may be spatial, and may even present spatial constraints to the movement of information. But agent-based systems are only rarely embedded in true geographic space (see, for example, [48]). Thus, agent-based concepts are important in developing and implementing decentralized spatial information systems (the focus of Chapter 7), as are other related concepts in software engineering, such as object-oriented programming (OOP) and cellular automata. Indeed, examples of all the algorithms presented in this book are available, implemented within an agent-based system (NetLogo) built using an object-oriented programming language (Java). However, such approaches do not tell us how to *design* decentralized spatial information systems; they can only help us develop and implement such systems in a manageable and modular fashion.

1.4 Ambient Spatial Intelligence

"Computing somewhere" provides the conceptual framework for this book and the problems of computing simultaneously in and about geographic space; geosensor networks are one important technology that motivates our interest in "computing somewhere"; decentralized spatial computing is the approach and technique for achieving computing somewhere in technologies such as geosensor networks. Our picture is completed by *ambient spatial intelligence*: the *application* of "computing somewhere" to real-world problems.

More specifically, ambient spatial intelligence (AmSI) is concerned with embedding into built and natural environments the intelligence to monitor geographical occurrences and respond to spatiotemporal queries. This idea of AmSI has its roots in the visions of *ubiquitous computing* [110] and, more recently, *ambient intelligence* (AmI) [32]. Ubiquitous computing is concerned with embedding unseen computing assistance in everyday objects. There are many different, related definitions of AmI, but all are similarly concerned with the idea of embedding context-sensitive computing devices in our environment to allow individuals to access information services through "natural" interactions with the environment. The key features of AmI include *ubiquitous* or *embedded computing* (unseen computing devices integrated with everyday objects and environments); *ease of interaction* (allowing users a range of natural ways to interact with information without display screens and keyboards); and *context-awareness* (automatically sensing the immediate environment, adapting to changes in that environment and to habits of users, and anticipating future user requirements) [1, 40].

AmSI differs in two important respects from the existing corpus on ambient intelligence and ubiquitous computing. First, AmSI is more narrowly focused on *spatial* intelligence. Many of the problems of AmI are not explicitly geospatial, such as how to monitor muscle activity or manage medication regimes in telemedicine [63, 68]. Second, AmSI is more broadly focused not only on communication of information to individual users in the environment, but also on integration with other spatial data sources and widespread dissemination of information to both local and remote users. Spatial information is by its nature a global framework with broad application to diverse domains. It is therefore to be expected that the information services for an individual will draw on wider spatial data sources, as well as be relevant to wider application domains.

Today's applications of geosensor networks cannot yet be termed AmSI; they lack important capabilities, such as sophisticated spatial and spatio-temporal information processing capabilities. (Indeed, helping remedy this issue is one of the broader objectives of the book.) Further, today's sensor node technology is not yet advanced enough to be truly "embeddable" in the environment. Chris Pfister has written about future sensor networks comprising a myriad individual nodes, each smaller than a pinhead, called "smart dust." But current technology is still too expensive, large and bulky, and energy-hungry to approximate Pfister's smart dust. Significant scientific, technological, and environmental hurdles bar the way. For example, in addition to the scientific issues of decentralized spatial computing and the technical issues of miniaturization and cost, the potential of many millions of low-cost, disposable sensor nodes to cause significant environmental pollution has yet to be adequately addressed.

In introducing ambient spatial intelligence, the objective is to outline a broader vision towards which the core topic of this book, decentralized spatial computing, contributes. This is not the place nor the time to specify the exact route to realize that vision (and if we could look into the future, we would almost certainly find the ultimate destination to be in important respects different from what we had envisioned). Instead, the vision can be illustrated with a specific example case study.

1.4.1 Case Study: Environmental Monitoring for Conservation Contracts

To illustrate the potential role of ambient spatial intelligence in real-world problems, we will look at one specific example within the domain of environmental monitoring and conservation contracts. *Conservation contracts* are contracts issued by governments remunerate private landholders for engaging in public-good environmental conservation activities. The range of conservation activities are potentially broad, such as improving biodiversity and encouraging native vegetation (for example, by excluding grazing by livestock or pests, or removing invasive non-native weeds); improving water quality (for example, by eliminating chemical pesticide or fertilizer usage); or increasing carbon sequestration (for example, by regeneration or revegetation of native trees on a degraded site). Figure 1.10 shows a typical Australian example of such a site, previously grazed and now undergoing revegetation as part of a conservation contract.

Fig. 1.10. An example of a typical Australian (Victorian) site of revegetation managed under a conservation contract

Conservation contracts have an increasingly important role to play in our management of the environment (e.g., [107]), particularly in the context of the international movement towards emissions trading schemes (ETSs). However, a critical limitation to the use of conservation contracts is informational. Objectively monitoring what conservation activities have occurred, and more importantly what effects have eventuated from those activities, presents a major problem for today's technology. Manual assessment, whether performed by a private landowner or by a government ecologist, can be a highly resource-intensive task.

Geosensor networks are therefore a natural technology to look towards for increasing the automation and efficiency of this task. Imagine the following somewhat futuristic scenario:

> *Following an online auction organized through a local government Web site, a conservation contract to reduce nitrogen leaching on a small parcel of private land is won by its owner, Charlie. Two days later a small box of 2,000 tiny geosensor nodes arrives by express post. The nodes are preconfigured and programmed with the capability to self-localize and monitor a range of relevant environmental parameters, including temperature, soil moisture, and light intensity. Charlie distributes the nodes by "sowing" small handfuls of sensor nodes around the site. The nodes activate, organize themselves into an ad hoc network, and localize themselves using a combination of ultrasound range-finding and low-power GPS. Over the following three years, the network monitors the environmental changes that result from Charlie's new management regime, involving the construction of new wetlands on the site. By monitoring the spatial changes that occur, including the emergence of the nitrogen "hot spots," and the location, area, and connectivity of new wetlands, the system is able to ensure that the conditions of the conservation contract are being met. Monthly progress summaries are automatically relayed to the government contract manager; Charlie can also monitor changes while in the field using a special application on her smartphone.*

There are substantial practical and theoretical hurdles to overcome before applications such as the one described above become more than mere flights of fancy. However, the example highlights the importance of "spatial" technologies and intelligence in overcoming these hurdles. More specifically, some of the ways in which *spatial* computation, and in particular *decentralized* spatial computation, can help us realize this sort of application include:

- *Monitoring where and when*: Most of the important phenomena we want to monitor and manage in the world are spatial and spatiotemporal. "Where" and "when" are central to every facet of environmental monitoring. The *location* of natural resources, the *connectivity* of wetlands, the *emergence* of nitrogen hot spots, and innumerable other examples fundamentally rely on (changes to) geographic space. And generating new knowledge about our changing environment requires intelligent ways of processing highly dynamic and distributed spatial data.

- *Information integration*: Space and time constitute a universal framework for integrating information from diverse sources. Such integration, across a range of scales, is vital to any monitoring application. For example, the application described above will need to relate monitored changes to larger-scale patterns. Droughts, cold snaps, or extreme weather may lessen the impact of a new management regime. Consequently, automated techniques for integrating spatial information both from across the network and from other centralized sources (e.g., weather stations, satellite remote sensing) are required.

- *Inferring meaning*: In monitoring, as in science more generally, there is frequently a gap between what can be measured and what we wish to know. The range of sensors available to applications, such as the temperature, soil moisture, and optical sensors in our example, will surely continue to expand. But, useful knowledge often cannot be purely objectively measured. Habitat types, such as wetlands, grasslands, or semi-arid woodland can at best be inferred from measured proxies, but can never be sensed directly. A vital mechanism for inferring new meaning from observed data is identifying patterns, and in particular spatial patterns. While an individual node may be able to say little about a habitat change, coordination between thousands of nodes may allow patterns to emerge that provide a "spatial signature" for a more meaningful change. Thus, spatial intelligence is needed to automate this inference process, even inside the monitoring system itself.

- *Reasoning under uncertainty*: Information about geographic space is inherently uncertain. All sensors, whether for environmental parameters or for location, are more or less unreliable. Even if equipped with GPS, the nodes in the example above may be unable to determine accurate and precise position due to signal attenuation from vegetation. Spatial intelligence, embedded in the network itself, is needed to ensure monitoring systems continue to operate in the presence of uncertainty. Akin to inferring meaning, identifying spatial patterns is important to identifying imperfections, such as sensor inaccuracies, in individual data items or data streams.

- *Interacting with people*: Finally, in most applications, a human will at some point need to use and understand the information generated by the network. In many of those cases, as in the example above, the person may be located in the environment itself, combining information from his or her own senses with that from the network. Spatial intelligence is required to ensure data is collated and presented in a form that can be easily accessed by the human users, ultimately helping to ensure better decision making.

Many of these important issues we return to in more detail in Part III of this book.

1.4.2 Alternative Technologies and Applications

This chapter focuses primarily on geosensor networks as the most important emerging technology for decentralized spatial computing, and on environmental monitoring as a key application for decentralized spatial computing. However, it is worth spending a moment to reflect on the fact that geosensor networks motivate, but do not define decentralized spatial computing. There are a range of alternative spatial technologies that are already beginning to be used to view the geographical world "decentrally." Any or all of these technologies may, in the future, be components of ambient spatial intelligence applications.

For example, transportation networks are highly instrumented spaces. Today's vehicles (such as cars and buses) and transportation infrastructure (e.g., roads, traffic signals, road signs) are increasingly able to sense information about their immediate geographic environment (such as weather, traffic speed, and road conditions in their vicinity), compute with that information, and communicate wirelessly with other nearby vehicles or with nearby infrastructure. The term *vehicle ad hoc networks* (VANETs) has been coined to refer to peer-to-peer networks of nearby vehicles. VANETs have the aim of supporting new applications that improve safety or efficiency in transportation, such as by warning road users in real time about adverse traffic conditions or traffic jams. VANETs do not present the same energy constraints as geosensor networks; vehicles are typically physically large and equipped with ample power supplies (engines) that can easily support wireless *wide area network* (WAN) communication. However, while energy constraints may not be an issue for VANETs, millions of vehicles moving rapidly around a network every day can readily lead to information overload and scalability problems. Further, VANETs can be regarded as sensor/actuator networks, with the highly dynamic nature of transportation requiring low latencies. As a result, researchers are already turning to decentralized models of computation in such environments (e.g., [112]).

Another example is social networking applications. Just as vehicles are increasingly instrumented and connected, so are people. The mobile handheld devices we all have in our pockets or bags, such as smartphones, PDAs, tablet computers, and media players, are increasingly connected via short-range personal area networks (PANs), WANs (wide area networks), or more often both. These devices may be able to sense many different parameters about our environment, using cameras, microphones, accelerometers, as well as determine their position, using GPS and other localization systems. Using these sensors, the devices may also be used to infer higher-level knowledge about the context of an individual, as when using the accelerometer or microphone to determine whether a person is in a vehicle, walking, or in a meeting (e.g., [91]). As with geosensor networks, the limited battery life of today's portable devices, combined with scalability issues, makes decentralization advantageous. But in a world where personal information is ever more accessible, a further advantage of decentralization is protecting privacy. A decentralized approach to social networking can ensure that personal data is only ever shared with a selected, small subset of system components, and never stored centrally. As already argued, this can potentially make the entire system more robust to attacks on personal privacy.

In summary, geosensor networks are but one technology for ambient spatial intelligence, and environmental applications but one motivation for studying decentralized spatial computing. However, there are many other technologies and applications for "computing somewhere" on and just beyond the horizon, from transportation and social networking to location-based services, emergency preparedness and response, elder care, defense, facilities management and logistics, and citizen science, to name but a few (see question 1.6). Figure 1.11 summarizes the relationships between geosensor networks, decentralized spatial computing, ambient spatial intelligence, and "computing somewhere."

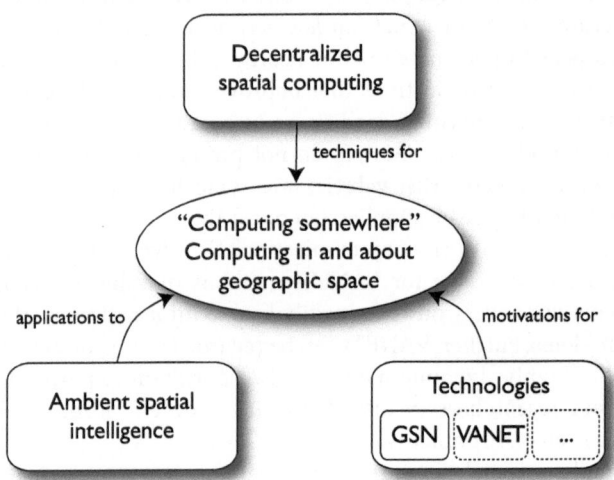

Fig. 1.11. Relationships of geosensor networks, decentralized spatial computing, and ambient spatial intelligence to "computing somewhere"

1.5 Chapter in a Nutshell

The problems of "computing somewhere" arise in spatial information systems where information is distributed throughout geographic space, and geography presents spatial constraints to the movement of that information. Geosensor networks provide a concrete example of a system for "computing somewhere." Nodes in a geosensor network generate information about their immediate geographic environment; limited system resources means that the energy costs associated with movement of information increase with distance.

Decentralized spatial computing is the study of decentralized techniques for computing with spatial information. Decentralized spatial computing presents both a challenge and an opportunity. Spatial computing usually requires the combination of information from distal geographic locations, a challenge in the context of spatial constraints on the movement of information. At the same time, the fundamentally autocorrelated structure of geographic space means that the data items required to satisfy a spatial query or perform a spatial operation are likely to be spatially proximally located.

Ambient spatial intelligence (AmSI) aims to apply technologies for "computing somewhere," such as geosensor networks, and decentralized spatial tools and techniques to problem domains such as environmental monitoring. The vision of AmSI is to augment built and natural environments with ubiquitous spatial computing capabilities.

Taken together, "computing somewhere" is the idea of computing in and about geographic space; geosensor networks constitute the technology that motivates our interest in "computing somewhere"; decentralized spatial computing is the technique we shall develop for enabling systems to achieve "computing somewhere"; and ambient spatial intelligence is the application of "computing somewhere" to real-world problems (Fig. 1.11).

Review questions

1.1 What is the difference between a *decentralized* and a *distributed* system? (Hint: see §1.3)
1.2 List and briefly explain three reasons why geosensor networks provide a unique spatial information source (Hint: see §1.2.1).
1.3 Every week new types of digital MEMS (microelectromechanical system) sensors are being designed and built. How many different kinds of MEMS sensors can you find out about? Try to classify each sensor you find into one of the six categories of sensor types in Table 1.1. (Hint: begin your research using a Web search).
1.4 Explain what is meant by the phrase "trade communication for computation" in the context of geosensor networks.

1.5 List and explain six reasons why decentralized spatial computing is a more appropriate model for computing in geosensor networks than centralized spatial computing.

1.6 Geosensor networks provide an important technological motivation for studying "computing somewhere." Transportation and social networking provide further motivations. But what other technologies (either existing today or more fanciful) might benefit from adopting the "computing somewhere" paradigm?

1.7 Our example of an AmSI application in §1.4 concerns environmental monitoring. What other applications of AmSI can you imagine, for example, in the domains of emergency response or location-based gaming?

Formal Foundations 2

Summary: This chapter sets out a precise model of decentralized spatial information systems, like geosensor networks. The model has three main levels, each of which builds on structures in the previous level. First, a minimal, neighborhood-based model of geosensor networks provides the most fundamental structures for sensing and communicating information about the environment. Next, an extended spatial model of static geosensor networks provides additional quantitative and qualitative structures to model spatial location. Third, a spatiotemporal model of dynamic geosensor networks provides the capability to precisely model spatial and environmental change over time. In addition, two auxiliary structures are also discussed: a model of the partial knowledge about the decentralized system available to each individual node; and the most common communication network structures, which constrain the movement of information in any decentralized spatial information system.

SETTING the scene for decentralized spatial computing, the previous chapter aimed to motivate our interest in the topic, frame the fundamental problems and concepts, and sketch the vision for the application of decentralized spatial computing to solving real-world problems. In this chapter, we come back down to earth (with a bump!) and examine the basic components of decentralized spatial computation by defining in a precise and abstract way the key computational elements of a geosensor network.

2.1 Introduction

A formal model of a geosensor network must provide assistance in representing and reasoning about the features of geosensor networks important in decentralized spatial computing. Equally important, a formal model also needs to suppress unnecessary detail about the geosensor network. For example, in domains such as network routing, hardware design, and operating systems,

M. Duckham, *Decentralized Spatial Computing*, DOI 10.1007/978-3-642-30853-6_2,
© Springer-Verlag Berlin Heidelberg 2013

many technical details of geosensor networks are important, including the specific details of:

- the characteristics and types of sensors available on the node;
- the processing speed, architecture, and capabilities of the microcontroller;
- the storage capacity and characteristics of the node's memory;
- the communications protocol used to exchange information between nodes; and
- the power requirements of specific node operations, such as the relative energy budgets required for computation, communication, and sensing.

However, in designing decentralized spatial algorithms it is helpful to abstract away from these technical details, especially as they are expected to develop rapidly, changing with future technological advances. Instead, the features of a geosensor network that are important to a decentralized spatial algorithm include:

- the set of nodes and the structure of the communication network connecting the nodes;
- the geographic locations of the nodes in their environment (either relative or absolute);
- the environmental parameters in each node's vicinity (i.e., what each node can sense); and
- the changes in all the above: nodes and the communication network, locations in geographic space, and environmental parameters.

This chapter builds up to a detailed model of a geosensor network in three stages. First we construct a basic model of a geosensor network which contains only minimal spatial information about *neighborhoods* in the communication network (§2.2). Then we add more sophisticated models of spatial location (§2.3). Finally, we add time to the mix, allowing modeling of not only where things are, but also of how they change (§2.4).

2.2 Neighborhood-Based Model

The most basic components of a geosensor network are:

- the nodes themselves, with their names or identities;
- the wireless communication network that connects and constrains the movement of information between nodes (i.e., the network neighborhood); and
- the values of environmental parameters that can be detected by sensors at each node.

Thus in the basic neighborhood-based model we ignore temporal issues, such as changes to the environment over time, and almost all spatial issues, such as the coordinate locations of the nodes.

At this point, we shall use some structures and syntax borrowed from discrete mathematics in order to precisely define the different components of a geosensor network. As a result, some familiarity with basic discrete mathematics structures (e.g., sets, relations, and functions) will be an advantage. Appendix A provides a brief discrete mathematics primer to these key concepts. However, for more information the reader is referred to any introductory discrete mathematics textbook (e.g., [53, 74]).

We can represent these basic components of a geosensor network formally as:

1. a *graph* $G = (V, E)$, which models the network and its connections;
2. a function $s : V \rightarrow C$, which models what a node can sense; and
3. a function *id* $: V \rightarrow \mathbb{N}$, which models the identities or names of each node in the network.

The set of vertices V of the graph represents nodes of the sensor network. The set of edges E in the graph represents direct, one-hop communication links between nodes. In this book we use the terms "node" and "link" to refer to sensor nodes and communication links, while "vertex" and "edge" are used to specifically refer to the formal, graph-based structures that are used to model sensor nodes and communication links. The graph itself is termed the *communication graph.* An undirected communication graph is used to model the situation where all communication links are bidirectional. A directed communication graph allows for the possibility that some communication links are unidirectional (i.e., where direct communication from v_1 to v_2 does not imply direct communication from v_2 to v_1). As we shall see in later chapters, for simplicity an undirected communication graph is often assumed; however, in reality communication between nodes may not be bidirectional.

The only spatial information available in the neighborhood-based model is that which is embedded in the communication graph (i.e., which nodes are adjacent to which other nodes). This information may be represented using the function $nbr : V \rightarrow 2^V$, where $nbr(v) \mapsto \{v' \in V | (v, v') \in E\}$. Note, however, that the neighborhood function nbr contains no information that is not already available from the communication graph $G-nbr$ is merely a notational convenience.

The function s, termed the *sensor function,* represents the data sensed by nodes in the geosensor network. The codomain C for the function s will depend on the sensors used. In many of our later discussions, the codomain C is assumed to be simply the real numbers $\mathbb{R}$. However, more sophisticated sensors and sensor arrays can also easily be represented in the neighborhood-based model. For example, imagine a sensor network where nodes are equipped with sensors for temperature (measured in degrees centigrade), humidity (measured from 0% to 100% relative humidity), and light intensity (normally measured in lux, varying for example, from 0 lux for total darkness to 100,000 lux or more for bright sunshine). This sensed data would be represented as the domain $C = \mathbb{R} \times [0, 100] \times \mathbb{R}^+$. For a particular node $v \in V$, writing $s(v) = (-4.3, 67, 500)$ would indicate that node v's sensors detected a temperature of -4.3°C (cold), relative humidity of 67% (comfortable), and light intensity of 500 lux (typical indoor conditions).

Finally, the function $id : V \to \mathbb{N}$, termed the *identifier function*, represents the identities or names of the nodes in the geosensor network. Without loss of generality, we model the set of names or identifiers as the natural numbers $\mathbb{N}$ (i.e., $\{1, 2, 3, \ldots\}$). For example, for some node $v \in V$, $id(v) = 5$ specifies that v's identifier is 5. In many (but not all) cases, we may wish model the situation where each node has a *unique* identifier. This uniqueness constraint can be represented by specifying that the *id* function is an *injection* (i.e., where each element in $\mathbb{N}$ is mapped to by at most one $v \in V$). In this book, we assume the *id* function is injective unless otherwise stated. Although it is important to distinguish between a node's identity and the node itself in many decentralized spatial algorithms, this book will occasionally refer to the node v (instead of the identity of the node $id(v)$) in contexts where no ambiguity exists.

Table 2.1 summarizes informally the different symbols used for a basic geosensor network.

2.2.1 Example Neighborhood-Based Model

To illustrate, consider the graph $G = (V, E)$ in Fig. 2.1. The communication graph G has 25 vertices, $V = \{a, b, \ldots, x, y\}$, and 42 edges, $E = \{\{a, b\}, \{a, f\}, \ldots, \{w, x\}, \{x, y\}\}$. We imagine that the nodes can sense a Boolean value, either *black* or *white* (indicated by the color of the nodes in Fig. 2.1). Thus, the sensor function may be represented as $s : V \to \{black, white\}$, where $s(x) \mapsto white$ if $x \in \{h, i, l, m, n, q\}$ and $s(x) \mapsto black$ otherwise. Finally, an (injective) identifier function could be constructed to map nodes to (unique) node identifiers 1, ..., 25 as $id : V \to \mathbb{N}$, where $id(a) = 1$, $id(b) = 2$, ..., $id(y) = 25$.

The only spatial information available about the network concerns the neighborhoods of nodes, a direct consequence of the communication graph. For example, $nbr(g) = \{b, f, h, l\}$ and $nbr(o) = \{j, n, t\}$.

2.3 Extended Spatial Model

Extending the neighborhood-based model of a geosensor network requires a mechanism to represent more sophisticated aspects of the node's spatial location, beyond basic neighborhoods in the communication graph. Unfortunately, this representation is complicated by the fact that there are many different types of location information an individual node may possess. The process of determining the locations of sensor nodes in a geosensor network is termed *localization*. "Location" in this context does not necessarily

Formal definition	Summary
$G = (V, E)$	Communication graph.
V	Set of nodes in a geosensor network. Lowercase letters v, v', v_1, etc. are used to refer to individual nodes in V.
E	Set of communication links between pairs of nodes in a geosensor network. A particular communication link l from v_1 to v_2 is written $l = (v_1, v_2)$ (unidirectional) or $l = \{v_1, v_2\}$ (bidirectional).
$nbr : V \to 2^V$	The neighborhood of a node, where $nbr(v)$ refers to the set of nodes that is in the *neighborhood* of $v \in V$ (i.e., those nodes within direct one-hop communication distance from v).
$s : V \to \mathbb{R}$	Sensor function sensing a particular environmental parameter. For example, $s(v) = 10.3$ indicates that the on-board sensor of node $v \in V$ senses environmental value of 10.3.
$id : V \to \mathbb{N}$	Identifier function for nodes. For example, $id(v) = 5$ indicates that a particular node $v \in V$ has identifier (or "name") 5.

Table 2.1. Summary of basic geosensor network model structures

imply the *coordinate location*. *Positioning* is the term for the special case of localization that generates *coordinate* location (i.e., position). However, localization may also involve less detailed quantitative information about the relative distances or directions (bearings) between nodes, or even qualitative information about a node's proximity to other nodes or known locations. Localization is a highly active area for current research. In general, localization techniques can be classified into *passive* or *active* techniques [119]. Active techniques rely on the active transmission of signals (such as radio frequency or ultrasound signals) from other nodes or beacons; passive techniques do not require active transmission and instead detect "naturally" occurring signals (cf. active and passive sensors in Table 1.1). Examples of active techniques include *lateration* (computing location based on distances to known locations) and *angulation* (computing location based on angles to known locations), as well as *proximity* systems (which determine the closest neighbors). GPS, for example, is a lateration-based, active positioning technique. Examples of passive techniques include *scene analysis* (determining location from analysis of digital camera images) and *dead reckoning* (determining displacement of

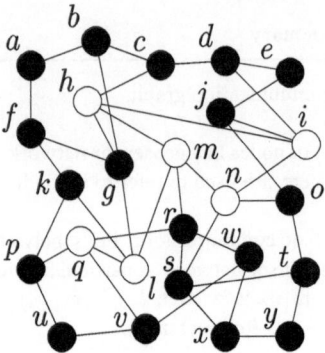

Fig. 2.1. Example neighborhood-based model, with graph $G = (V, E)$, where $V = \{a, b, \ldots, x, y\}$ and $E = \{\{a, b\}, \{a, f\}, \ldots, \{w, x\}, \{x, y\}\}$, sensor function $s : V \rightarrow \{black, white\}$, and identifier $id : V \rightarrow \mathbb{N}$, where $id(a) = 1$, $id(b) = 2, \ldots$

mobile nodes, for example, through inertial tracking). For more information about localization, see [54, 60, 119].

From the perspective of the spatial model of geosensor networks, there are two top-level classes of spatial information generated by localization:

- *absolute location* is spatial information about the location of a sensor node that is referenced to some external system. This most common external reference system is a geodetic framework, where coordinate location can be determined, termed a *positioning system*. However, other external reference systems are possible, such as referencing to known locations within a transportation network (such as known intersections or kilometer posts, a process known as *stationing*).
- *relative location* is spatial information about the location of a sensor node that is referenced to the locations of other nearby sensor nodes.

Table 2.2 summarizes five of the most common types of location information available to sensor nodes in a geosensor network, giving their formal definitions alongside an informal summary. Absolute location information generally concerns either coordinate location (for example, in two or even three dimensions), or close proximity to some known locations, termed "anchor locations" (such as intersections in a transportation network). GPS, for example, is potentially able to provide absolute coordinate location for nodes. Alternatively, RFID (radio frequency identification) tags, attached to mobile nodes, can provide absolute location in terms of proximity to RFID readers at known locations.

Relative location information might include information about the distances from a node to its neighbors; the quantitative bearings from a node to its neighbors; or qualitative bearings of a node's neighbors, in terms of the (counter)clockwise sequence of neighbors around a node (termed "cyclic ordering"). Figure 2.2 summarizes diagrammatically these different types of location information, in addition to relative neighborhood location, the most basic type of spatial information assumed to be available to all geosensor networks (see §2.2).

Type	Method	Definition	Summary
Relative	Neighborhood distance	$dist : E \to \mathbb{R}$	$dist(v, v')$ refers to distance between a node v and its neighbor v', where $(v, v') \in E$ (Fig. 2.2d).
	Neighborhood bearing	$bear : E \to \mathbb{R}$	$bear(v, v')$ refers to the bearing of node v' from node v, where $(v, v') \in E$ (Fig. 2.2e).
	Cyclic ordering	$cyc : E \to V$	$cyc(v, v')$ refers to the next neighbor of v in an anticlockwise direction from v' (Fig. 2.2f).
Absolute	Coordinate location (position)	$p : V \to \mathbb{R}^d$	$p(v)$ refers to the coordinate location of node $v \in V$ (Fig. 2.2c). In most cases in this book, we assume planar coordinates (i.e., $d = 2$).
	Anchor location	$anch : V \to A$	$anch(v)$ refers to the *anchor* location of a node (i.e., the nearest anchor). A is a set of anchors at known locations (which may or may not include nodes in V, Fig. 2.2b).

Table 2.2. Examples of common types of location information in geosensor networks (cf. Fig. 2.2)

As might be expected with spatial information, the different types of location information listed in Table 2.2 and Fig. 2.2 are interconnected. Information about the absolute (coordinate) locations of nodes and the neighborhoods of a node can be combined to compute (using standard geometry) information about the neighborhood distances and bearings of nodes. Further, information about the neighborhood bearings can be used to compute the cyclic ordering of nodes (the cyclic ordering is a less precise, qualitative version of the quantitative neighborhood bearing).

2.3.1 Example Spatial Model

Building on the neighborhood-based model example in §2.2.1, consider now the graph $G = (V, E)$ in Fig. 2.3. The communication graph, sensor function, and identifier function in Fig. 2.3 are identical to those in Fig. 2.1. However, the

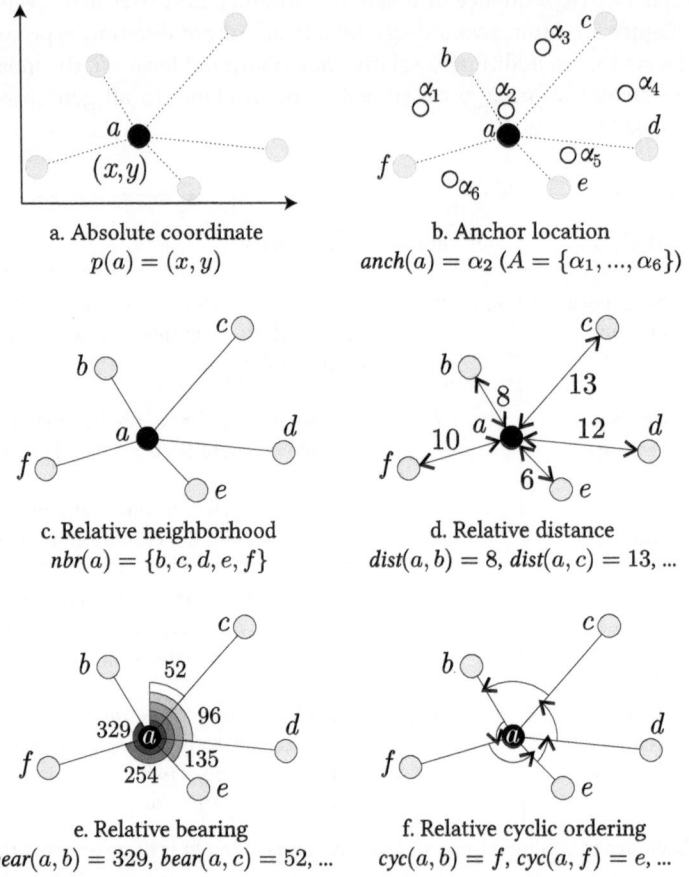

a. Absolute coordinate
$p(a) = (x, y)$

b. Anchor location
$anch(a) = \alpha_2$ $(A = \{\alpha_1, ..., \alpha_6\})$

c. Relative neighborhood
$nbr(a) = \{b, c, d, e, f\}$

d. Relative distance
$dist(a, b) = 8$, $dist(a, c) = 13$, ...

e. Relative bearing
$bear(a, b) = 329$, $bear(a, c) = 52$, ...

f. Relative cyclic ordering
$cyc(a, b) = f$, $cyc(a, f) = e$, ...

Fig. 2.2. Summary of common types of location information available to a node

nodes are no longer arbitrarily placed, but are arranged in a grid to emphasize their spatial locations.

In a spatial model, we might have access to any or all the levels of spatial information given in Table 2.2 (and potentially more). For example, assume each edge is has length 1 unit. It then follows that:

- the neighborhood distance $dist : E \rightarrow \mathbb{R}$ has assignment mapping $dist(v_1, v_2) \mapsto 1$;
- the neighborhood bearing $bear : E \rightarrow \mathbb{R}$ has assignment mapping:

$$bear(v_1, v_2) \mapsto \begin{cases} 0 \text{ if } id(v_2) < id(v_1) - 1 \\ 90 \text{ if } id(v_2) = id(v_1) + 1 \\ 180 \text{ if } id(v_2) = id(v_1) - 1 \\ 270 \text{ if } id(v_2) > id(v_1) + 1 \end{cases}$$

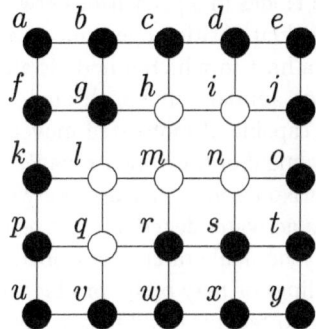

Fig. 2.3. Example spatial model, assuming an identical communication graph to that in Fig. 2.1, but with planar coordinate embedding such that all edges have length 1 and node u is located at the origin $(0, 0)$

- the (anticlockwise) cyclic ordering $cyc : E \to V$ has assignment mapping $cyc(a, b) = f$, $cyc(a, f) = b$, $cyc(b, a) = g$, $cyc(b, g) = c$, $cyc(b, c) = a$, and so forth.

If we additionally add the assumption of a planar coordinate system, with node u at the origin, we may further specify the position of each node as:

- the planar positioning function $p : V \to \mathbb{R}^2$ has assignment mapping $p(a) = (0, 5)$, $p(b) = (1, 5)$, $p(c) = (2, 5)$, ..., $p(y) = (5, 0)$.

Note that as long as we have the highest level of information, the planar positioning function p, all the other levels of relative location information may be deduced from p.

2.4 Spatiotemporal Model

The two models discussed so far have been *atemporal*, in the sense that they are static and make no reference to change over time. As we shall see in later chapters, in some situations it reasonable to use such simplified models. However, in general, geosensor networks monitor changing phenomena and may themselves be subject to change. It is possible to identify three main types of change that are important to decentralized spatial computing:

- *Environmental dynamism*: Geosensor networks are usually tasked with monitoring geographic environments that change; static environments are both rare in the world (try to think of a place that does not change) and by definition not especially interesting to monitor with geosensor networks (a single survey or a photograph would be enough to record a static environment).

- *Node mobility*: Nodes in a geosensor network may not be static, and may instead change their position over time. Node mobility may occur through movement of a host to which a node is attached (e.g., nodes attached to people, vehicles, animals, etc.) or through more purposeful mobility (e.g., robotic nodes capable of motorized movement or opportunistic mobility using the movement of the wind or waves).
- *Node volatility*: Nodes in a geosensor network may be *volatile* in the sense that they may activate, deactivate, or reactivate over time. Changes in activation may be deliberately controlled by the nodes or the network (e.g., duty cycling) or may be by-products of other processes (e.g., nodes deactivate due to technical failures or depletion of energy resources, or nodes are activated by new nodes being introduced into the network). Mobility can be seen as a special case of volatility, where volatile nodes may deactivate but immediately reactivate in a new location.

As a result, in most cases, it is important additionally to be able to model changes in the environment and the geosensor network. Informally, this is achieved by extending each of the structures described in the previous two sections, covering the neighborhood-based and extended spatial models, with time-varying capabilities. Table 2.3 summarizes the extended time-varying structures. Environmental dynamism can be modeled using a time-varying sensor function ($s : V \times T \to \mathbb{R}$) which represents the different data generated by each node's sensors over time. The only structure we assume cannot change is the identifier function; node identities are assumed to persist through time.

2.4.1 Example Spatiotemporal Model

Figure 2.4 shows an example of a spatiotemporal model of a geosensor network, building on the structures in §2.2.1 and §2.3.1. In this simple example we model only environmental dynamism: the communication graph and nodes are assumed to be static. Thus, the sensed value for node c, for example, changes from time t_2 to t_3: $s(c, t_1) = black$, $s(c, t_2) = black$, $s(c, t_3) = white$.

More sophisticated spatiotemporal models must additionally deal with change over time in the absolute or relative locations of nodes (mobility) and/or in the active set of nodes (volatility). For example, Fig. 2.5 shows a spatiotemporal model with two mobile nodes, with absolute locations referenced to intersections in a traffic network (anchors). Thus, in contrast to earlier figures, the network in Fig. 2.5 is the *transportation* network, not the communication network. The changing location of the mobile node v_1, for example, is given by: $anch(v_1, t_1) = n_9$, $anch(v_1, t_2) = n_{10}$, $anch(v_1, t_3) = n_{15}$.

Atemporal structure	Temporal extension	Informal summary
$s : V \to \mathbb{R}$	$s : V \times T \to \mathbb{R}$	$s(v, t)$ refers to the value sensed by node v at time t.
$p : V \to \mathbb{R}^n$	$p : V \times T \to \mathbb{R}^n$	$p(v, t)$ refers to the coordinate location of node v at time t.
$anch : V \to V_{anch}$	$anch : V \times T \to V_{anch}$	$anch(v, t)$ refers to anchor location of node v at time t.
$nbr : V \to V$	$nbr : V \times T \to V$	$nbr(v, t)$ refers to the set of neighbors of node v at time t.
$dist : E \to \mathbb{R}$	$dist : E \times T \to \mathbb{R}$	$dist(v, v', t)$ refers to the distance between v and its neighbor v' at time t.
$bear : E \to \mathbb{R}$	$bear : E \times T \to \mathbb{R}$	$bear(v, v', t)$ refers to the bearing of neighbor v' from v at time t.
$cyc : E \to nbr(v)$	$cyc : E \times T \to V$	$cyc(v, v', t)$ refers to the next neighbor of v in an anticlockwise direction from v' at time t.
$G = (V, E)$	$G(t) = (V, E(t))$	For mobile (but not volatile) geosensor networks, $G(t)$ refers to the communication graph at time t, $E(t)$ refers to the set of links in G at time t.
$G = (V, E)$	$G(t) = (V(t), E(t))$	For volatile (including volatile and mobile) geosensor networks, $G(t)$ refers to the communication graph at time t, $E(t)$ refers to the set of links in G at time t, and $V(t)$ refers to set of nodes at time t.

Table 2.3. Extending static definitions to allow for change over a totally ordered set T of discrete times (including informal descriptions for particular time $t \in T$)

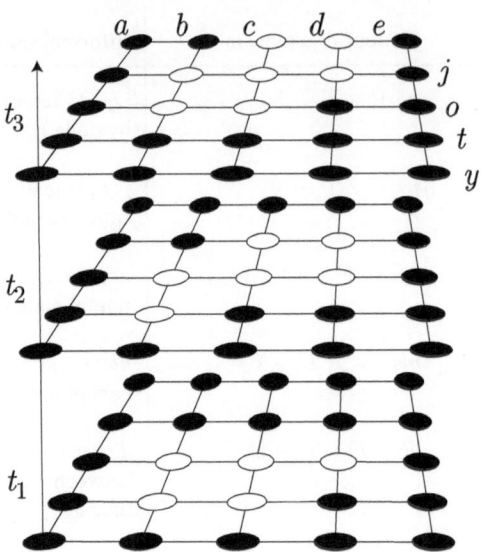

Fig. 2.4. Example spatiotemporal model for modeling environmental dynamism ($s : V \times T \rightarrow$ {*black, white*}) using a static communication graph, $G = (V, E)$, for times $T = \{t_1, t_2, t_3\}$

2.5 Partial Knowledge

The formal structures introduced above provide an overarching framework for representing geosensor networks with a range of different characteristics. However, let us now turn away from overarching frameworks and instead consider the individual nodes. The information available to a particular node will typically differ from that in the structures already introduced in two very important respects:

1. *Partial knowledge*: By default, a node in a geosensor network is expected to have information only about its *own* state. Any information about the state of other nodes in the system must be explicitly communicated to that node. Thus, we must be able to represent the situation where nodes have only *partial* knowledge of the state of the entire network.
2. *Uncertain knowledge*: Uncertainty is an endemic feature of geospatial information. Thus, we must also be able to represent the situation where a node's knowledge of its geographic environment may in some ways diverge from the "ideal."

The issue of uncertain knowledge is vital, but will be held in abeyance until Part III, Chapter 8, when we shall address this issue directly. Until then we shall make the simplifying assumption that nodes have ideal location and environmental sensors which can make perfect observations of geographic space.

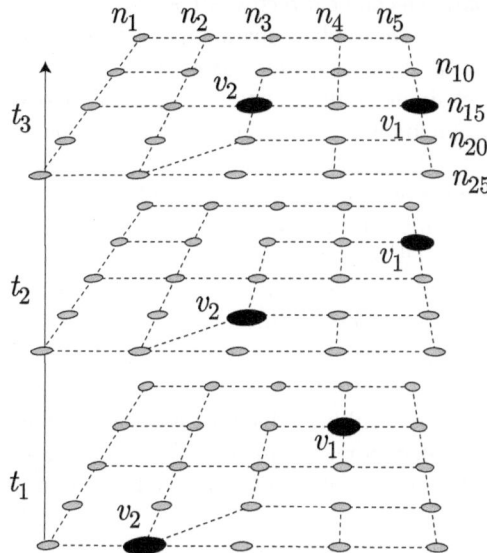

Fig. 2.5. Example spatiotemporal model of mobile nodes ($anch : V \times T \to A$), using the static nodes $A = \{n_1, \ldots, n_{25}\}$ of a transportation network as anchors, and times $T = \{t_1, t_2, t_3\}$

The issue of partial knowledge, however, does need to be addressed directly at this point. Some of the most frequent mistakes in decentralized spatial algorithm design arise from mistakes in distinguishing between information that is *local* to an individual node and that which is *global* to the entire system. Only information local to a node (such as that node's sensor data or its location) can be *directly* accessed by that node. All other data must be explicitly communicated to a node. Thus, one of the keys to successful decentralized algorithm design is carefully and rigorously to identify a node's *local* information—its partial knowledge of the entire system.

The formal model set out above makes no distinction between local and global information. Functions are defined at the global level: for example, the sensor function $s : V \to C$ is defined globally, across all nodes in the network. An individual node $v \in V$ can only access a small part of that function directly, the piece of information that relates to its own sensor function, $s(v)$.

To distinguish between a global function and a node's local knowledge about that function requires some additional notation. For some global function $f : V \to C$ we will use the overdot notation, $\mathring{f}$ (read "my" f or "local" f), to refer to an arbitrary (unspecified) node's local knowledge of that function. Thus, for any function f with domain V, $\mathring{f}$ is interpreted to mean $f(\circ)$ for some node $\circ \in V$ that is clear from the context.

This approach can be directly extended to those functions with a domain that is a product set (such as spatiotemporal algorithms). Any function with a product set as its domain, such as $g : V \times T \to C$, can be rewritten as an equivalent chain of single-argument functions, such as $g : V \to (T \to C)$ (a transformation known as "Currying"). Thus, we may write $\mathring{g}(t)$ ("my g of t") in place of $g(\circ, t)$ for some arbitrary $\circ \in V$ and time $t \in T^1$.

For example, where we might write globally $s(\circ) = 10$, inside a decentralized algorithm we instead write $\mathring{s} = 10$ ("my s is 10"). Similarly, in a spatiotemporal algorithm, instead of writing the global statement "$s(\circ, t) = 10$" to indicate that node $\circ \in V$ senses the value 10 at time t, we will use the corresponding local function inside a protocol, $\mathring{s}(t) = 10$. Finally, in a spatial algorithm $\mathring{dist}(v) = 12$ may be used in place of $dist(\circ, v) = 12$

In this way, direct references to global functions can be avoided in the specification of algorithms, which helps to ensure that only local information is used in the algorithm. Information that is not part of the local knowledge of a node must be explicitly communicated to that node before it can be used.

2.6 Neighborhood Structure

One final important component of our formal foundations remains only partially defined: the precise neighborhood structure of the communication graph. In "computing somewhere" the neighborhood structure is a key constraint on the movement of information. The communication graph $G = (V, E)$ specifies for each node $v \in V$ those nodes that are are capable of direct communication with v (given by $nbr(v)$, defined as the set $\{v' | \{v, v'\} \in E\}$; see §2.2). This section examines more closely some of the most common neighborhood structures that can constrain communication in a geosensor network, or indeed in any decentralized spatial information system.

2.6.1 Unit Disk Graph (UDG)

In a geosensor network the constraints on the movement of information are imposed by the communication network structure, which in turn is constrained by the relative locations of the nodes. Our model has so far considered the spatial locations of the nodes and the structure of the network separately. However, because of the limited communication range of nodes (see §1.2.2), the structure of the communication network connecting

---

[1] More specifically, given the function $g : V \to (T \to C)$, the result of applying the function g to $\circ$, $g(\circ)$, yields a function $h : T \to C$. In turn, this function h can be applied to some time t, $h(t)$. Informally, Currying ensures that any functions with multiple arguments can be decomposed into a chain of single argument functions, each applied in sequence.

geosensors is in practice inherently dependent on the spatial location of the nodes.

The most basic network structure used to represent this dependency is the *unit disk graph* (UDG). The UDG begins by assuming nodes have a uniform communication range, c (the "unit distance"). The UDG is then the communication graph that is formed by creating edges to connect all nodes that are within unit distance c (i.e., that are within direct communication range). Figure 2.6 illustrates an example UDG for a set of nodes and an arbitrary unit distance c.

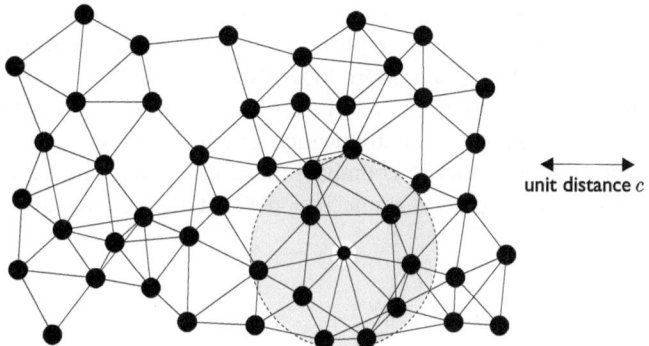

unit distance c

Fig. 2.6. Example unit disk graph (UDG), highlighting the unit distance communication range of one particular node (gray circle)

For a given set of nodes V, node positions $p : V \to \mathbb{R}^2$, and unit distance c, the UDG is formally defined as:

$$\text{UDG} = (V, \{\{u, v\}|(u, v) \in V \text{ and } 0 < \delta(p(u), p(v)) \leq c\})$$

where $\delta(a, b)$ is the Euclidean distance between two points a and b.

It is important to note at this point that the UDG is a model of the communication graph that results from the *physical* constraints of limited power radio frequency communication. Consequently, the UDG is often assumed as the default structure for a network, even in cases where nodes have no capability for localization or other knowledge of their coordinate location (i.e., the *locator* function $p : V \to \mathbb{R}$ is not known to the nodes). In other words, the UDG is assumed to result from the nodes' actual position in geographic space, even in cases where they have no knowledge of that position.

Two implicit assumptions behind the UDG are:

* communication is bidirectional; and
* the communication range is uniform across the network.

In practice, neither of these assumptions may hold. Environmental conditions, as well as varying energy resources of nodes, will often result in different nodes in a network having different communication distances. In turn, this leads to unidirectional communication, where the fact that a node can transmit directly to another does not imply that it can receive messages communicated from that node. Of course, where the varying communication ranges of different nodes are known or can be estimated, a (directed) communication graph can still be constructed to represent that situation.

The UDG represents a common baseline assumption about communication in a geosensor network, and is used as the foundation of several other neighborhood structures, discussed further in the following sections. All of these other network structures are *spanning subgraphs* of the UDG. A spanning subgraph is a graph that contains all the vertices but a subset of edges of another graph. More formally, for a graph $G = (V, E)$ a spanning subgraph is a graph $G' = (V, E')$ such that $E' \subseteq E$. This property is important because a communication graph that is *not* a spanning subgraph of the UDG necessarily contains one or more edges that are not contained within the UDG (and so connects nodes that are too far apart to communicate directly).

Topology control

A natural question that follows the discussion of the UDG is "What happens if we adjust the unit distance?" In the extreme case where the unit distance c is very large, we obtain the *complete* graph of the vertices V (the graph where an edge exists between every pair of nodes). As already discussed, such situations are not especially interesting in the context of decentralized spatial computing, since they present no constraints on the movement of information (any node can communicate directly with any other node). Further, the fundamental resource constraints of geosensor networks means that they are also in practice not commonly encountered.

Moving to the other extreme, for any graph there exists a unique minimum unit distance c such that the communication graph remains connected (i.e., there exists a path connecting any pair of nodes, and as a result it would be possible to multi-hop information from any node to any other node). In this case, the unique minimum unit distance is termed the critical transmission range (CTR). The UDG resulting from using the CTR as the unit distance is called the minimum radius graph (MRG). The MRG for the node set in Fig. 2.6 is illustrated in Fig. 2.7. Any further reduction in the unit distance c will lead the graph to become disconnected.

As a result of the direct relationship between the distance of communication and the power required to communicate (see §1.2.2), topology control has clear practical implications for geosensor networks. At least in theory, the MRG provides a mechanism for setting globally a minimum energy

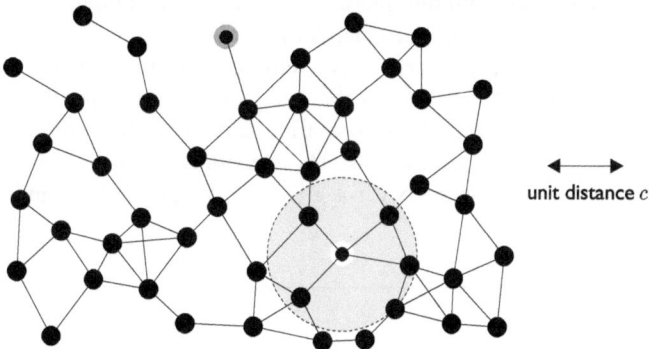

unit distance c

Fig. 2.7. Minimum radius graph (MRG) for the nodes from Fig. 2.6. The node highlighted in gray will become disconnected by any further reductions in the unit distance c

level for wireless transceivers that still ensures nodes can engage in (multi-hop) communication. In practice, the simplifying assumptions of the UDG (discussed above), and the desire for improved robustness in communication by allowing for some redundancy in communication paths, means that a network would normally aim to operate with a unit distance c that is some way above the CTR.

2.6.2 Plane and Planar Graphs

The structure of the UDG is determined spatially, based on the coordinate locations of the nodes and the unit distance. However, many of the algorithms we will look at in later chapters require further spatial structure. In particular, a common requirement concerns the *planarity* of the communication graph. A *planar* graph is a graph that can be embedded in the plane in such a way that no edges cross (i.e., edges only intersect at nodes). A *plane* graph is a planar graph plus a particular planar embedding the ensures that no edges cross. In short, a planar graph *could* be drawn on a sheet of paper without any crossing edges; a plane graph (or a *planar map*) is a specific drawing of a planar graph that has no crossing edges. The graphs in Figs. 2.6 and 2.7 are clearly not plane (they have several crossing edges), and in fact are non-planar (i.e., it is impossible to redraw the figure in a way that preserves all nodes and connections but removes crossing edges).

Four of the most important plane graph structures commonly used with geosensor networks are explored in the following subsections. As already highlighted, each of these structures is a spanning subgraph of the UDG.

Delaunay triangulations

A triangulation is a *maximal plane graph*: a plane graph where no edge can be added between existing nodes without making the graph non-planar.

There are many different possible maximal plane graphs for a set of vertices in the plane. One special maximal plane graph, the *Delaunay triangulation*, has the defining property that a circumcircle through the vertices of any triangle contains no other vertices. Figure 2.8 illustrates this property. More intuitively, this property ensures the Delaunay triangulation has the "fattest" triangles of all triangulations. The Delaunay triangulation is an important structure in GIScience, with a great many applications and articles in the literature.

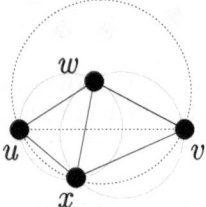

Fig. 2.8. The Delaunay triangulation has the defining property that a circumcircle through the vertices of any triangle contains no other vertices. The dashed circumcircle contains vertex w, and so the edge uv is not in the Delaunay triangulation

Unfortunately, the Delaunay triangulation (or indeed any maximal plane subgraph) is not guaranteed to be a subgraph of the UDG. Any maximal plane graph may have arbitrarily long edges. (Imagine, for example, a set of vertices in the shape of a crescent: any triangulation of these points will necessarily require an edge between the two extreme tips of the crescent). In the context of geosensor network, arbitrarily long edges translate into arbitrarily large amounts of power required to transmit directly between two nodes.

For a given set of vertices, the unit Delaunay triangulation (UDT) is the intersection of the UDG and the Delaunay triangulation (i.e., the graph formed by the edges that are contained in both the UDG and the Delaunay triangulation of the vertices). Figure 2.9 shows the UDT derived from the UDG in Fig. 2.6. Note that there are many possible edges that could be added to the graph in Fig. 2.9 without leading to a non-planar result: in fact the UDT is not a triangulation at all!

Gabriel graph

The Gabriel graph (GG) has the property that an edge exists between two nodes $u, v \in V$ only if there is no node $w \in V$ such that the angle formed by uwv is greater than $90°$. Equivalently, this condition is more easily framed using Pythagoras' theorem rather than trigonometry, as follows (where E_{UDG} is the set of edges in the UDG for a given unit distance):

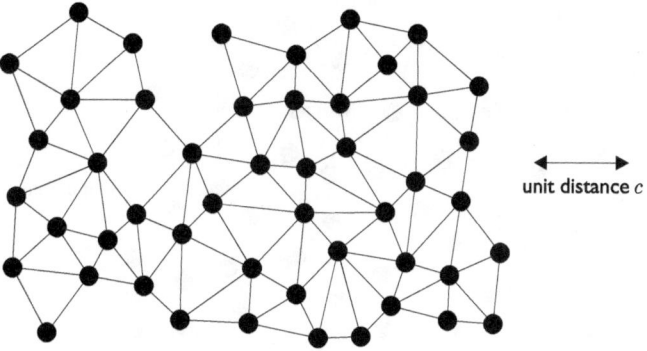

Fig. 2.9. Example unit Delaunay triangulation (UDT) (cf. Fig. 2.6)

$$\text{GG} = (V, \{(u,v) \in E_{\text{UDG}}| \text{ for all } w \in V,$$
$$\delta(p(u), p(w))^2 + \delta(p(w), p(v))^2 \geq \delta(p(u), p(v))^2\}).$$

Figure 2.10 illustrates this definition graphically. Figure 2.11 shows the GG derived from the UDG in Fig. 2.6.

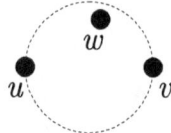

Fig. 2.10. The Gabriel graph has the property that a pair of nodes u and v may only be connected by an edge if there exists no other node w such that the angle formed by uvw is greater than $90°$ (i.e., there exists no w located in the dashed circle)

Relative neighborhood graph

The relative neighborhood graph (RNG) has the property that two nodes $u, v \in V$ will not be connected by an edge if there exists a node $w \in V$ where w is closer to u and to v than u and v are to each other. Formally,

$$\text{RNG} = (V, \{(u,v) \in E_{\text{UDG}}| \text{ for all } w \in V,$$
$$\delta(p(u), p(w)) \geq \delta(p(u), p(v) \text{ or } \delta(p(w), p(v)) \geq \delta(p(u), p(v)\})$$

where E_{UDG} is the set of edges in the UDG for a given unit distance. Figure 2.12 illustrates this definition graphically. Like the GG, the RNG is a

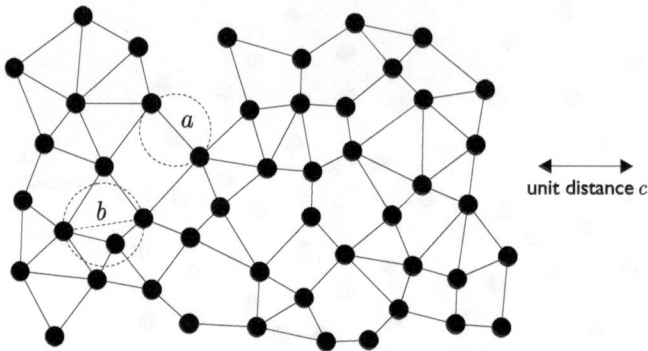

Fig. 2.11. Example Gabriel graph (GG), highlighting examples of (*a*) an edge maintained from the UDG; and (*b*) an edge removed from the UDG (cf. Fig. 2.6)

spanning subgraph of the UDG. As could be deduced by comparing Figs. 2.10 and 2.12, the RNG is also a spanning subgraph of the GG (since any edge excluded by the GG condition will also be excluded by the more restrictive RNG condition). Figure 2.13 shows the RNG derived from the UDG in Fig. 2.6.

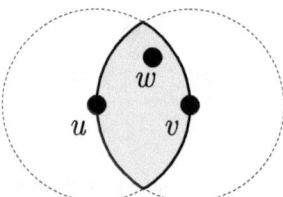

Fig. 2.12. The relative neighborhood graph has the property that any pair of nodes *u* and *v* will not be connected by an edge if there exists a node *w* that is closer to both *u* and *v* than *u* and *v* are to each other (i.e., there exists no *w* located in the highlighted lune)

2.6.3 Trees

A *tree* is is a graph where any pair of vertices is connected by exactly one path (sequence of adjacent nodes). Figure 2.14 shows an example of a tree that is a spanning subgraph (subtree) of the UDG. For a graph such as the UDG in Fig. 2.6 there are many possible spanning subtrees (approximately 1×10^{30}—a "nonillion"—to be more precise; the exact number can be computed using a result in graph theory called *matrix tree theorem*). In fact, the tree shown in Fig. 2.14 is a special, uniquely defined tree, called the *minimum spanning tree* (MST). The minimum spanning tree has the property that it

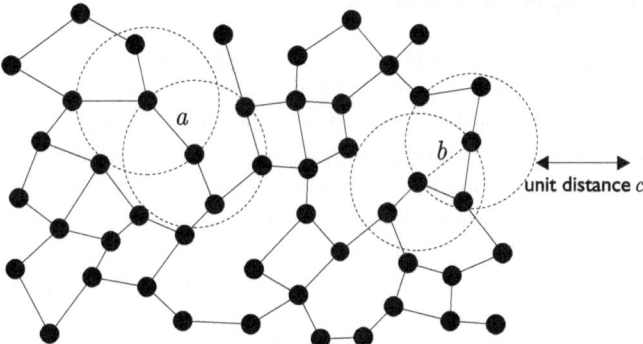

Fig. 2.13. Example relative neighborhood graph (RNG), highlighting examples of a, an edge maintained from the UDG; and b an edge removed from the GG (cf. Figs. 2.6 and 2.11)

is the spanning tree in the graph that has the minimum possible total edge length (i.e., adding up the lengths of all the edges in the graph in Fig. 2.14 results in a total length less than any other spanning tree of the UDG in Fig. 2.6). The minimum spanning tree also happens to be a spanning subtree of the RNG (and by implication the GG).

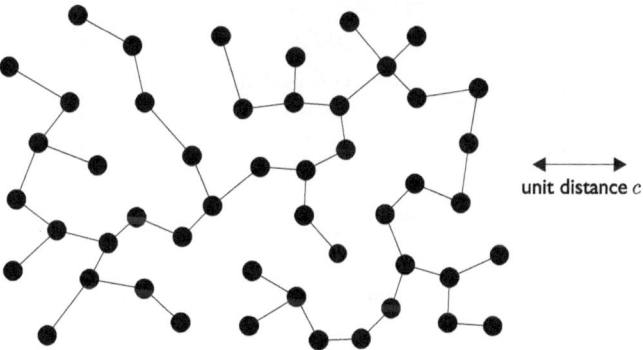

Fig. 2.14. Example minimum spanning tree (MST) (cf. Fig. 2.6)

Trees are commonly used because they provide a convenient neighborhood structure for wireless sensor networks. A *rooted tree* has one designated node as its *root*. Because a tree by definition contains no cycles (since every pair of nodes is connected by exactly one path), edges in a rooted tree have a natural orientation (towards or away from the root). As we shall see, this feature can be helpful in decentralized computation, as it provides a natural structure for aggregating and centralizing data.

2.7 Chapter in a Nutshell

The formal foundations set out in this chapter are founded on three inter-connected models (Fig. 2.15). At the core, the formal foundations depend on the neighborhood-based model of nodes, their sensors, and their communication network. Building on this core, the extended spatial model depends on the neighborhood-based model, and introduces more sophisticated representations of the spatial locations of nodes. In turn, the spatiotemporal model depends on the neighborhood-based or the extended spatial model, further extending the core by representing change in the network and in the environment over time.

In addition, two auxiliary models are added. First, we define a model of the partial knowledge of a node, restricting the information available at each node to that which it senses directly or which is explicitly communicated to that node. Second, a neighborhood model provides more details about the different neighborhood structures that may be used to constrain communication between nodes. A third auxiliary model, allowing for uncertainty in the information sensed by a node, will be added later (Chapter 8).

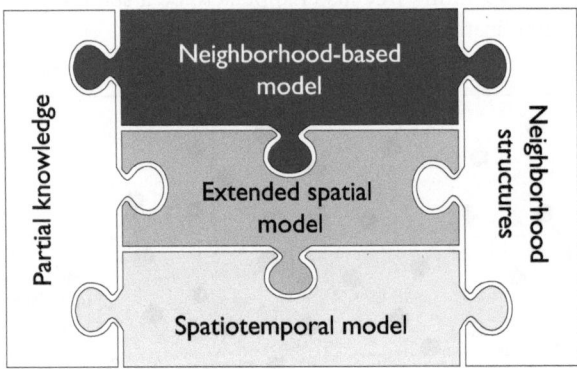

Fig. 2.15. Summary of formal model structure

These foundations deliberately suppress the technical or application details of sensor networks. The aim is to provide a basis for discussing decentralized spatial computing independently of changing node architectures and software, sensor technologies, batteries and energy harvesting, routing protocols, and application requirements.

Importantly, the formal foundations are the basis for all subsequent chapters in the book. Not all these structures are required at all times. Indeed, the models are designed to be deployed in stages, supporting increasingly sophisticated design and analysis of decentralized spatial algorithms. In some decentralized spatial information systems, only the extended spatial model (and by implication the neighborhood-based model) are required. But, based on the extended spatial model, any combination of the remaining two core models can be "plugged in" depending on the specific application requirements.

It is worth highlighting that there are several features our basic model has not yet addressed, including:

- *Node heterogeneity*, where nodes possessing different spatial and environmental sensing capabilities are combined in the same network. Homogeneous networks are assumed in Parts I and II of this book, with an exploration of the important issues connected with node heterogeneity discussed in Chapter 9.
- *Three-dimensional networks*, such as might be deployed for monitoring marine environments, vegetation structure, or the atmosphere. Although it is straightforward to include 3D coordinate locations in the positioning of nodes, many of the later structures in this chapter (such as planar graphs) rely on a two-dimensional representation of geographic space. Extending algorithms to monitor environments in three spatial dimensions (for example, additionally incorporating information about the vertical profile of changes in sea temperature and salinity across a coral reef) remains an important challenge for future work.
- *Asynchronous networks*, where different sensors cannot be relied upon to sample at the same time or frequency. Extended models of spatiotemporal networks that can allow for asynchronous sensing are discussed in Chapter 9.

Armed with our formal foundations, we are almost ready to start exploring the design and analysis of decentralized spatial algorithms, capable of processing spatial information with no centralized control. However, before that, we must first turn our attention to the computational characteristics of decentralized spatial computing. These algorithmic foundations are the topic of the next chapter.

Review questions

2.1 Under a certain condition, it will always be the case that for a communication graph $G = (V, E)$, $v \in nbr(v')$ assuming $v' \in nbr(v)$. What is this condition? (Hint: see §2.2).

2.2 Table 2.2 provides five examples of common methods for specifying location in geosensor networks. What other examples of different methods for specifying relative and absolute location information can you think of?

2.3 Using a Web search, try to find different technologies for localization that are capable of generating the information required for the different methods for specifying location in Table 2.2 and in your answer to question 2.2 above. (Hint: to start you off, GPS is one example of a technology for generating coordinate location—but there are others!)

2.4 In §2.3, the different methods for specifying location are categorized into two types: absolute and relative. However, one might also classify location information as quantitative (concerned with the *measurable* aspects of space) and qualitative (referring to discrete and meaningful *categories* or "qualities"). Reclassify the different methods for specifying location in Table 2.2 and in your answer to question 2.2 above into quantitative and qualitative location information.

2.5 Node mobility can be seen as a special case of node volatility. Explain in your own words why this is so. (Hint: see §2.4).

2.6 The Yao graph is another graph structure. In a Yao graph, the space around each node is partitioned into sectors of fixed opening angle (e.g., $60°$). Each node is then connected to its the unique nearest neighbor (if any) in each of these sectors [120]. Is the Yao graph a spanning subgraph of the UDG? Check your answer by sketching a Yao graph for the nodes in Fig. 2.6.

2.7 We have stated that the relative neighborhood graph is a (spanning) subgraph of the Gabriel graph. Construct a proof (or find and understand an existing proof) that this must always be the case (i.e., that the RNG is necessarily a subgraph of the GG).

Algorithmic Foundations

3

Summary: This chapter presents a technique for designing decentralized spatial computing algorithms, based on the combination and extension of some conventional algorithm design tools. First, the chapter sets out a language for specifying decentralized spatial algorithms, based on the specification of local protocols for individual node behavior, rather than on global system behavior. Next, the approach defines a structured analysis technique for understanding and identifying any failures in the global system behavior that emerge from the local protocols. The technique is iterative, successively identifying and then correcting faults, refining the algorithm until it performs as expected.

CHAPTER 2 armed us with a mechanism for precisely specifying *what* the important components of a decentralized spatial information system are. Moving up a level, this chapter focuses on specifying *how* a decentralized spatial system captures, transforms, and processes information.

Specifying how a decentralized spatial information system works means designing *algorithms*: precise specifications of computational procedures for solving a problem. However, decentralized spatial algorithm design is rather different from centralized spatial algorithm design. In both decentralized and centralized algorithms the *problems* to be solved are typically the same. They are usually specified at the *global* level, generating new information about the state of the system as a whole (e.g., the areas of or topological relationships between spatial regions, the identification of coordinated movement patterns amongst mobile entities, the detection of salient changes amongst a set of spatial fields). In a centralized spatial algorithm, the procedure to be specified is assumed to have *global* access to all the data necessary to solve this problem. However, in a decentralized spatial algorithm, the computational procedures are only specified at the level of *individual* system components, each of which only has access to *local* information in its immediate geographic

M. Duckham, *Decentralized Spatial Computing*, DOI 10.1007/978-3-642-30853-6_3,
© Springer-Verlag Berlin Heidelberg 2013

neighborhood. Only through the interaction between individual components does the solution to the problem emerge. This characteristic, going from local specifications to global solutions, is the central challenge of decentralized spatial algorithm design (and indeed the central theme of this book).

The decentralized spatial algorithm design process used in this book iterates over two distinct phases (Fig. 3.1). First, the *specification phase* sets out in a structured way the computational procedure and interactions each individual system component will engage in. Second, the *analysis phase* critically examines the specification, identifying problems and limitations, faults and weaknesses. Depending on the results of the analysis phase, it may be necessary to revise the specification for further analysis. The phases iterate until no further changes are needed, and we have confidence that the algorithm adequately solves the problem.

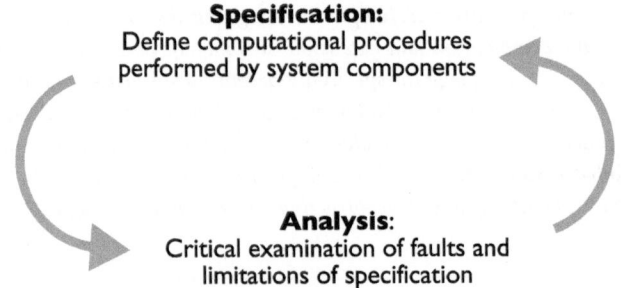

Specification:
Define computational procedures
performed by system components

Analysis:
Critical examination of faults and
limitations of specification

Fig. 3.1. Specification and analysis in decentralized spatial algorithm design

3.1 Algorithm Specification

This section introduces the algorithm specification style used throughout the remainder of this book. The style adopted is based on an established technique for decentralized algorithm design developed by Nicola Santoro [99] (who also wrote the foreword to this book). In addition to being well used and understood in the distributed systems community, the technique is relatively simple and intuitive. But why did we choose this specific style? What were the alternatives?

Conventional centralized algorithm specification techniques are not well suited to designing decentralized spatial algorithms. Because these techniques focus on the global system state, in a decentralized context it is difficult to analyze the interactions between nodes and represent the local knowledge of each individual node. In turn, this makes detecting and correcting faults in an algorithm much harder.

There are, however, a number of existing formal tools for modeling and analyzing *decentralized* systems that one might use for algorithm design. An influential formal model that deals explicitly with the interactions between different processes is Robin Milner's CCS (calculus of communicating systems), and its more recent relations, the pi-calculus and bigraphs [81–83]. These are powerful formalisms that are helping us address fundamental problems in geographic information science (e.g., [116]), but are relatively complex for non-mathematicians to use confidently in more applied domains such as algorithm design. Similarly, related models such as IOA (input-output automata, [77]) have been applied to decentralized spatial algorithms (e.g., [36]), but at the cost of high complexity. So, while these alternative models have the advantage of more formal rigor, their substantial additional complexity makes them much harder to learn and use. Hence, the less formal, more intuitive techniques from Nicola Santoro [99] were chosen for this book as a practical compromise between complexity and rigor.

3.1.1 Components of Decentralized Spatial Algorithms

There are four main components of all the decentralized algorithms explored in this book:

1. *restrictions* on the environment in which the algorithm operates;
2. *events* that occur at nodes, such as the receipt of a message from a neighbor;
3. *actions* that a node can perform in response to the different events that occur; and
4. *states* of a node, which allow nodes to retain knowledge of previous interactions.

Restrictions

The *restrictions* for any decentralized spatial algorithm encompass the assumptions that are made about the environment in which the algorithm will be operating. These restrictions relate closely to the five key components of our formal foundation, introduced in the previous chapter (see Fig. 2.15), specifically:

- *Non-spatial restrictions* on what data can be sensed by nodes;
- *Spatial restrictions* on whether nodes are assumed to be able to sense their absolute or relative location, neighborhood, distance, direction, cyclic ordering with respect to nearby nodes (see §2.3);
- *Temporal restrictions* on whether nodes are assumed to be static, volatile, or mobile, and on whether the sensed environment changes over time;
- *Uncertainty restrictions* on how reliable communication, sensing, positioning, and computation are assumed to be; and

- *Network restrictions* on whether the graph is planar or non-planar, on whether UDG, RNG, GG, or DT are assumed for the communication graph, and on what other network connectivity or topology restrictions exist.

Using fewer restrictions is desirable from an application perspective, as this tends to lead to more flexible algorithms with fewer assumptions about the environment in which the algorithm will be operating. However, using more restrictions is desirable from a design perspective, as it tends to lead to simpler algorithms that are easier to specify and analyze. One of the marks of a well-designed algorithm is that it can strike a good balance between being simple while at the same time as requiring relatively few, or at least realistic, restrictions.

As a concrete example, several spatial algorithms can be greatly simplified by restricting the algorithm to a maximal plane graph as the communication network (since there are no "holes" in the network). However, as we have already seen, for an arbitrary set of nodes and unit distance, the UDG may in general have no maximal plane subgraph (see §2.6.2). Consequently, while using such a restriction will simplify the design, it will also limit the practical usefulness of any algorithm that relies on it.

Events

Modeling the interaction between nodes in a decentralized system is fundamental to decentralized spatial algorithm design. As already discussed, interaction between system elements is a distinguishing characteristic of decentralized spatial algorithms, when compared with centralized spatial algorithms. Interaction is modeled using *events*. Nodes in a decentralized spatial system are *reactive*, responding to three different types of events that can occur at a node (cf. [99]):

1. the *receipt* of a message, sent by a neighboring node;
2. a *triggered* event, such as a scheduled alarm or periodic sensor reading; or
3. a *spontaneous* impulse, external to the system.

Using only these three types of system events, we can construct a remarkable range of different algorithms and behaviors (including all the algorithms in this book).

Actions

Actions are procedures that specify how a node will respond to a particular event. Actions must be "atomic" procedures, in the sense that they must be finite sequences of operations that must terminate within a finite amount of time (e.g., without infinite loops), and must execute without interruption (i.e., without needing to wait for, or being altered by the arrival of further events). Requiring actions to be atomic in this way substantially simplifies the process of algorithm design.

States

For any nontrivial decentralized spatial algorithm, nodes usually need to retain some knowledge of previous interactions in order to respond correctly to subsequent events. *States* are used to maintain this knowledge, allowing nodes to perform different actions in response to the same event, depending on previous interactions. An algorithm may specify any finite number of states. Every node must be in exactly one state at any time. An algorithm must therefore specify the starting states of all the nodes with which the network is initialized.

The action a node takes in response to a particular event will depend on that node's state. For every state and event, the algorithm must define a unique action, specifying what a node in a particular state will do in response to a particular event. Sometimes, to be concise, one or more events (and by implication actions) for a particular state are omitted. In these cases, the action is assumed to be empty (i.e., a node in that state does nothing in response to the event). Actions often cause a node to *transition* to a new state, preparing it for new responses to later events.

A convenient way of graphically summarizing states and transitions is by using a *state diagram*. A state diagram depicts the different node states as circles, with arrows connecting states where a transition between two states is possible. In some cases it is also convenient to annotate transitions with information about the events and/or conditions associated with the transition.

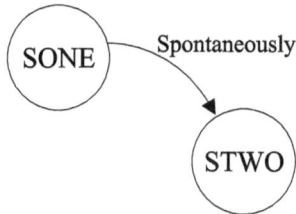

Fig. 3.2. Simple state diagram with two states and one transition (cf. Fig. 3.3)

Figure 3.2 illustrates the syntax of state diagrams, showing two states (SONE, STWO) and one transition between those two states (from SONE to STWO). The state diagram can be formally represented as a directed graph, called a *state transition system*. The state transition system comprises a set of states (e.g., {SONE, STWO}, as in Fig. 3.2) and transitions between its members (e.g., {(SONE, STWO)}). Using state diagrams to depict state transition systems provides a convenient and intuitive graphical syntax for summarizing the high-level behavior of an algorithm.

3.1.2 Protocols

In this book, decentralized algorithms are specified through the definition of *protocols*. The terms "algorithm" and "protocol" are near-synonyms in the context of distributed systems. However, while "algorithm" emphasizes the sequence of actions required by the global system to complete some task, "protocol" emphasizes the rules that govern interactions between individual system components. Given the importance of interaction in decentralized computing, it should be no surprise that protocols are at the core of our decentralized algorithm specifications.

Figure 3.3 illustrates the outline structure used for all the protocols in this book, after [99]. The structure begins with a "header," which specifies the global context in which the algorithm executes. First comes a list of the restrictions, in addition to the state transition system (State Trans. Sys.) and the initial states for all nodes.

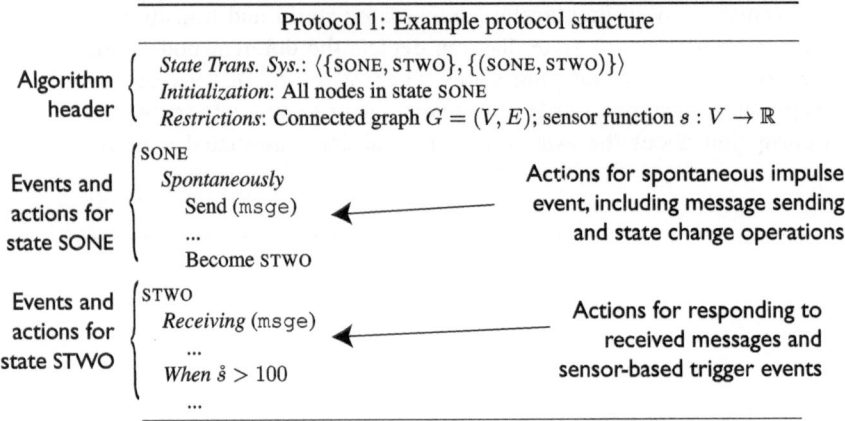

Fig. 3.3. Restrictions, states, events, and actions in decentralized spatial algorithms

The protocol proper then specifies what actions a node takes in response to different events and in different states. The protocol itself is structured hierarchically into different blocks. Within this structure the details of the actions take a back seat to the states and events. Top-level blocks specify the different states (as defined in the state transition system). Any actions and events within the block for a state apply only to nodes in that state. We adopt the convention throughout this book that states are named using four-letter abbreviations, in small capital letters.

At the next level come one or more events. The event must be identified using one of three keywords (corresponding to the three different types of events in §3.1.1): *Receiving*, for receiving a message; *Spontaneously*, for a spontaneous impulse; and *When*, for a scheduled or triggered event. The three event types are always typeset in italics.

At the lowest level, protocols take on a more conventional algorithmic form, specifying the sequence of atomic operations required to complete the action. These operations can include:

- processing local information about the node, such as location, sensed data, or neighbors;
- processing, storing, or accessing local information about other nodes in the network, communicated in previously received messages or computed and stored from earlier actions;
- changing the state of the node (using the **become** keyword);
- sending messages to all (using the **broadcast** keyword) or one (using the **send** keyword, termed *unicast*) neighboring node;
- other operations, including logical tests and control flow (e.g., "if-then-else" statements).

Thus, in an action, a node may perform most of the operations that one would expect in any (spatial) algorithm, with the exceptions that 1. the action must be atomic and cannot be interrupted by any other events; and 2. the node can only process local information or information it has previously received and processed or stored.

With respect to this second issue, protocols can only ever access partial knowledge of a global function, in the form of *local* functions (as already discussed in §2.5). Thus, even though the restrictions to an algorithm may define a global function, such as a sensor function $s : V \to D$, within the protocol a node is only allowed to access local versions of this function, such as $\mathring{s}$ (i.e., $s(\circ)$, where $\circ$ is the node currently under consideration and actually executing the protocol).

In summary, the most important feature of every protocol is that it specifies how each *individual* node will operate, but not how the system as a whole will operate. The structure of protocols (requiring atomic operations for actions; prohibiting access to global functions within actions; focusing first and foremost on the states and events that govern iterations between individual nodes) is designed to provide a clear line between what the rules governing individual system elements are and what the behavior of the system as a whole is.

Summary of protocol syntax

To recap, in addition to the formal structures defined in Chapter 2 the main syntactical conventions used in specifying algorithms are:

- States: Formatted in small capitals, e.g., INIT, DONE;
- Events: Formatted in italics, i.e., *Receiving*, *When*, or *Spontaneously*;
- Actions: A number of standard operations, formatted in bold, including **become** (change state), **send** (unicast to a specified subset of neighbors), **broadcast** (send to all network neighbors), **one-of** (arbitrarily select one element from a list or set), **set** (update the value of a local variable defined in the header, available to all actions of a node), and **let** (create and update the value of a temporary variable, not defined in the header and only available within the scope of a particular action); and
- String literals: Words and names typeset in a typewriter font, in particular the names of messages, such as msge.

3.1.3 Example: Specifying the "Crowd!" Algorithm

On p. 4, we introduced an informal, decentralized algorithm for determining whether a group of people constitutes a crowd. The algorithm operated by passing a piece of paper around to crowd members, each individual adding a tally mark as the message progressed.

The first step in more precisely specifying this algorithm is to identify the restrictions, events, and actions required. Starting with restrictions: the crowd membership must remain static (no crowd members leave or join). The crowd must also be "connected" in the sense that there must exist some route (path) for passing a message, hand to hand, through the crowd such that a message sent from any crowd member can potentially (eventually) be received by any other crowd member. Finally, message passing must be reliable, in the sense that no one drops, loses, or spoils the piece of paper.

Turning to the protocol, there are two events that are required by our algorithm. First, someone has to start the process, by placing a tally mark on a piece of paper and passing it on. This is an example of a spontaneous impulse, an event that happens to initiate the computation. Second, crowd members must respond to being passed a piece of paper; thus, receiving a message is another event.

In order to respond to events in the correct way, we define three states for nodes: INIT for the node that starts the process by sending the first message; IDLE for nodes that have not yet received any message; and DONE for nodes that have previously received a message. There are then also three actions. The first action is spontaneously to send the first message. This action is performed by a node in the INIT state responding to a spontaneous impulse. After sending the message, this node then transitions into a visited DONE state. The second action is performed by any node in the IDLE state when it receives the message. This action specifies a node must add its tally mark to the piece of paper, and check whether the tally is greater than 1,000 (the size of our crowd). If it is, the node shouts "Crowd!"; otherwise the message is given to an arbitrarily chosen neighbor. The third and final action, performed by any node in a DONE state when it receives the message, simply forwards the message on in search of more tally marks.

The algorithm can be more concisely presented as in Protocol 3.1. The structure for Protocol 3.1 is exactly as discussed in the previous section (cf. Fig. 3.3). The algorithm starts with header information about the restrictions (static, connected graph); the allowed states (INIT, IDLE, DONE) and state transitions (only transitions from INIT or IDLE to DONE are defined); and the nodes' initial states (all nodes must be in state IDLE, except for one designated node that is in the INIT state); Then for each state, Protocol 3.1 specifies the actions that correspond to the different events for that state. Each action must correspond to a unique event/state pair. Where events are missing for a state (e.g., no *Receiving* event for the INIT state), actions for event/state pairs are empty (i.e., no operations are performed). (In fact, in the case of our algorithm, as long as the system is correctly initialized with just one node in state INIT, it is not possible for a node in the INIT state to receive a message.) The state diagram for Protocol 3.1 is shown in Fig. 3.4.

Protocol 3.1. Finding the crowd (cf. p. 5)

Restrictions: Connected graph $G = (V, E)$, node neighborhood $nbr : V \rightarrow 2^V$
State Trans. Sys.: $(\{\text{INIT}, \text{IDLE}, \text{DONE}\}, \{(\text{INIT}, \text{DONE}), (\text{IDLE}, \text{DONE})\})$
Initialization: One node in state INIT, all other nodes in state IDLE

INIT
  Spontaneously   #*Spontaneous impulse event*
    send (msge, 1) to **one-of** $n\mathring{b}r$ #*Pass initial tally to neighbor*
    become DONE   #*Transition to visited state*

IDLE
  Receiving (msge, n)   #*Receiving message event*
    if $n \geq 999$ **then**   #*Check if this will be 1,000th tally*
      shout "Crowd!"   #*Crowd found*
    else
      send (msge, $n + 1$) to **one-of** $n\mathring{b}r$ #*Pass message to neighbor*
      become DONE   #*Transition to* DONE *state*

DONE
  Receiving (msge, n)   #*Receiving message event*
    send (msge, n) to **one-of** $n\mathring{b}r$ #*Pass message to neighbor*

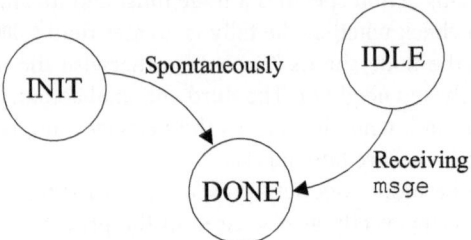

Fig. 3.4. State diagram for Protocol 3.1

3.2 Algorithm Analysis

Despite out having specified an algorithm using the tools presented in the previous section, our work is not yet done! Algorithm design is typically an iterative process, whereby algorithms are gradually refined, by our step-by-step eliminating errors and better understanding limitations. An analysis of the algorithm must provide the structure for this refinement process.

Analysis is especially important in decentralized spatial algorithms because of the need to specify *local* procedures to solve *global* problems. For example, although we have proposed a protocol that specifies how *individuals* will act and collaborate (Protocol 3.1), it is not immediately clear exactly how (or even if) this protocol will solve the problem of detecting a crowd. Analysis aims to support improved understanding and progressive refinement of the protocol, ultimately leading to an algorithm in which we can have more confidence, and which better addresses the problem at hand. In this book we use three main tools to analyze a protocol:

- Adversarial analysis, which aims to uncover problems and understand limitations of the algorithm through critical examination of the "worst cases" facing the algorithm;
- Computational analysis, which describes the computational characteristics of the algorithm, with particular emphasis on the computational efficiency; and
- Sequence diagrams, which provide insights into the dynamic interactions between nodes and the emergent behavior of the network.

3.2.1 Adversarial Analysis

Adversarial analysis is the most fundamental technique used to analyze decentralized spatial algorithms. In an adversarial analysis, the designer turns from building to destroying the algorithm. The designer adopts the role of an adversary, whose objective is to find as many scenarios as possible where the algorithm fails in some way, including generating incorrect results, executing inefficiently, failing to terminate (executing indefinitely without generating the desired result), and deadlock (reaching a non-terminal state where no further actions can be performed). The primary ways an adversary can achieve this are by:

- selecting unfavorable initialization scenarios where the protocol fails; and
- identifying unfavorable interaction sequences for groups of nodes executing the protocol which lead the algorithm to fail.

"Unfavorable" in this context means "allowed by the algorithm specification, but not well behaved or otherwise undesirable." More specifically, in the context of decentralized spatial algorithms, the common ways in which an adversary may cause an algorithm to fail include:

- choosing unfavorable network topologies;
- choosing unfavorable spatial distributions of nodes;
- choosing unfavorable sequences or scheduling of events or interactions between nodes;
- choosing unfavorable spatial or spatiotemporal configurations of the geographic environment;
- introducing uncertainty into nodes' location or environmental sensors; and
- introducing unreliability into the communication between nodes.

Importantly, the designer need only find *one* scenario where the algorithm fails, no matter how unlikely, to successfully "beat" the algorithm, necessitating another iteration of the design process. The ways in which the designer can remedy the failure, and proceed with the next design iteration, are discussed in §3.3. However, in this section we focus purely on the adversarial process: finding faults.

Example adversarial analysis

Adopting the adversary's position, there are several ways in which our "Crowd!" algorithm might fail.

- Because a node randomly selects a neighbor to forward the tally message to, there exists the possibility that a message will be caught in an indefinite loop. In such a scenario, a message could be forever passed between nodes in the DONE state, which have previously received the message. If this occurs, the tally will then never be incremented, and so the algorithm will never terminate. It does not matter how unlikely this is to occur—while it remains a *possibility*, the adversary is correct to identify this as a potential algorithm failure.

- Similarly, if the number of nodes in the graph is smaller than the size of the crowd being sought, the algorithm will always fail to terminate, since the tally can never exceed to the total number of nodes in the graph. Thus, the algorithm may not distinguish between a negative answer (i.e., there is no crowd) and a positive answer that is simply taking a long time to compute.

- Even in cases where the cardinality of the set of nodes is greater than the size of the crowd, and the algorithm does terminate, it may require an arbitrarily large number of messages for termination to occur. As a result, the algorithm is not expected to be very efficient. The issue of efficiency is critical in resource-constrained decentralized spatial computing environments such as geosensor networks, and is explored in more detail in §3.2.2.

- If the communication in the network is unreliable, and a message is lost (for example, a node fails to correctly receive the tally message), the algorithm will fail. Issues connected with communication reliability are amongst those explored further in Chapter 8.

3.2.2 Computational Analysis

The next tool in our analysis arsenal is computational analysis. Computational analysis aims to understand how efficient the algorithm is in its use of limited computational resources. The basic tools of computational analysis will be familiar to students of computer science (indeed, such students may want to skim the following section on "Computational complexity"). However, as we shall see, the application of computational analysis to decentralized spatial computing is a little different from that to conventional centralized spatial algorithms.

Computational complexity

The essence of computational analysis is to understand how the resources consumed by an algorithm (such as information processing, data storage, or data communication resources) are expected to vary as a function of the size of the input. More efficient algorithms are judged to be those that are more *scalable*: where the demand for resources increases less rapidly as a function of the input size. For geosensor networks, the most important input is the size of the network itself, in terms of the number of nodes in the network. A particularly scalable algorithm will consume only marginally more computational resources as we add more and more sensor nodes to a network.

We could assess the efficiency of an algorithm by implementing and testing it, measuring how quickly it executes, and how many resources it consumes. Such *empirical* evaluations provide important insights into an algorithm (discussed in much greater detail in Part III of this book), but are also subject to limitations. In particular, it is difficult to control the effects of individual programming skills and the range of possible environments an algorithm will face using empirical evaluations. Perhaps an algorithm performs badly just because it happened to be inefficiently implemented. Or perhaps it performs well because it happened to be tested in a range of favorable environments. Further, in the case of geosensor networks, implementing an algorithm and testing it in the field are typically complex and time-consuming activities. Ideally, we want to be able to have high confidence about the expected performance of an algorithm before investing the time and resources required for a field deployment, or even before simulating it. By contrast, an analysis of computational complexity aims to provide a pen-and-paper procedure for investigating and comparing the theoretical efficiency of different algorithms in an abstract way, independent of any specific implementation, system architecture, or technology.

Figure 3.5 shows some typical examples of how the resources consumed by an algorithm can grow as a function of input size. These response curves can be ranked according to their asymptotic orders of growth (i.e., how the resources consumed compare for arbitrarily large input sizes). The most efficient algorithms have *constant* orders of growth. In these cases, the resources required by an algorithm are independent of the size of the input, n. Next come *logarithmic, linear,* and *linarithmic* orders, where the resources required grow as a function of $\log n$, n, and $n \log n$ respectively, where n is the input size. The resources required by *polynomial* and lastly *exponential* order algorithms scale as a function of n^k and k^n respectively, where k is some constant (independent of n).

It is worth emphasizing that when comparing algorithms it is the *asymptotic* scalability that matters, irrespective of the level of computational resources required for a *particular* input size. For example, constant order computational complexity algorithms are the most scalable, even though for the smallest input sizes the constant order growth curve in Fig. 3.5 requires *more* resources than other orders of growth.

Big-oh notation is used to concisely denote these (and other) orders of computational complexity. Specifically, the statement that "an algorithm is in $O(f(n))$" means that the computational resources required by that algorithm are bounded from above by some constant multiple of the function $f(n)$ for the inputs n. For example, if an algorithm has logarithmic computational complexity, we say the algorithm is in $O(\log n)$. This means that the resources consumed by the algorithm increase at most by $k \log n$ as inputs n increase. Big-oh notated orders are included on the response curves in Fig. 3.5.

It is sometimes also useful to be able to discuss the *minimum* rather than *maximum* computational complexity of an algorithm. Similarly to big-oh notation, big-omega notation, $\Omega(f(n))$, indicates that the computational resources required by an algorithm are bounded from *below* by some constant multiple of the function $f(n)$. So, if an algorithm is in $\Omega(n)$, it means that it

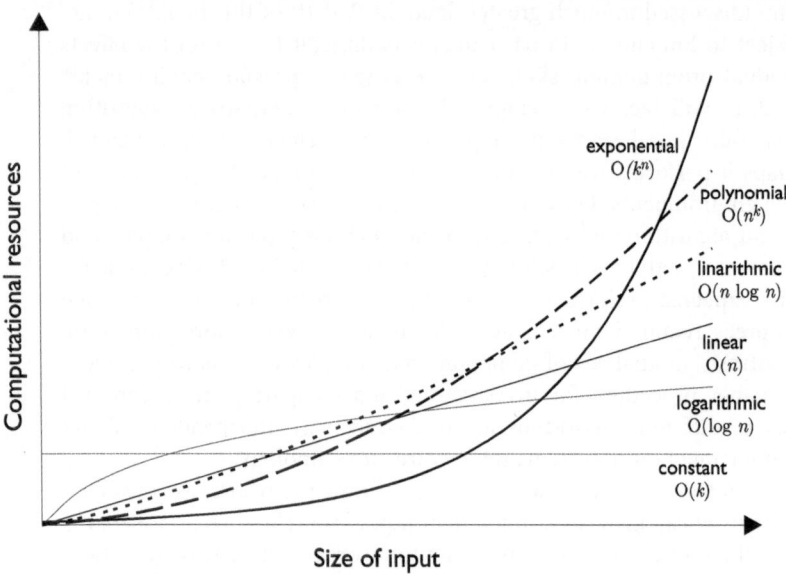

Fig. 3.5. Computational complexity: typical response curves for computational resources as a function of input data set size n

requires computational resources of size *at least* kn as inputs n increase. To complete the set, big-theta notation indicates that an algorithm is bounded from above *and* below by some order of growth. Thus, if an algorithm is in both $O(n^2)$ and $\Omega(n^2)$ we can more concisely write that it is in $\Theta(n^2)$.

Communication complexity

In centralized spatial algorithms, the most important computational resources are typically the amount of processing *time* required to complete an algorithm (termed *time complexity*) and the amount of storage *space* needed to maintain the data structures required for an algorithm (termed *space complexity*). Because of the limited processing and storage capacities of today's geosensor nodes, both these indicators are still important in decentralized spatial algorithms. However, in a decentralized spatial information system, by far the most important resource limitation is *communication complexity*: the amount of communication required by an algorithm as a function of network size. As we have already seen, in a geosensor network energy is the most scarce resource, and communication is the most critical operation in a node's energy budget.

Important measures of communication complexity include the numbers of messages *sent* and *received*. The number of messages sent and received are usually not equivalent, as a consequence of both broadcast (where one message sent from a node may be received by multiple neighbors) and more practical communication considerations (where, for example, in practice multiple messages sent at the same time may collide and never be received). We might also be interested in other measures of the communication complexity, such as the *length* (in numbers of bits) of messages sent and received, or the *latency* of messages (i.e., the amount of time required for a message to be delivered as a function of input size).

The specific details of how much energy is required to transmit or receive a certain number or length of messages vary substantially depending on the specific hardware and communication protocols used. In some hardware, the transceiver characteristics are such that receiving a message is almost as expensive as transmitting a message; in others, message receipt may be associated with lower energy budgets than transmission. Similarly, packet-based communication protocols may mean message length is not *directly* related to the energy required to transmit or receive a message. By abstracting away from these specific details, communication complexity analysis instead aims to capture the underlying communication resources consumed by an algorithm. This information can then be used to compare algorithms even as technologies evolve, and can even identify which technologies are best suited to a particular algorithm.

Table 3.1 compares the communication complexity of some basic decentralized algorithms, many of which we shall investigate in more detail in later chapters.

Computational complexity	Big-oh	Example algorithm
Constant	$O(k)$	Single broadcast: Number of sent messages required for a single broadcast "ping" message from a subset of k nodes from a network of fixed size.
Logarithmic	$O(\log n)$	Single multi-hop message in a tree: Number of messages sent and received in transmitting a single message from a source node to the sink (root) node in a routing tree.
Linear	$O(n)$	Boundary detection: Number of messages sent in order to detect boundaries, where neighboring nodes sense values on either side of some threshold (Protocol 4.14).
Linarithmic	$O(n \log n)$	Leader election in a ring: Number of messages sent and received in order to elect a single leader from amongst a set of nodes, with the network structures as a ring (Protocol 4.12).
Polynomial	$O(n^2)$	Random routing: Number of messages sent and received by an algorithm that randomly routes a unicast message around a planar network until every node has received the message at least once (e.g., Protocol 3.1, assuming planar communication network).

Table 3.1. Some decentralized operations and algorithms, and their communication complexities

One final point is worth noting. Normally in analysis of computational complexity, any constant factors are ignored (so, for example, $O(2n) = O(4n) = O(8n) = \cdots = O(n)$). While the order is the most important indicator of efficiency, in analyzing decentralized spatial algorithms the constant factors may also be important when comparing algorithms of the same order. For example, an algorithm that requires $2n$ messages to be sent is of the same order of scalability as one that requires $4n$ messages. However, because of the unique resource constraints in decentralized spatial information systems such as geosensor networks, the difference between an algorithm that requires $2n$ and one that requires $4n$ messages is still significant: all other things being equal, the former algorithm will operate for twice as long as the latter before the nodes in the network start to die. As a result, in some cases we will abuse the big-oh notation slightly, and distinguish between constant factors, for example, highlighting the difference in efficiency of algorithms that are in $O(2n)$ and $O(4n)$.

Example: Communication complexity of the "Crowd!" algorithm

Turning back to our "Crowd!" algorithm (Protocol 3.1), we can quickly see that it is potentially very inefficient in terms of messages sent. As already identified in the adversarial analysis in §3.2.1, because each node randomly selects a neighbor, there is always the possibility that two nodes happen to pass a message backwards and forwards in an infinite loop, resulting in an infinite number of messages sent. Even if we ignore this possibility, groupings comprising fewer than 1,000 people will also result in an infinite number of messages sent.

Leaving aside both these possibilities briefly, the total number of messages sent is related to a well-studied problem in mathematics, termed the *cover time* for a graph. The cover time for a graph is the *expected* number of steps taken by a random walk[1] that visits all nodes in the network. The number of steps in a random walk through the graph translates directly into the number of messages sent and received by our algorithm.

It is a well-known result of graph theory that the cover time for a graph is at best $O(n \log n)$ and at worst $O(n^3)$, where n is the number of nodes in the graph. The best cases relate to highly connected graphs (such as complete graphs—recall that a complete graph is a graph where every pair of vertices is connected by an edge). The worst cases relate to sparser graphs with fewer connections between nodes. As a result, the cover time in decentralized spatial computing applications is certain to be amongst the worst rather than the best cases (cf. §2.6.1).

Happily, things are not quite as bad as that! If we can ensure the network is planar (for example, using the Gabriel graph or relative neighborhood graph; see §2.6.2) then the worst case reduces to $O(n^2)$ [59]. Further, the

---

[1] "Random walk" is the mathematical term for the path that our message will take, where at each node the next step is randomly selected.

analysis above assumes that the size of the group of people operating the algorithm is equal to the size of the crowd we are looking for (e.g., we are looking amongst a group of 1,000 people for a crowd, defined to be 1,000 people or more). In such cases it is essential that the message visit *every* person in the crowd. In cases where the group of people is much larger than our definition of a crowd (for example, where we are looking for a crowd of 1,000 amongst a group of 10,000 people), the expected communication complexity is lower, since the algorithm can terminate without having to visit all n people in the group.

Unsurprisingly, this variation of the cover time for a graph is also well studied in graph theory, and is termed the *partial cover time* of a graph (the cover time for a proportion of nodes from a graph). It has been shown that in many cases the partial cover time for graphs with direct analogy to geosensor networks can be $O(n \log n)$ [8].

Summary: Best, expected, and worst case complexities

So, what can we learn from all this? The details of cover time are not so important to the central topics in this book (although they are admittedly an active research topic in the domain of wireless sensor networks more broadly). However, there are two "take home" messages highlighted by this computational analysis. First, the discussion provides an example of the many different factors that can affect the efficiency of a decentralized spatial algorithm. The exact communication complexity depends on the number of nodes in the graph, but also on the structure of the graph and even to an extent on the size of the crowd we are looking for relative to the total number of nodes in the graph.

Second, the analysis demonstrates the importance of the *best, expected,* and *worst* case computational complexities. Computer scientists are by nature a pessimistic bunch, and are typically most interested in the *worst* case complexity. Knowing the worst case provides an lower bound for the performance of an algorithm, and a guarantee that the algorithm in practice will perform at least as well if not better than that lower bound (so perhaps that means computer scientists are in fact an *optimistic* bunch). However, the average or expected case computational complexity is also an important guide to the efficiency of an algorithm. This is especially so in the context of decentralized spatial algorithms where (as we shall see in Part III of this book) uncertainty is an endemic issue, and algorithms that generate *approximate* answers, as well as those that generate *exact* answers, can be very useful. Finally, the best case complexity is the least interesting of the different measures (not least since Murphy's Law predicts it will never eventuate). However, in some cases it can still provide useful information about the algorithm, adding to our understanding of what is possible with our algorithm and what is not.

Load balance

The discussion of computational complexity above concerns itself solely with the question of how the communication complexity scales for the network as a whole. While this provides an important guide to the *overall* resource consumption of an algorithm, it can obscure important details about individual node resource usage, and about how resource usage is distributed amongst nodes in the network (termed *load balance*).

Load balance is an especially important consideration in geosensor networks. Where nodes rely on battery power, and cannot recharge or harvest energy from the environment (see §1.2.2), once a node's energy resources are depleted it will die and no longer be able to operate as part of the network. In turn, this will lead to increased load on other remaining nodes in the network, a decrease in the spatial granularity of the network, and depending on the network structure may even cause the network to become disconnected and ultimately the algorithm to fail. Balancing the computational load evenly across the network can help to ensure that a homogeneous network continues to operate as long as possible within the overall energy constraints imposed on its nodes. Conversely, if the load is not balanced, it may be possible to identify a smaller subset of nodes that will bear the greater load, and augment those with additional resources (such as extra battery packs), again helping the network as a whole to operate for longer.

As an example, consider again the "Crowd!" algorithm, where the network is structured as a planar graph. As we have seen, in this case the algorithm has an overall communication complexity of $O(n^2)$. However, the total number of messages sent/received by an individual node is expected to be much lower. Because the message path traces a random walk through the graph, ignoring edge effects, every node is equally likely to be located at a self-intersection in the path. Thus, the number of messages sent/received by an *individual* node is expected to have a communication complexity of $O(n)$. This means the algorithm has an optimal expected load balance given the overall complexity: the workload of the algorithm is spread equally amongst all nodes in the network.

By contrast, consider an adaptation of the "Crowd!" protocol, as set out in Protocol 3.2. This revised protocol uses no decentralized processing at all, and is restricted to scenarios where the communications network is structured as a rooted tree. Each node sends a single message to its *parent*, the next node along the unique path to the root node (recall from §2.6.3, that a tree is a graph with exactly one path between every pair of nodes; and a rooted tree is a tree with a designated root node). Knowledge of each node's parent node is represented using the function $parent : V \rightarrow V \cup \{\varnothing\}$ (i.e., $parent(v) = v'$ indicates that the parent for node v is v'; $parent(v) = \varnothing$ indicates that node v is the root). Non-root nodes forward any received messages unaltered to their parent, the next node on the path to the root. The root node simply counts all the messages it receives, and shouts "Crowd!" when the 999th message is received (i.e., 1,000 nodes including the root).

Protocol 3.2. Finding the crowd II

Restrictions: Rooted tree $G = (V, E)$, *parent* $: V \to V \cup \{\varnothing\}$
State Trans. Sys.: $(\{\text{ROOT}, \text{INIT}, \text{DONE}\}, \{(\text{INIT}, \text{DONE})\})$
Initialization: One node in state ROOT, all other nodes in state INIT
Local data: Integer n count of messages received, initialized $n := 0$

INIT

   Spontaneously   #*Spontaneous impulse event*
      become DONE   #*Transition to* DONE *state*
      send (msge) to *parent*   #*Pass initial message to parent*
   Receiving (msge)   #*Receiving message event*
      send (msge) to *parent*   #*Forward message to unique parent*

DONE

   Receiving (msge)   #*Receiving message event*
      send (msge) to *parent*   #*Forward message to parent*

ROOT

   Receiving (msge)   #*Receiving message event*
      set $n := n + 1$ #*Store total number of messages received*
      if $n = 999$ **then**   #*Check if 1,000 messages received*
         shout "Crowd!"   #*Crowd found*

The communication complexity of Protocol 3.2 is $O(\sum_{v \in V} depth(v))$, where $depth(v)$ is the length of the path from v to the root. In the best case, where the length of the longest path in the tree (termed the *diameter*) is 2, this translates into an overall communication complexity of $O(n)$. In the worst case, where the diameter of the tree is $|V| - 1$, this results in a communication complexity of $O(n^2)$. Thus, depending on the structure of the tree, the communication complexity of Protocol 3.2 is at least as good as or better than the random walk algorithm in Protocol 3.1 ($O(n^2)$ for the expected case assuming a planar graph).

Despite this, up to $O(n)$ messages may be received by a single node (the root) in the worst case using Protocol 3.2. This is equivalent to the maximum expected number of messages for any node using Protocol 3.1, as discussed above. Thus, Protocol 3.2 exhibits greater imbalance of load than the Protocol 3.1, where all nodes share the load equally. In homogeneous networks where all nodes have access to the same resources, such load imbalance is important information as it has direct implications on network longevity, since the network becomes disconnected as nodes die from resource depletion.

Load balance and overall communication complexity

Although the load balance and the overall communication complexity contain different and important information, the two measures of efficiency are related. Clearly, at least in the worst case scenario, the load balance can never exceed the overall communication complexity (i.e., the total communication required for the entire network cannot be less than the communication required at a single node). Further, optimal load balance is achieved where every node takes an equal share of the overall communication. Thus, communication complexity can never be more than n times the load balance (e.g., if the worst case load balance for an algorithm is in $O(n)$, then the overall communication complexity cannot exceed $O(n^2)$).

3.2.3 Sequence Diagrams

One final tool used for understanding the dynamic behavior of a decentralized spatial algorithm is a sequence diagram. A sequence diagram provides a specific example of how nodes using a protocol can interact, highlighting the sequence of events and state transitions that occur. The sequence diagrams used in this book are based on standard graphical syntax used in system development, although adapted slightly to the requirements of decentralized spatial algorithm analysis.

A sequence diagram is composed of two or more vertical "lifelines," each of which is labeled with its corresponding node. Each lifeline is annotated with a sequence of events and state transitions that occur at that node, using the convention that preceding events and states are placed above subsequent ones. The initial state for each node is the first annotation on every lifeline. Labeled horizontal arrows between lifelines are used to depict messages sent and received. Triggered or spontaneous events are depicted using a labeled involutory arrow (an arrow that turns in on its own lifeline).

Figure 3.6 shows an example sequence diagram corresponding to an execution of Protocol 3.1. The network has six nodes, with node v_4 the only node in the INIT state (all other nodes IDLE, as per Protocol 3.1). The resulting spontaneous event initiates a chain of subsequent events, involving in turn most of the other nodes in the network.

Sequence diagrams can provide important insights into the dynamic behavior associated with a protocol, enabling the designer to explore the precise sequence and interactions between events. However, basic sequence diagrams, such as that in Fig. 3.6, can only represent the most basic of network topologies, for example, where immediate horizontal neighbors in the diagram are assumed to be network neighbors. As a result, sequence diagrams can provide only limited assistance in understanding the more complex spatial and spatiotemporal behaviors of protocols. It is also possible to depict both space and time together using a *spatial sequence diagram*. Figure 3.7 gives an example of a spatial sequence diagram, again for an execution of the

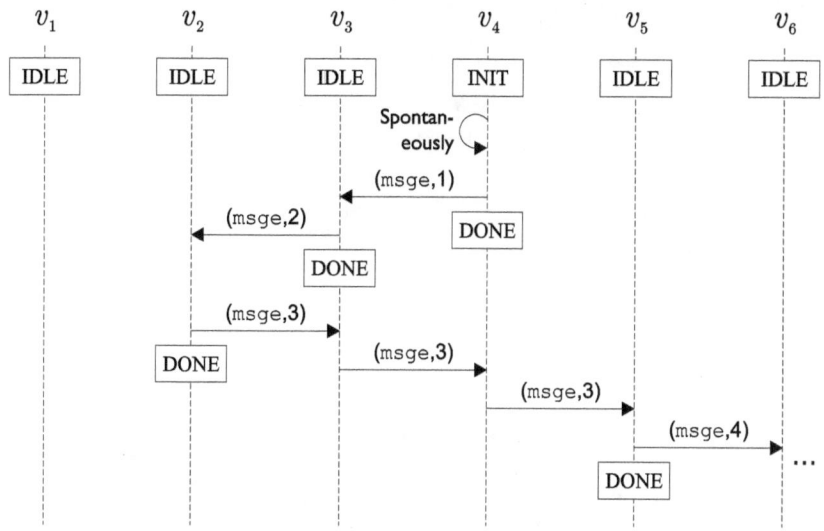

Fig. 3.6. Sequence diagram, showing example events and transitions for six nodes using the "Crowd!" protocol (cf. Protocol 3.1)

"Crowd!" protocol on p. 65. The limitations of drawing two spatial and one temporal dimension on two-dimensional paper mean that spatial sequence diagrams can quickly become cluttered and difficult to understand. Where basic sequence diagrams suppress spatial characteristics in the interest of increased ease of interpretation, spatial sequence diagrams depict complex spatiotemporal protocol behaviors, but at the cost of decreased clarity. Thus, in this book we rely primarily on basic sequence diagrams, using spatial sequence diagrams only when feasible and when a protocol demands the depiction of more complex spatiotemporal behaviors.

3.3 Design Iteration

If a thorough adversarial analysis reveals no problems with a protocol, then we can have high confidence that the protocol will operate as expected in practice, assuming the specified restrictions hold. Using our design technique it is never possible to be *certain* that a protocol will perform as expected; regardless of how thorough an adversarial analysis may have been, it is always possible that we may have missed some combination of unfavorable circumstances that will lead to algorithm failure. Formal methods for proving certain properties of decentralized algorithms do exist, but as already discussed these tools are mathematically more demanding and harder to apply to practical algorithm design problems (see §3.1). Thus, where adversarial

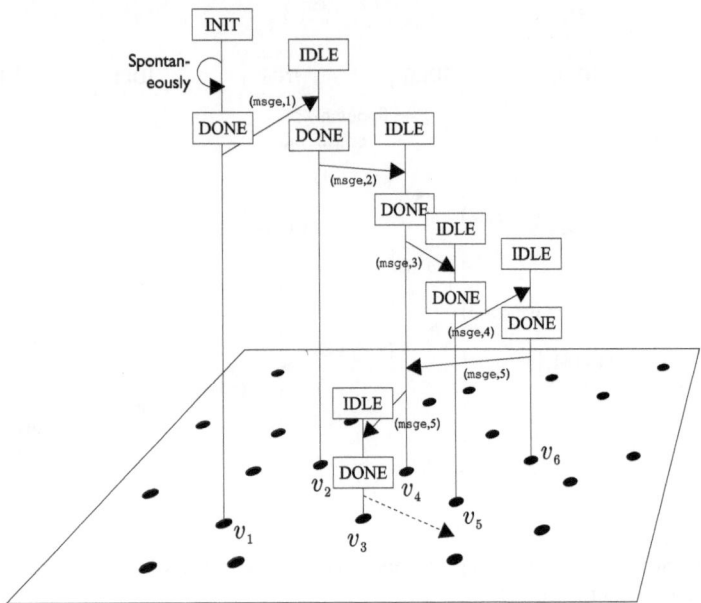

Fig. 3.7. Spatial sequence diagram, showing planar spatial configuration for nodes using the "Crowd!" protocol (cf. Protocol 3.1)

analysis results in no further problems, the next stage is to move to implementation, usually first in a simulation system and ultimately in a practical deployment (discussed further in Chapter 7).

However, at least for early iterations, it is to be expected in most cases that the adversarial analysis will reveal some problems that must be addressed (such as those identified for the "Crowd!" algorithm at the beginning of this chapter, in §3.2.1). In these cases, the design must be revised, adapting the specification and repeating the adversarial analysis with the new protocol. The next iteration may proceed following changes to the protocol and/or changes to the restrictions to resolve the problems identified.

For example, in light of the high communication complexity of the "Crowd!" algorithm, one possible adaptation is to store information about which neighbors a node has previously forwarded a message to. Then nodes could be instructed to only forward a message to each neighbor at most once. Protocol 3.3 provides this revised protocol, with the addition of a locally stored set of visited neighbors N. Nodes in the DONE state will only forward messages to unvisited neighbors, to whom that node has not previously communicated the tally. Once a node has forwarded messages to all its neighbors it transitions to state DEAD, and performs no further actions in response to any events.

Protocol 3.3. Finding the crowd III

Restrictions: Connected graph $G = (V, E)$, node neighborhood $nbr : V \to 2^V$
State Trans. Sys.: $(\{\text{INIT}, \text{IDLE}, \text{DONE}, \text{DEAD}\}, \{(\text{INIT}, \text{DONE}), (\text{IDLE}, \text{DONE}), (\text{DONE}, \text{DEAD})\})$
Initialization: One node in state INIT, all other nodes in state IDLE
Local data: Set N of visited neighbors, initialized $N := \overset{\circ}{nbr}$

INIT
   Spontaneously   *#Spontaneous impulse event*
      become DONE   *#Transition to visited state*
      let $v := $ **one-of** N
      send ($\mathtt{msge}$, 1) **to** v *#Pass initial tally to neighbor*
      set $N := N - \{v\}$

IDLE
   Receiving ($\mathtt{msge}$, n)   *#Receiving message event from node v*
      if $n \geq 999$ **then**   *#Check if this will be 1,000th tally*
         shout "Crowd!"   *#Crowd found*
      else
         become DONE   *#Transition to* DONE *state*
         let $v := $ **one-of** N
         send ($\mathtt{msge}$, $n + 1$) **to** v *#Pass message to unvisited neighbor*
         set $N := N - \{v\}$

DONE
   Receiving ($\mathtt{msge}$, n)   *#Receiving message event*
      let $v := $ **one-of** N
      send ($\mathtt{msge}$, n) **to** v *#Pass message to neighbor*
      set $N := N - \{v\}$
      if $N = \varnothing$ **then**
         become DEAD

This revised protocol represents a substantial improvement in terms of computational complexity, because the total number of messages sent will be in the worst case equal to the number of edges in the graph G (for a pair of nodes v_1, v_2 at most two messages will be sent along the edge $\{v_1, v_2\}$, one from v_1 to v_2, one from v_2 to v_1). Thus, the worst case overall communication complexity of the algorithm is $O(|V|^2)$ (since there are at most $|V|^2$ edges in a complete graph). Indeed, because of the relative sparseness of geosensor networks, the average case is likely to be much better. Additionally restricting the graph to be planar further reduces the communication complexity to $O(|V|)$, since the number of edges in a planar graph is linearly related to the number of vertices in the graph (by the Euler-Poincaré formula).

However, despite this apparent improvement, an adversarial analysis of the revised protocol in Protocol 3.3 reveals a fundamental flaw, illustrated by the sequence diagram in Fig. 3.8. Only forwarding to neighbors once can mean the algorithm fails when a message arrives at a node with no unvisited neighbors, since a node in the DEAD state will no longer forward messages.

Even if we were to further adapt the algorithm to ensure that messages were never passed to a DEAD node, it would still be possible for the network to become effectively disconnected when one or more DEAD nodes formed a group where all paths between the message and unvisited nodes pass through them.

Thus, having identified flaws in our revised algorithm, we would need to again iterate our design process, either testing more sophisticated protocols to address the flaws identified by the adversary, or abandoning this line of adaptation altogether. This example highlights the importance of retesting a revised algorithm, and completing the design iteration. It is never adequate to propose a revision to an algorithm without "closing the loop" and performing an adversarial analysis on the revision.

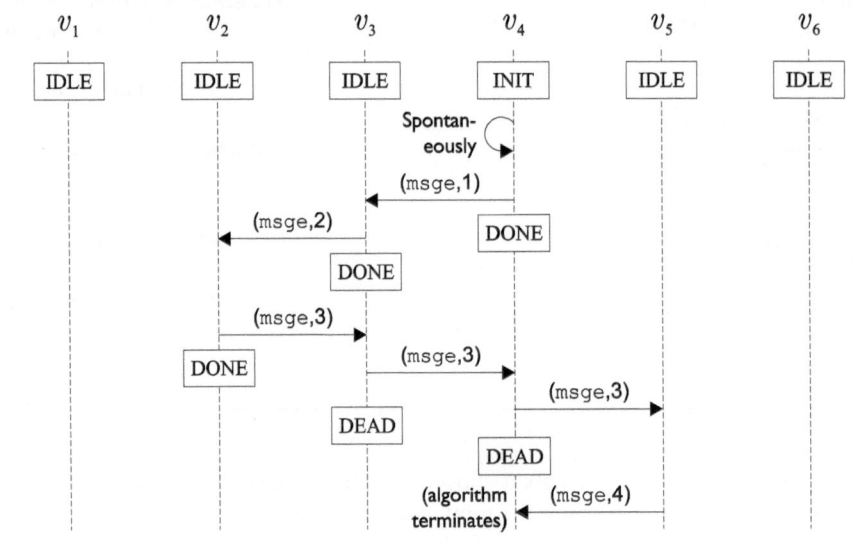

Fig. 3.8. Unfavorable sequence diagram for Protocol 3.3

3.4 Chapter in a Nutshell

Building on the formal model constructed in Chapter 2, this chapter has set out a computational approach to designing decentralized spatial algorithms. The approach is founded on the combination of established (non-spatial) models of distributed and decentralized computing, with the addition of some uniquely spatial tools, such as spatial sequence diagrams. The two stages of the technique, specification and analysis, provide the tools for designing and iteratively refining protocols for decentralized spatial algorithms.

The key feature of our technique is that it provides the mechanism for defining local protocols that exhibit desired global behaviors. Specifications are *local*, defined at the level of individual nodes; the analysis is both local and *global*, focusing on the overall properties and behaviors of the protocol executing amongst groups of interacting nodes. The challenge of moving from local to global is one of the central themes in this book.

Together, the ideas presented in Chapters 1–3 provide (almost!) everything we need to develop our own decentralized spatial algorithms. Thus, this chapter concludes Part I of this book. In Part II, the focus turns to applying the tools, models, and concepts developed in Part I to increasingly sophisticated spatial and spatiotemporal problems.

Review questions

3.1 What are the five classes of restriction that can be placed on any protocol?

3.2 You are shown a protocol where a *Receiving* action begins processing a message, but waits for a second message to be received before completing the processing. What, specifically, is the feature that makes this protocol malformed?

3.3 Given a protocol based on the state transition system ($\{\text{INIT}, \text{IDLE}, \text{ROOT}\}$, $\{(\text{ROOT}, \text{IDLE}), (\text{INIT}, \text{IDLE}), (\text{IDLE}, \text{ROOT})\}$), how many *distinct* (i.e., that transition to a different state) **become** statements would you expect to find in that protocol?

3.4 You are shown a protocol that contains only message receipt (*Receiving*) events, and no spontaneous (*Spontaneously*) or triggered (*When*) events. What does this protocol do?

3.5 A protocol is known to have overall communication complexity $O(\sqrt{n})$? What is the optimal load balance for this protocol?

Part II

Algorithms for Decentralized Spatial Computing

Part II of this book applies the concepts and techniques developed in Part I to the development of foundational algorithms for decentralized spatial computing. In three chapters, the book steps through a suite of increasingly sophisticated algorithms, from algorithms with minimal spatial information about their neighborhoods (Chapter 4); to algorithms with access to more detailed spatial information, such as direction, distance, or coordinate location (Chapter 5); to truly spatiotemporal algorithms, which monitor environments that are dynamic, even using networks that are volatile or mobile (Chapter 6).

Underlying the material in Part II of this book is the question: "What makes decentralized spatial computing different from centralized spatial computing?" Many of the decentralized spatial algorithms in Part II have direct centralized counterparts, and may already be well understood in the context of GISs and spatial databases. Representing a spatial region and computing its area, for example, are basic functions of any GIS or spatial database. Yet, such basic capabilities must be revisited in the context of decentralized spatial computing. Conversely, as the book builds towards spatiotemporal algorithms, for which GISs and spatial databases are notoriously ill equipped, we shall see examples of decentralized spatial algorithms for operations and queries that can challenge even today's well-established centralized spatial information systems.

Neighborhood-Based Algorithms

4

Summary: This chapter introduces the most basic type of decentralized spatial algorithm: neighborhood-based algorithms. In a neighborhood-based algorithm, nodes are assumed to have access only to minimal spatial information: the identities of their neighbors. Despite this simplicity, neighborhood-based algorithms can provide a range of important functionality, including the dissemination and routing of information, the construction of network topologies, and some basic qualitative spatial functionality, including operations on region boundaries.

NEIGHBORHOOD is the most basic type of spatial information generated by a decentralized spatial information system (see §2.3). A node's neighborhood is composed of all those nearby neighbors that can engage in direct (one-hop) communication with that node. This chapter introduces the most basic of decentralized spatial algorithms, termed *neighborhood-based* algorithms. Neighborhood-based algorithms do not use any spatial information other than information about a node's neighborhood. The "Crowd!" algorithm, used as a running example in Part I of this book, is one example of a neighborhood-based algorithm.

Neighborhood-based algorithms rely solely on the instantaneous state of a node and its immediate one-hop neighbors. Consequently, the restrictions listed in the header of a neighborhood-based algorithm typically include the minimal restrictions, such as the communication graph and its associated neighborhood function (*nbr*; see Fig. 2.2), the sensor function, and the identifier function. Despite this simplicity, there is a range of important and fundamental neighborhood-based algorithms.

This chapter begins by looking at examples of the simplest neighborhood-based algorithms: flooding and gossiping. Flooding and gossiping are widely used to disseminate information throughout a decentralized spatial information system, such as a geosensor network.

M. Duckham, *Decentralized Spatial Computing*, DOI 10.1007/978-3-642-30853-6_4,
© Springer-Verlag Berlin Heidelberg 2013

4.1 Flooding

When one node holds some important new information, it may often be desirable for the entire network to be updated with this information. There are many scenarios where information needs to be disseminated in this way. For example, suppose we wish to reprogram an active network to perform a new query or task (termed *retasking*). In this situation, it may be convenient to inform just *one* node of the new task, and then have that node initiate the dissemination of the new task to the entire network (often referred to as *injecting* a query into a network).

4.1.1 Information Diffusion: A First Pass

Flooding is a process that aims to disseminate information throughout the network, from one source node to every other node in the network. But how should we design an algorithm to achieve this? A first pass would be to require the node that holds the new information to tell all its neighbors; then require all the other nodes in turn to tell all *their* neighbors when they are told something. Protocol 4.1 provides a precise specification of this approach in the style of algorithms that we have already encountered.

Protocol 4.1. Flooding #1: Information diffusion.

Restrictions: Reliable communication; connected, bidirected communication graph $G = (V, E)$; neighborhood function $nbr : V \rightarrow 2^V$
State Trans. Sys.: $(\{\text{INIT}, \text{IDLE}\}, \{(\text{INIT}, \text{IDLE})\})$
Initialization: One node in state INIT, all other nodes in state IDLE

INIT
  Spontaneously
    broadcast (msge)   #*Broadcast msge to neighbors*
    become IDLE

IDLE
  Receiving (msge)
    broadcast (msge)   #*Rebroadcast msge to neighbors*

The restrictions to Protocol 4.1 are that we assume:

1. reliable communication, where messages sent will always arrive within a finite amount of time (although necessarily in the order in which they were sent); and
2. a connected bidirected communication graph, $G = (V, E)$, with associated neighborhood function $nbr : V \rightarrow 2^V$.

These restrictions are so basic to all our neighborhood-based algorithms, that we will refer to them in future with a special mnemonic, $\mathcal{NB}$ (calligraphic NB for "neighborhood-based" restrictions).

An analysis of this algorithm quickly reveals, however, an important flaw in this approach: the algorithm will never terminate. If all nodes rebroadcast all the messages they receive, nodes can expect to rebroadcast the same message over and over and over again. As a result, we can expect Protocol 4.1 to result in unbounded communication overheads, which in turn will rapidly deplete the network of all its energy and bandwidth resources.

4.1.2 Basic Flooding

We might fix this issue by changing the rules in the protocol slightly. Instead of requiring a node to blindly tell all its neighbors, the protocol could be adapted to require a node to "tell all my neighbors only *once* when I am told something." This adaptation is shown in Protocol 4.2, where a new state, DONE, has been introduced. A node that has broadcast a message transitions into the new state, DONE. Because the actions in the DONE state are empty, nodes in the DONE state that subsequently receive a message will simply ignore it.

Protocol 4.2. Flooding #2: Basic flooding algorithm

Restrictions: $\mathcal{NB}$ (reliable communication; connected, bidirected communication graph $G = (V, E)$; neighborhood function $nbr : V \to 2^V$)
State Trans. Sys.: $(\{\text{INIT}, \text{IDLE}, \text{DONE}\}, \{(\text{INIT}, \text{DONE}), (\text{IDLE}, \text{DONE})\})$
Initialization: One node in state INIT, all other nodes in state IDLE

INIT
    Spontaneously
        broadcast (msge)   *#Broadcast msge to neighbors*
        become DONE

IDLE
    Receiving (msge)
        broadcast (msge)   *#Rebroadcast msge to neighbors*
        become DONE

In terms of messages sent, the version of flooding in Protocol 4.2 improves dramatically on Protocol 4.1, requiring that each node send in total exactly one message. More precisely, this translates into a load balance of $\Theta(1)$ and an overall communication complexity of $\Theta(|V|)$ for a connected communication graph $G = (V, E)$ (recall: Θ asserts that an algorithm is bounded above and below by some order of growth). Thus the load balance for messages received is optimal (see §3.2.2).

Basic flooding appears slightly less efficient in terms of messages *received* than in terms of messages sent. A message must be received by each node from all its neighbors. Thus, the total communication complexity for messages received is $O(|E|)$. In the worst case of a complete graph, $|E|$ is proportional $|V|^2$. However, as we have already seen, in decentralized spatial information systems it is safe to assume relatively sparse graphs, where the communication distance is substantially smaller than the network extent, and consequently $|E|$ is approximately linearly proportional to $|V|$.

The load balance for messages received is also then necessarily less efficient than that for messages sent. Each node v is required to receive $O(deg(v))$ messages, where $deg(v)$ is the degree of node v (the number of edges incident with v; see Appendix A).

Fig. 4.1 provides an example execution of the basic flooding algorithm (Protocol 4.2), showing how each node will broadcast a received message only once, before transitioning into the DONE state.

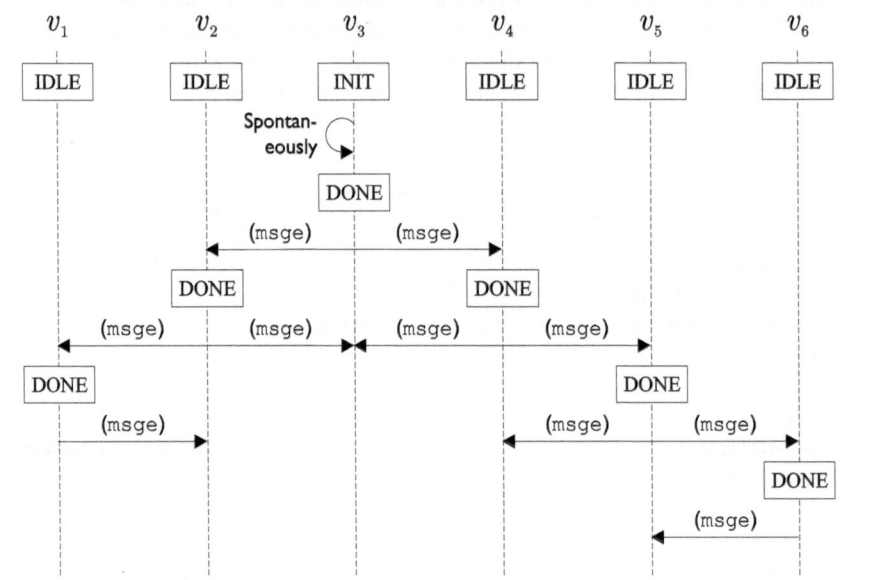

Fig. 4.1. Sequence diagram for example execution of Protocol 4.2

Basic flooding can therefore be relatively efficient in cases where a message needs to be communicated to the entire network, although potentially much less efficient in cases where we need to communicate a message to a specific node or subset of nodes.

However, an analysis of the basic flooding algorithm in Protocol 4.2 reveals that it will work only when one message must be sent. Nodes that transition into the DONE state can no longer respond to subsequent new messages. In cases where a network must respond to multiple messages (e.g., in response to multiple environmental events), it will not usually be known in advance how many messages might need to be flooded. Consequently, further adaptations are required to deal with flooding of multiple messages.

4.1.3 Hop-Count Flooding

An alternative approach to enable multiple messages to be handled, while avoiding infinite retransmission of messages, is to count the number of times a message is retransmitted. In *hop-count* flooding, Protocol 4.3, every time a message is forwarded, the hop count h for that message is incremented. Messages are only retransmitted up to a certain predetermined number of hops (in Protocol 4.3, the maximum hop count is 10). In effect, the algorithm now ensures messages *degrade* as they move through the network. A node will tell its neighbors about a message only as long other nodes have not previously retold this message a predefined number of times. In Protocol 4.3, new messages are generated whenever a node locally detects a high sensor value (in our example a sensed value greater than 100, $\overset{\circ}{s} > 100$).

Protocol 4.3. Flooding #3: Hop-count flooding, with sensor value trigger

---

Restrictions: $\mathcal{NB}$; sensor function $s : V \rightarrow \mathbb{R}$
State Trans. Sys.: $(\{\text{IDLE}\}, \varnothing\})$
Initialization: All nodes in state IDLE

IDLE
   When $\overset{\circ}{s} > 100$ *#Check for sensor value over threshold 100*
      broadcast (msge, 0)   *#Broadcast msge to neighbors*
   Receiving (msge, h)
     if $h < 10$ **then**   *#Maximum hop count of 10*
      broadcast (msge, $h + 1$)   *#Broadcast msge to neighbors*

The hop-count flooding in Protocol 4.3 does not suffer from the single-use problem of basic flooding in Protocol 4.2 (where only one message can be flooded). However, hop-count flooding, as well as many other forms of flooding, can suffer from high levels of redundancy, where the same message may be received or retransmitted by a node multiple times.

The term *implosion* is sometimes used to describe this situation, where a node receives the same message multiple times. With an overall communication complexity of $O(h|V|)$, in the worst case with high hop count thresholds (e.g., $h \approx |V|$), implosion can lead hop-count flooding to become just as inefficient as our first pass at information diffusion.

Let us now adopt an adversarial position: how could an adversary cause hop-count flooding to fail? One way is to note that *every* node that detects a high sensor value will generate a new message with an initial hop count of 0. Thus, if an adversary places the network in an environment with universally high temperature, new messages will be generated by every node. Coupling this with high hop count threshold, in the worst case each node might have to forward h messages from every node, leading to overall communication complexity of $O(h|V|^2)$.

A subtle adversary might also construct the network to have a *diameter* of $h + 1$, and then assert that an environmental change occurs at the periphery of the network (i.e., at one of the nodes with an eccentricity equal to h). In this case, Protocol 4.3 cannot guarantee delivery of a message throughout the network. A message sent from the periphery of the network will reach its maximum hop count before traversing the entire network. The diameter of the network is global information that will not normally be known to individual nodes in a decentralized algorithm.

4.1.4 Surprise Flooding

To further address the problem of implosion, one final adaptation of the flooding approach would be to ask each node to tell its neighbors only when it is told something that it has never heard before. In this case, nodes are required to *remember* what they have heard, compare received messages with those stored in memory, and only forward new information (i.e., information that "surprises" them).

Protocol 4.4 provides the formal specification of such an algorithm. Each message generated in response to a sensed environmental change now contains information about the sensed value, and the identifier of the node that initiated that message (hence the restrictions in Protocol 4.4 additionally require that nodes have access to local versions of the *id* function). Further, all nodes that receive a message must store that message as *local data*. Local data is akin to variables in a programming language: in comprises are registers in the local memory of a node that can store information generated by one action for access by subsequent actions. The local data in Protocol 4.4 is a set S of sensed value/identifier pairs, $\mathbb{R} \times \mathbb{N}$. When a node receives a message, it first compares the payload of that message to the local data stored in S. Message payloads that do not appear in S are new ("Surprise!") and can be rebroadcast after they are is stored in S. Message payloads that are already stored in S can safely be ignored, and not rebroadcast.

Computational analysis of the surprise flooding algorithm will reveal that a node must broadcast exactly one message for each environmental change detected (load balance $\Theta(1)$ and overall communication complexity $\Theta(|V|)$ for messages sent). In the worst case, where an adversary makes every node in the network detect an environmental change, this leads to an overall communication complexity $\Theta(|V|^2)$.

This efficiency is comparable to that of the basic flooding algorithm discussed earlier, at least in cases where only one environmental change occurs. Turning to the adversarial analysis, where there are large numbers of *distinct* environmental changes, the amount of storage required for local data (S) will increase in proportion to the number of distinct changes. We might

---

Protocol 4.4. Flooding #4: Surprise flooding algorithm

Restrictions: $\mathcal{NB}$, $s : V \to \mathbb{R}$, $id : V \to \mathbb{N}$
State Trans. Sys.: $(\{\text{IDLE}\}, \varnothing)$
Initialization: All nodes in state IDLE
Local data: Set $S \subset \mathbb{R} \times \mathbb{N}$ of received payloads, initialized to $S := \varnothing$

IDLE
When $\overset{\circ}{s} > 100$ #*Check for sensor value over threshold 100*
    broadcast (msge, $\overset{\circ}{s}$, id)   #*Broadcast msge, sensed value, and identifier to neighbors*
Receiving (msge, s, i)
    if $(s, i) \notin S$ then   #*Check if the received sensed value and id pair are surprising*
        set $S := S \cup \{(s, i)\}$ #*Store message payload*
        broadcast (msge, s, i)   #*Broadcast msge, sensed value, and identifier to neighbors*

---

attempt to remedy this by in a variety of ways, for example, by discarding the oldest messages when storage is exhausted, replacing our set S with a FIFO list of fixed length. But such approaches come with their own problems (for example, when a delayed message *is* rebroadcast by a node because it discarded its memory of previously broadcasting that message).

4.1.5 Summary

What a lot of flooding! Flooding is a basic technique in wireless sensor networks, and often used as a "fallback" procedure when other, more intelligent and efficient, mechanisms fail. However, it is already well studied in the literature (see question 4.1), and only of marginal interest to our core theme of what makes decentralized *spatial* computing special. Our main objective in discussing flooding in such detail is to illustrate several important features of the decentralized algorithm design process. Specifically, three general points, already identified in Chapter 3, are worth re-emphasizing about this process:

- The design process is iterative. While we may start with a simple solution, a thorough computational and adversarial analysis should reveal any shortcomings of that solution. In turn, these shortcomings may necessitate adaptation or increased sophistication in the protocol. Further, there may be no "silver bullet" that will work in all situations; the limitations of a protocol may be appropriate in some scenarios, but not in others. In such cases, it is vital to make explicit through the analysis process all the restrictions on, and limitations of, a particular protocol.
- Adversaries may posit a wide range of adverse conditions to defeat an algorithm. We have seen adversarial changes to the structure of the network; to the parameters of an algorithm (e.g., the interaction between network diameter and hop count in hop-count flooding); and to the environment (e.g., the adverse interaction between unfavorable environments

that reduce efficiency in hop-count and surprise flooding). Later analyses
will uncover other adverse conditions.

- There often exist trade-offs between different aspects of efficiency. For
  example, in surprise flooding it was possible to reduce communication
  complexity, but only at the cost of increased storage complexity.

As we progress to more sophisticated algorithms in later sections, the
level of detail in the analysis process will be gradually reduced. But, it is
worth keeping flooding in mind as an exemplar of the full design process.

4.2 Gossiping and Rumor Routing

The immediate objective of this section is to introduce two commonly encountered information dissemination protocols: *gossiping* and *rumor routing*.
However, proceeding from the discussion above, the broader context of this
section is to incrementally introduce the reader to the next level of complexity
of algorithm design. Thus, gossiping is introduced as a stepping stone to
rumor routing, which uses gossiping as one component. Rumor routing will
be the first "nontrivial" protocol encountered in this book. All the protocols so
far encountered, up to and including gossiping, can be summarized as a single
sentence (such as "tell your neighbor what you hear," or "tell your neighbor
what you hear, if it is news to you"). As we shall see, rumor routing involves
a number of interacting strategies, and so cannot be so simply explained.

4.2.1 Gossiping

Gossiping is a set of neighborhood-based algorithms closely related to flooding that aim to reduce the redundancy associated with implosion by randomly selecting a subset of messages to send.

The term "gossiping" is occasionally used to refer to algorithms where a
node forwards a message to only one (or some fixed number of) randomly
selected neighbor(s). This approach is, in essence, a generalization of the
basic "Crowd!" algorithm we have already seen on p. 65, and is more properly
termed an *epidemic* algorithm [27] (or simply a random walk in the special
case where each node forwards to only one neighbor). Usually, "gossiping"
refers to a generalization of flooding where each node randomly decides
whether to rebroadcast a message with some probability g.

Protocol 4.5 specifies the procedure for a basic gossiping algorithm. Comparing Protocol 4.5 with Protocol 4.2 (basic flooding), it should be clear that
the two protocols are almost identical, with the single difference that IDLE
nodes are not guaranteed to rebroadcast received messages. Nodes receiving
a message randomly decide whether to rebroadcast it, with the probability of
selecting rebroadcasting equal to g.

Protocol 4.5. Gossiping

Restrictions: $\mathcal{NB}$; gossip probability $g \in [0, 1]$
State Trans. Sys.: $(\{\text{INIT}, \text{IDLE}, \text{DONE}\}, \{(\text{INIT}, \text{DONE}), (\text{IDLE}, \text{DONE})\})$
Initialization: One node in state INIT, all other nodes in state IDLE

INIT

    Spontaneously
        broadcast (msge)   *#Broadcast msge to all neighbors*
        become DONE

IDLE

    Receiving (msge)
        Randomly select a number $n \in \{0, 1\}$ with probability $P(n = 1) = g$
        if $n = 1$ **then**
            broadcast (msge)   *#Rebroadcast message with probability g*
        become DONE

If the probability that a node will broadcast a message $g = 1$ then the nodes will always rebroadcast, and the gossiping algorithm becomes equivalent to the basic flooding we saw earlier in Protocol 4.2. Thus, the computational analysis of gossiping is similar to that for flooding. However, in cases where $g < 1$, some nodes will likely not rebroadcast the message.

At certain levels of g, the message may still be received by all nodes because of implosion (i.e., even when some neighbors don't rebroadcast a message, a node may still receive a message from another neighbor). In experiments, reductions of the order of 30% fewer messages transmitted have been observed using gossiping, when compared with flooding [52].

At lower levels of g, the gossip may die out completely and never reach nodes in some parts of the network. Whether a particular gossip probability will lead to a gossip reaching the entire network or dying out depends both on random chance and on the topology of the network. In any case, gossiping cannot *guarantee* delivery of a message, except in the special case where $g = 1$. For example, we might set the gossip probability to 0.999 recurring. But, an adversary might still set up the network topology such that there exists a node with only one neighbor that (by chance) happens to randomly choose *not* to forward the message it receives.

As for flooding, there are several variants of gossiping which improve on the performance and reliability of gossiping. For example, where source nodes have low connectivity, gossiping can be adapted to initially increase the gossip probability to avoid the possibility of gossip messages dying prematurely (see question 4.2).

4.2.2 Rumor Routing

Rumor routing [16] is a well-known algorithm that exhibits aspects of all three of the neighborhood-based algorithms already seen: flooding, gossiping, and random walk. A version of the rumor routing algorithm is shown in Protocol 4.6. As foreshadowed at the beginning of this section, the rumor routing protocol is noticeably more complex than the simpler algorithms we have so far seen. So, we will take a little time to explore its behavior in detail. Through exposure to these more sophisticated algorithms over the course of this book, you can expect to become more proficient at understanding how they work, and why they are structured the way they are. While Protocol 4.6 looks daunting, it can be easily broken down into manageable, understandable pieces by approaching it systematically using the structures we have already encountered.

The objective of rumor routing is to enable information about meaningful environmental changes (such as the appearance of a "hot spot") to be routed from nodes that detect those changes (termed *sources*) to nodes that have an interest in those changes (termed *sinks*). These meaningful changes in the environment are typically referred to as "events." At this point, it is important to note the potential for confusion between environmental "events" and *system* events happening at individual nodes (i.e., spontaneous events, triggered events, or message received events). Where there is possible ambiguity, the text will make explicit which type of event is being discussed.

In rumor routing, messages from the source and from the sink travel on random walks through the network. The message from the source is termed a *resource* (rsrc); that from the sink is termed a *request* (rqst). On their random walk, these messages leave a trail of "breadcrumbs," where each node that receives a message stores the identity of the predecessor node from which it received the message. When request and resource messages meet at a node, that node can then use the breadcrumbs to route information directly between the source and the sink. In Protocol 4.6, the algorithm terminates when the request has been routed to the source. However, at this point the network now contains all the information required to engage in further information routing, such as routing new information from the source directly back to the sink (see question 4.3). Figure 4.2 summarizes rumor routing graphically. Resource and request messages traverse random walks through the graph until they meet.

Examining Protocol 4.6 in more detail, we can decompose the algorithm into its key components using the structures already identified at the beginning of Chapter 3:

- *Restrictions*: Like all the protocols in this section, Protocol 4.6 requires the neighborhood-based algorithm restrictions $\mathcal{NB}$ (reliable communication and a connected, bidirectional communication graph G with associated neighborhood function *nbr*); a sensor function s; an identifier function *id*; and a gossip probability g, which ensures that only a small proportion of the nodes detecting an environmental event will become sources.
- *States*: Protocol 4.6 has three states: IDLE for nodes that have not heard about a resource; RSRC for nodes that have detected or received informa-

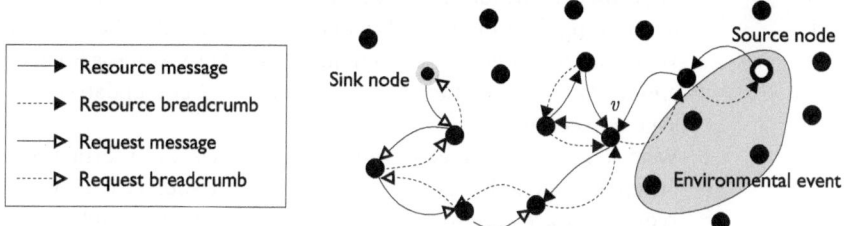

Fig. 4.2. Summary of sinks, sources, resources, requests, and breadcrumbs in rumor routing, highlighting route optimization at node v

tion about an environmental event; and DONE for a source node that has received a request for information about its environmental event.

- *(System) Events*: The sink and the source are associated with two spontaneous system events, external to the system: a query for information about an environmental event, initiated by a sink node; and the detection of an environmental event by a source node. For example, the request at the sink node might result from a user injecting a query into the network; the resource at the source node might be triggered by sensing a value over some threshold. Correspondingly, there are two types of messages transmitted by the protocol: a request for information about an environmental event (rqst message), and a resource advertising information about that environmental event (rsrc message).

- *Actions*: In Protocol 4.6, DONE is a terminating state where no actions are defined. Thus we would expect a total of two states (IDLE, RSRC) × four system events (two external triggers, two message types) = eight actions. In fact, we have only *seven* actions, since we ignore the possibility that a node that has already heard about a resource (i.e., in state RSRC) subsequently locally detects an environmental event. This is a reasonable simplification as we are restricted in the first chapters in this part of the book to static environments; however, in later chapters we shall also look at more complex spatiotemporal situations, monitoring dynamic environments. In brief, each of these actions have simple atomic tasks:

  1. When an IDLE node detects an environmental event, it gossips a resource message with gossip probability g to a randomly selected neighbor (a random walk);
  2. When an IDLE node receives an external request for information about an environmental event, it forwards the request message to a randomly selected neighbor;
  3. When an IDLE node receives a resource message, it transitions to state RSRC, indicating it now has knowledge of the resource, and stores a breadcrumb leading to the neighbor that sent the message. The

node then forwards the resource message to an arbitrary neighbor (continuing the random walk), updating the message hop count;

4. When an IDLE node receives a request message, it stores a breadcrumb, and forwards the request to an arbitrary neighbor, updating the message hop count;

5. When an RSRC node receives an external request for information about an environmental event, if it is not itself a source, it forwards the request towards the source using its resource breadcrumb;

6. When an RSRC node receives a resource message, it updates its breadcrumb only in cases where the message hop count indicates a shorter route to the source, and then forwards the resource to an arbitrary neighbor, updating the message hop count;

7. When an RSRC node receives a request message, if the node is a source the algorithm terminates; otherwise the message is forwarded towards the source using the stored breadcrumb.

Step 6 is particularly interesting, since it enables the system to incrementally optimize routes to the source. Nodes in state RSRC that receive information about more direct routes to another source (indicated by smaller hop counts for incoming rsrc messages) can update their breadcrumbs accordingly (or vice versa avoid updating breadcrumbs for incoming rsrc messages that indicate larger hop counts to the source). Figure 4.2 provides an illustration of this process, where node v does not update its breadcrumb the second time it receives a rsrc message.

Figure 4.3 shows a spatial sequence diagram for an example execution of the rumor routing protocol. The figure illustrates the sequence of messages and state changes resulting from a single source node (v_9) and sink node (v_1), with request and resource messages meeting at v_5.

At first glance, rumor routing does not appear to be an especially promising approach. It combines features of random walks, flooding, and gossiping, but is as a result open to many of the drawbacks of each of these approaches. The use of gossiping to initiate resource messages means that some (small) environmental events may result in no rsrc messages being generated. For example, if the gossip probability is set to 0.1 and an important environmental event is only detected by five nodes, then there is a 59% chance (0.9^5) that *none* of the nodes that detect the environmental event will initiate a resource message. In such cases, the version of the algorithm presented in Protocol 4.6 should still *eventually* connect sink and source, but only through the blind random walk taken by the rqst message (and with all the properties and problems of random walks discussed in §3.2.2).

Despite these drawbacks, rumor routing can offer an advantage over flooding either resource requests or environmental event notifications throughout the network. In cases where many more environmental events are occurring than the expected number of queries for information, it will be more efficient to simply flood requests from sinks throughout the network. Similarly,

---

Protocol 4.6. Rumor routing

Restrictions: $\mathcal{NB}$; $s : V \to R$; $id : V \to \mathbb{N}$; gossip probability g

State Trans. Sys.: $(\{\text{IDLE}, \text{RSRC}, \text{DONE}\}, \{(\text{IDLE}, \text{RSRC}), (\text{RSRC}, \text{DONE})\})$

Initialization: All nodes in state IDLE

Local data: Resource predecessor id $i_s := 0$; Resource hop count $h_s := 0$; Request predecessor id $i_q := 0$

IDLE

   When $\overset{\circ}{s} > 100$ *#Check for sensor value over threshold 100*

      if $n = 1$, where n is randomly selected with probability $P(n = 1) = g$ then

         send (rsrc, 1, $\overset{\circ}{id}$) to one-of $\overset{\circ}{nbr}$ *#Gossip resource to random neighbor*

         become RSRC

   When External request for environmental event initiated

      send (rqst, $\overset{\circ}{id}$) to one-of $\overset{\circ}{nbr}$ *#Unicast request to random neighbor*

   Receiving (rsrc, h, i)

      set $h_s := h$ and $i_s := i$ *#Store hop count, breadcrumb to resource predecessor*

      send (rsrc, $h + 1$, $\overset{\circ}{id}$) to one-of $\overset{\circ}{nbr}$ *#Unicast rsrc to random neighbor*

      become RSRC   *#Receiving resource message*

   Receiving (rqst, i)   *#Receiving request message*

      set $i_q := i$ *#Store breadcrumb to request predecessor*

      send (rqst, $\overset{\circ}{id}$) to one-of $\overset{\circ}{nbr}$ *#Unicast rqst to random neighbor*

RSRC

   When External request for environmental event initiated

      if $h_s = 0$ then   *#Check if source reached*

         become DONE

      else

         send (rqst, $\overset{\circ}{id}$) to node $v \in \overset{\circ}{nbr}$ with identifier i_s *#Follow request breadcrumb*

   Receiving (rsrc, h, i)   *#Receiving resource message*

      if $h < h_s$ then   *#Check for shorter route to resource*

         set $h_s := h$ and $i_s := i$ *#Store breadcrumb to resource predecessor*

         send (rsrc, $h_s + 1$, $\overset{\circ}{id}$) to one-of $\overset{\circ}{nbr}$ *#Unicast rsrc to random neighbor*

   Receiving (rqst, i)   *#Receiving request message*

      if $h_s = 0$ then

         become DONE   *#Request has located source*

      else

         set $i_q := i$ *#Store breadcrumb to request predecessor*

         send (rqst, $\overset{\circ}{id}$) to node $v \in \overset{\circ}{nbr}$ with identifier i_s *#Follow resource breadcrumb*

---

where a relatively large number of requests is expected, but environmental events are rare, it will instead be more efficient to flood information about resources. However, in intermediate cases, where the levels of queries for information and environmental event notifications are expected to be roughly comparable, there will be an advantage to using rumor routing, as it is able to balance the effort expended by the network between resources and requests. In experiments, rumor routing has been shown in some cases to more than halve the total number of messages required to connect sink and

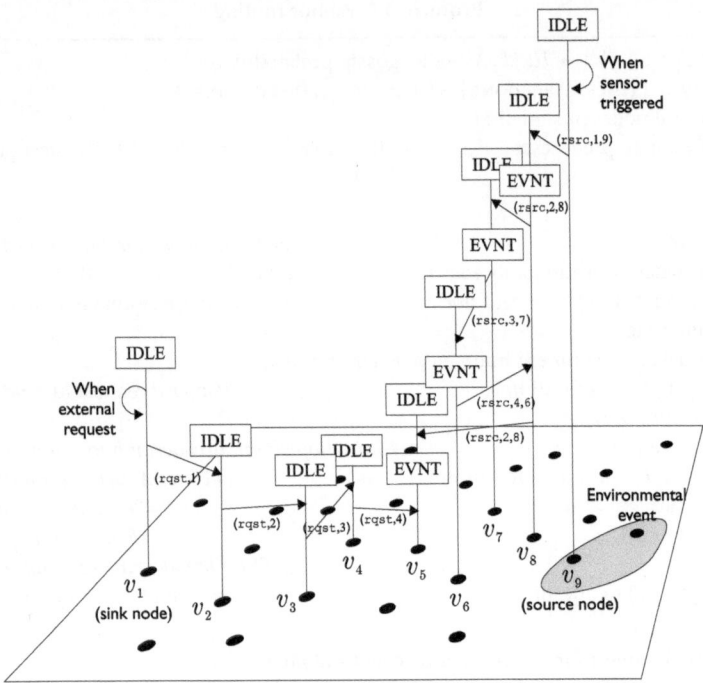

Fig. 4.3. Rumor routing: Spatial sequence diagram highlighting source node (v_9), sink node (v_1) and using the shortest known path to the source (v_8)

source where the balance of requests and resources makes the choice between environmental event or user query flooding difficult (see [16]).

While rumor routing does assume prior knowledge of the expected balance of requests and resources for an application, it assumes no knowledge about any relationship between the spatial locations of source and sink nodes. Random walks are used by request and resource messages to traverse the network—all neighbors of a node are regarded as equally likely to be a step closer to the source/sink. In later chapters, we shall improve on this situation given the inherent structure of geographic space (near things are more related to each other).

4.2.3 Summary

There are two important lessons that can be taken from this discussion of gossiping and rumor routing with broader relevance to the design of decentralized spatial algorithms.

- More sophisticated algorithms, tailored to specific scenarios or with improved efficiency, can often be constructed from the combination of simpler algorithms. For example, rumor routing makes use of aspects of random walks, flooding, and gossiping. As we progress through this and later chapters we shall see many more examples of this "divide-and-conquer" approach, leading ultimately to a discussion of modularization of algorithms.

- When attempting to understand or construct a complex protocol such as rumor routing, it is important to approach the protocol systematically, breaking the complexity down into its constituent parts: the restrictions, states, system events, and finally actions for each state/event pair. In most of the algorithms in this book, individual *actions* are trivial: actions only rarely involve more than four or five lines of pseudocode. What is non-trivial are the *interactions* between neighboring nodes. These interactions can be understood by studying the system states, transitions, and events.

4.3 Tree Structures

Trees are a fundamental neighborhood structure in communication networks, and in computer science more generally. We have already seen one example of a tree, the minimum spanning tree in §2.6.3. Recall that in a tree there exists exactly one path through the network between any pair of nodes.

Trees are an important example of *overlay network* for decentralized spatial computing. "Overlay network" is the term used for a logical network structure that is built on another logical or physical network. In the case of geosensor networks, the underlying physical network structure is defined by the potential for direct, one-hop communication between nearby nodes. As already discussed, we normally use the UDG as (a model of) the structure of this underlying physical network (see §1.2.2). An overlay network for a geosensor network is typically a spanning subgraph of this underlying physical network structure (see discussion of spanning subgraphs on p. 48). Overlay networks can help to restrict the potential for communication in a way that improves the efficiency or organization of computation in the network. There are many existing decentralized algorithms that rely on a tree structure for routing information between nodes.

The remainder of this section looks first at the construction of tree-based overlay networks, and then shows how these overlay networks can assist in efficient computation.

4.3.1 Establishing a Tree with a Known Root

The question of how to establish a tree overlay network is relatively simple to answer in the special case where we have a unique node that is designated as the initiator for the algorithm (or equivalently as the root of the tree). The unique root can simply broadcast a "tree" message with its identity to all its neighbors. When a neighbor hears a tree message for the first time, it stores the identity of the sending node. This sending node then becomes the unique "parent" for that node: the next node on the (unique) path to the root. The receiving node then rebroadcasts the tree message with its own identifier. Nodes that hear a second or subsequent tree message simply ignore it. Protocol 4.7 provides a decentralized algorithm for establishing a rooted tree.

Protocol 4.7. Establishing a rooted tree

Restrictions: $\mathcal{NB}$; identifier function $id : V \to \mathbb{N}$
State Trans. Sys.: $(\{\text{ROOT}, \text{IDLE}, \text{DONE}\}, \{(\text{ROOT}, \text{DONE}), (\text{IDLE}, \text{DONE})\})$
Initialization: One node in state ROOT, all other nodes in state IDLE
Local data: $parent : V \to \mathbb{N} \cup \{-1\}$, initialized $parent := -1$

ROOT
   Spontaneously
      broadcast $(\texttt{tree}, \overset{\circ}{id})$ *#Broadcast root identifier*
      become DONE

IDLE
   Receiving $(\texttt{tree}, i)$
      set $\overset{\circ}{parent} := i$ *#Store parent id*
      broadcast $(\texttt{tree}, \overset{\circ}{id})$ *#Broadcast node identifier*
      become DONE

The protocol requires one designated root node, initialized to state ROOT. The local data of the algorithm is a function $parent : V \to \mathbb{N} \cup \{-1\}$, initialized for each node to $parent := -1$. Thus, for a node v, $parent(v) < 0$ indicates that the node has no parent (i.e., it is either the root, or has not yet received a tree message). For a non-root node v, following the receipt of a tree message from a node v', the node will store $parent(v) := i$ where $i = id(v')$.

Protocol 4.7 has overall communication complexity $\Theta(|V|)$ and optimal load balance $\Theta(1)$ for messages both sent and received. With the exception of the root node (which sends one message and receives zero messages), each node sends exactly one message and receives exactly one message. Thus, the basic algorithm for establishing a rooted tree is relatively efficient. Figure 4.4 provides an example of a spatial sequence diagram for Protocol 4.7.

From an adversarial perspective, Protocol 4.7 can make no guarantees about the structure of the overlay network beyond being a tree. Because nodes adopt as their parent the *first* neighbor to send a tree message, it is likely that delays in sending, receiving, and processing messages will lead to relatively long routes (in terms of both number of hops and total geographic distance over which nodes are transmitted) being established between some nodes and the root when compared with the shortest path. In the worst case, an adversary may structure the network and communication to result

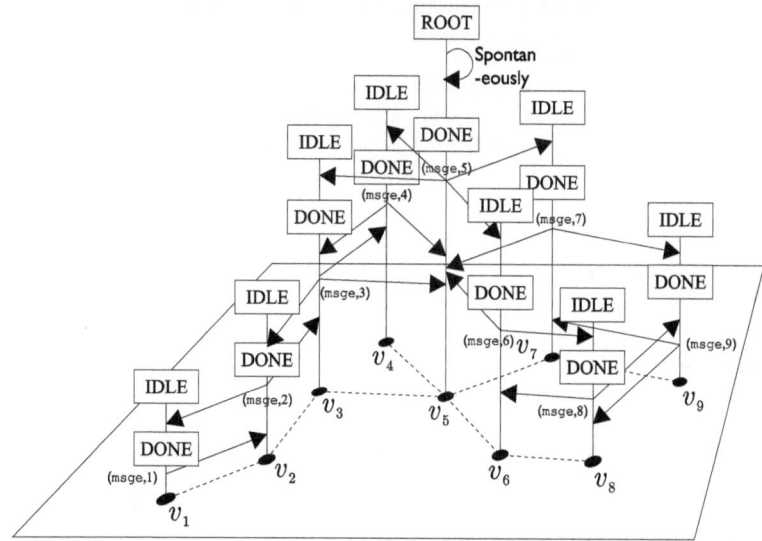

Fig. 4.4. Establishing a rooted tree: spatial sequence diagram for Protocol 4.7 with root node v_5

in a tree with maximum depth (i.e., diameter) $|V|$. As we have already seen, such degenerate network structures may have adverse effects on the computational characteristics of some algorithms (see §3.2.2 on load balance).

4.3.2 Establishing a Shortest Path Tree with a Known Root

It is possible to design a revision of Protocol 4.7 that can deal with the criticism above: the paths between nodes and the root may in some cases be substantially longer than the shortest path. In the extension to Protocol 4.7, shown in Protocol 4.8, the algorithm starts similarly to that of the basic rooted tree, except that a hop count is also broadcast along with node identifiers. Further, nodes in Protocol 4.8 have an additional action for responding to tree message events after having transitioned to a DONE state (i.e., after they have already received one tree message). This additional action enables a node that subsequently hears about a shorter route to the root to update their parent in the overlay network.

Protocol 4.8 generates an overlay network that is more efficient, in the sense that the path between a node and the root will be at least as short (in terms of number of hops) as, and frequently shorter than, that resulting from Protocol 4.7. However, the communication complexity of Protocol 4.8 is much less efficient than that of Protocol 4.7. Because nodes can continue to listen for, and respond to tree messages after transitioning to a DONE state, it is possible that an individual node may send and receive more than one message. In the worst case, an adversary may design a network and execution

---

Protocol 4.8. Establishing a rooted shortest path tree

Restrictions: $\mathcal{NB}$; identifier function $id : V \to \mathbb{N}$
State Trans. Sys.: $(\{\text{ROOT}, \text{IDLE}, \text{DONE}\}, \{(\text{ROOT}, \text{DONE}), (\text{IDLE}, \text{DONE})\})$
Initialization: One node in state ROOT, all other nodes in state IDLE
Local data: $parent : V \to \mathbb{N} \cup \{-1\}$, initialized $parent := -1$; hop count to root r, initialized $r := 0$

ROOT
    Spontaneously
        broadcast $(\texttt{tree}, id, 0)$ *#Broadcast root identifier and hop count of zero*
        become DONE

IDLE
    Receiving $(\texttt{tree}, i, h)$
        set $parent := i$ *#Store parent id*
        set $r := h + 1$ *#Store hop count to root*
        broadcast $(\texttt{tree}, id, r)$ *#Broadcast node identifier incrementing hop count*
        become DONE

DONE
    Receiving $(\texttt{tree}, i, h)$
        if $h + 1 < r$ **then**
            set $parent := i$ *#Store new parent id*
            set $r := h + 1$ *#Store hop new count to root*
            broadcast $(\texttt{tree}, id, r)$ *#Broadcast node identifier incrementing hop count*

---

of Protocol 4.8 such that a node receives and sends $|V| - 1$ messages (see question 4.5). Thus, the worst case load balance is $O(|V|)$, and the overall communication complexity is $O(|V|^2)$. In practice, such a situation is highly unlikely to eventuate, and on average the communication complexity of Protocol 4.8 may be closer to that of Protocol 4.7 (a property investigated further in Part III of this book).

4.3.3 Establishing a Bidirected Tree with a Known Root

The information generated by Protocols 4.7 and 4.8 is in the form of the *parent* function. Thus, after execution has terminated, each node will have gained information about its unique parent in the overlay network. For some algorithms (including the TAG algorithm in the following section), individual nodes may additionally require information about their *children*.

The algorithm in Protocol 4.9 provides a simple extension to Protocol 4.7 that can generate both the parents and the children for each node. The idea behind Protocol 4.9 is that the tree message contains information about a node and its parent. Nodes in the DONE state (those which have already found their parents, and broadcast their tree message) may then listen in for any *neighbor* which announces in a tree message who their parent is. If a node v

overhears a neighbor v' announcing that v is its parent, then v can store the identifier of v' in its list of children, $children : V \rightarrow 2^{\mathbb{N}}$.

Protocol 4.9. Establishing a bidirected rooted tree

Restrictions: $\mathcal{NB}$; identifier function $id : V \rightarrow \mathbb{N}$
State Trans. Sys.: $(\{\text{ROOT}, \text{IDLE}, \text{DONE}\}, \{(\text{ROOT}, \text{DONE}), (\text{IDLE}, \text{DONE})\})$
Initialization: One node in state ROOT, all other nodes in state IDLE
Local data: $children : V \rightarrow 2^{\mathbb{N}}$, initialized to $children := \varnothing$; $parent : V \rightarrow \mathbb{N} \cup \{-1\}$, initialized to $parent := -1$

ROOT
   Spontaneously
      broadcast (tree, id, -1)   #*Broadcast root identifier*
      become DONE

IDLE
   Receiving (tree, i, j)
      set $parent := i$ #*Store parent id*
      broadcast (tree, id, i)   #*Broadcast node identifier*
      become DONE

DONE
   Receiving (tree, i, j)
      if $j = id$ **then**   #*Internal node overhears a child*
         set $children := children \cup \{i\}$ #*Store child identifier*

Protocol 4.9 again has optimal $\Theta(1)$ load balance and $\Theta(|V|)$ messages sent, since each node must again send exactly one message. However, nodes may need to receive multiple messages. Protocol 4.9 may require each node to receive a tree message from each of its neighbors, leading to load balance of $O(deg(v))$ and overall communication complexity of $O(|E|)$ messages received.

4.3.4 Aggregation in Trees

One of the best known examples of an algorithm that utilizes a tree overlay network is TAG (Tiny aggregation, named after the wireless sensor network operating system TinyOS), first explored in [78]. In TAG, each node senses some data about its environment, as in Fig. 4.5. TAG assists in computing aggregate functions (such as count, min, max, sum, average) over this data. Centralized computation of aggregate statistics such as "What is the maximum value in the network?" requires that every node's data be forwarded to the root, where it can be processed using a conventional centralized algorithm (Fig. 4.5a). Instead, TAG partially processes the query in the network itself. Instead of nodes forwarding all data, each node locally computes the maximum value and forwards only that value (Fig. 4.5b).

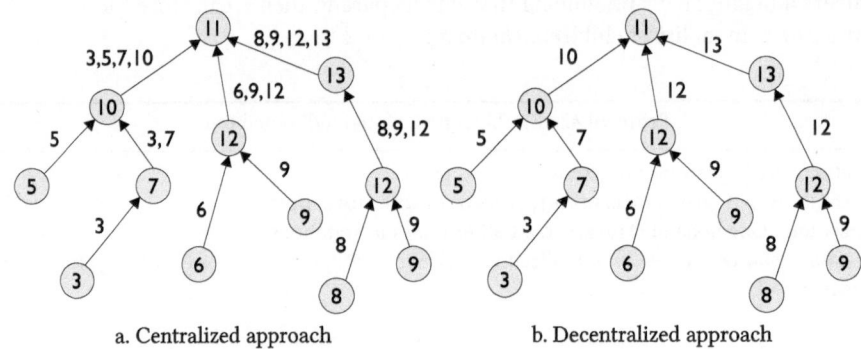

a. Centralized approach   b. Decentralized approach

Fig. 4.5. Trading communication for computation: computing the max value in TAG

TAG assumes as a restriction that there exists a bidirected rooted tree as an overlay network. Thus, listed in the restrictions in Protocol 4.10 are the *parent* and *children* functions that were computed in Protocol 4.9. In short, TAG assumes that we have previously computed the appropriate overlay network using some algorithm like Protocol 4.9.

TAG begins with each leaf node (a node that has no children) forwarding its sensed values in a `tagx` message to its unique parent. When a parent has received `tagx` messages from all of its children, it performs the required computation (finding the maximum) on the set of received sensed values, along with its own sensed value. The result of this computation is then forwarded to that node's own parent, and the computation cascades upwards through the network to its root.

Figure 4.6 shows a sequence diagram for an example execution of the tag algorithm. In Protocol 4.10, each non-root node must transmit exactly one message, and each non-leaf node must receive exactly one message. Thus the overall communication complexity is $\Theta(|V|)$, with optimal load balance $\Theta(1)$. By contrast, the alternative of forwarding all sensed values to the root requires $O(\sum_{v \in V} depth(v))$, where $depth(v)$ is the length of the path from v to the root (the depth of v). In the best case this translates into an overall communication complexity of $O(|V|)$ (the same as that of the TAG algorithm); in the worst case, where the network has diameter $|V| - 1$ this results in a communication complexity of $O(|V|^2)$. We might try to adapt this naïve approach and instead send the *set* of sensed values from a node and its children as a *single* message. While this approach guarantees the *number* of messages sent is $O(|V|)$, in the worst case the total *length* of messages sent is still proportional to $O(|V|^2)$. Thus, TAG performs at least as well, and usually better than, these naïve alternatives.

TAG relates to a broader class of problems known as *distributed function evaluation*: computing a function with specific characteristics in a decentral-

---

Protocol 4.10. Computing the maximum value using tiny aggregation (TAG, cf. [78])

Restrictions: $\mathcal{NB}$; $s : V \to \mathbb{R}$; $children : V \to 2^{\mathbb{N}}$; $parent : V \to \mathbb{N} \cup \{-1\}$ (see Protocol 4.9)
State Trans. Sys.: $(\{\text{IDLE}, \text{LSTN}\}, \{(\text{IDLE}, \text{LSTN})\})$
Initialization: All nodes in state IDLE
Local data: Maximum value m, received message count r initialized to $r := 0$

IDLE

   Spontaneously

      set $m := \overset{\circ}{s}$ *#Initialize m*

      if $children = \varnothing$ then   *#Check for leaf node*

         send (tagx, m) to $\overset{\circ}{parent}$ *#Send sensed value to parent node*

      become LSTN

LSTN

   Receiving (tagx, d)

      set $r := r + 1$ *#Increment received message count*

      set $m := \max(d, m)$ *#Store the largest known data point*

      if $|\overset{\circ}{children}| = r$ then   *#Check if all children have forwarded data*

         if $\overset{\circ}{parent} = -1$ then   *#Check if this is a root node*

            Maximum value in network is m *#Max value found*

         else

            send (tagx, m) to $\overset{\circ}{parent}$ *#Send sensed value to parent node*

---

ized network. The "specific characteristics" required for TAG are that the functions are associative and commutative binary operations.[1] Such functions are known algebraically as a *commutative semigroup* [99]. For example, the function $\max : \mathbb{R} \times \mathbb{R} \to \mathbb{R}$ is commutative—$\max(n_1, n_2) = \max(n_2, n_1)$—and associative—$\max(\max(n_1, n_2), n_3) = \max(n_1, \max(n_2, n_3))$. Many other common binary operations, such as minimum and summation, similarly form commutative semigroups.

Crucially, most *spatial* functions *cannot* be treated as associative and commutative functions over the set of all nodes, because they involve spatial relationships between nodes and spatial entities. Even in cases where we may need to evaluate an associative and commutative function for a spatial entity (e.g., the minimum of all values in a spatial region) simple tree overlay networks may not provide efficient answers because there may be no direct relation between the proximity of nodes in space and the proximity of nodes in the tree (discussed further in the summary below).

4.3.5 Summary

There are two further important points to note in the discussion of decentralized computing with tree structures.

---

[1] A binary operation is a function of the form $f : X \times X \to X$. Refer to Appendix A for more information.

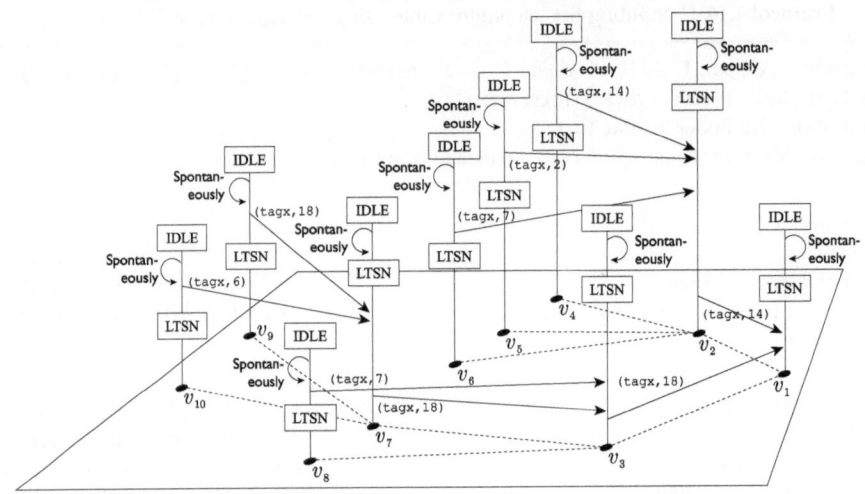

Fig. 4.6. TAG: Spatial sequence diagram for Protocol 4.10

- Trees are one of the most common overlay network structures. A range of other overlay network structures will be discussed in later algorithms. However, most of these structures share the same basic aims: they constrain the opportunities for communication, but in return result in more efficient, simpler, or more robust computation. As a result, there is often a balance to be struck between the resources expended in generating the overlay network (for example, in building the tree structure using Protocols 4.7–4.9) and the resources saved in using improved algorithms employing that overlay network (for example, in computing the maximum value using the TAG algorithm, Protocol 4.10).

- The TAG algorithm in Protocol 4.10 provides a further example of how to construct more sophisticated algorithms from simpler components. In particular, Protocol 4.10 illustrates the strategy of including in the restrictions to a protocol data that has previous been computed using another algorithm. In the case of Protocol 4.10, the restrictions require that a tree overlay network has been previously computed, for example, using Protocol 4.9.

One further point about trees more generally, rather than about algorithms for trees specifically, is worth highlighting. Trees are a helpful structure for designing decentralized algorithms because of the property that there exists exactly one path through the network between any pair of nodes. This property can help, for example, in avoiding the problem of implosion inherent in flooding or gossiping, where a node may receive the same message multiple times (§4.1). However, trees also have important disadvantages, especially in the context of monitoring spatial phenomena.

The first disadvantage of trees is computational. Trees are defined by their global properties (every pair of nodes is connected by a single path). As a result, building and maintaining trees requires algorithms that traverse the entire network. This can be especially expensive in terms of communication complexity in cases where node neighbors may change over time, for example, as a result of node mobility. Second, and more subtly, trees often have rather poor spatial characteristics. Two nodes that are topologically neighbors in a tree (i.e., one hop neighbors) can be expected to be spatially close (given the assumption of limited direct communication range). However, the converse is not true: two nodes that are spatially close can be arbitrarily far apart in terms of their topological distance in the tree. For example, in the tree in Fig. 4.7, spatially nearby nodes v_1 and v_2 are topologically rather distant, only connected by a path of nine intermediate nodes. Using this tree neighborhood structure means that algorithms that rely on collaboration between these nodes, for example, in cases where both nodes are monitoring the same geographic region, are likely to incur additional complexity and efficiency costs when compared with some other neighborhood structures.

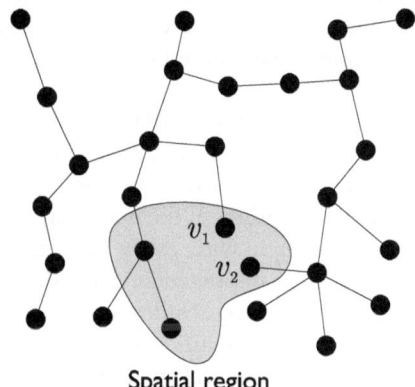

Spatial region

Fig. 4.7. Trees may offer poor spatial properties, with spatially proximal nodes (e.g., v_1, v_2) being distal in the tree

4.4 Leader Election

The previous section introduced several algorithms for computing an overlay network tree. However, all of these assumed as part of their initialization that exactly one node initiated the algorithm (in state ROOT in Protocols 4.7–4.9). This may be reasonable in some cases, for example, where a designated node is tasked with relaying information through a WAN (wide area network) gateway or where an external system has injected a query into the network at a specific node. However, in other cases, this assumption may be too strong. If there is no obvious node to initiate computation, how can we construct the tree?

In fact, this question relates closely to an important class of problems in distributed computing called *leader election*. Leader election concerns the mechanisms that can be employed by a distributed system to choose, from amongst the system nodes, a single coordinator. In recent years, many ingenious leader election algorithms have been developed for different types of network. Although we shall not attempt to give a thorough exposition of this important topic (see [99]), the need to elect a leader to coordinate a computational process, even if temporarily, is common in many spatial algorithms. As a consequence, this section will introduce some of the key ideas and approaches behind leader election.

4.4.1 Building a Tree Using Multiple Initiators

Turning back to our specific example of how to generate a tree when there is no designated initiator, Protocol 4.11 gives an algorithm for generating a tree when there may be multiple initiators for the algorithm. Its key difference from the algorithms already encountered in §4.3 is that in Protocol 4.11 there is no single designated node in the INIT state; instead all nodes begin in the INIT state.

The approach taken in Protocol 4.11 is similar to that in the shortest path tree algorithm, Protocol 4.8. However, instead of broadcasting the number of hops that a message has traveled, tree messages in Protocol 4.11 contain information about the identifier of the node that initiated the construction of that particular tree. To begin with, every node initiates the construction of a tree, broadcasting a tree message along with its own identifier, which becomes the identifier for that tree. Nodes receiving a tree message first check to see if the incoming tree's identifier is smaller than that of the tree they are currently part of. If it is, then that node will update its parent accordingly and propagate the received message. If it is not, then the receiving node will simply ignore the message. In this way, the trees rooted in nodes with larger identifiers are progressively "killed" by nodes that have joined trees rooted in nodes with smaller identifiers. Figure 4.8 shows an example of the tree construction process resulting from Protocol 4.11.

A critical requirement of this approach is that each node must have a *unique* identifier. Without this, an adversary could place two nodes with the same minimum identifier in the network. In turn, this would lead to an overlay network structure that is not a tree (in fact it would result in two distinct trees—a set of trees is termed a *forest*). This requirement of distinct identifiers or other unique values for nodes is fundamental to all leader election algorithms. In cases where it is possible that two nodes are

---

Protocol 4.11. Establishing a tree with multiple initiators

Restrictions: $\mathcal{NB}$; *injective* function $id : V \to \mathbb{N}$

State Trans. Sys.: $(\{\text{INIT}, \text{IDLE}\}, \{(\text{INIT}, \text{IDLE})\})$

Initialization: All nodes in state INIT

Local data: $parent : V \to \mathbb{N} \cup \{-1\}$, initialized $parent := -1$; smallest root identifier m, initialized
   to $m := id$

INIT

   Spontaneously
      broadcast $(\texttt{tree}, \overset{\circ}{id}, \overset{\circ}{id})$ *#Broadcast root identifier and tree identifier*
      become IDLE

IDLE

   Receiving $(\texttt{tree}, i, t)$
      if $t < m$ **then**   *#Check if smaller tree identifier received*
         set $parent := i$ *#Store new parent id*
         set $m := t$ *#Store id of new root*
         broadcast $(\texttt{tree}, \overset{\circ}{id}, t)$ *#Broadcast node identifier and tree identifier*

---

indistinguishable (i.e., they have the same state, including identifiers), any leader election algorithm cannot be relied upon to correctly identify a leader. In the case of Protocol 4.11, the requirement of unique identifiers is listed in the restrictions, which require that the identifier function *id* be an injection (i.e., every node has a distinct identifier).

Unfortunately, like Protocol 4.8 on which it is based, Protocol 4.11 is not especially efficient. In the worst case, nodes may approach having to forward $|V|$ messages (one from each node with a lower identifier), leading to an overall communication complexity of $O(|V|^2)$ and a load balance of $O(|V|)$.

Despite this worst case efficiency, executing the algorithm until no further messages are generated will result in a tree where every node knows its parent and the identifier of the root node. The root will be the only node without a parent. Thus, although Protocol 4.11 has ostensibly constructed a rooted tree with multiple initiators, it has also provided a (not especially efficient) algorithm for electing a leader in a network.

Aside from the computational issues, one further problem might be raised by an adversary: how does a (root) node know that the algorithm has terminated, and that no further messages will be generated somewhere in the network? Unfortunately, the answer to this question is: "it doesn't!" Using Protocol 4.11, no matter how much time a root node allows for the algorithm to terminate, there is always a chance that it will subsequently receive (severely delayed) messages from another root with a lower identifier. In order to confirm that a root really is the only root requires an extended protocol that checks the entire network to ensure that no other trees are still under construction. This process of checking will further decrease the efficiency of the algorithm (see question 4.8, and, later, §8.1.4).

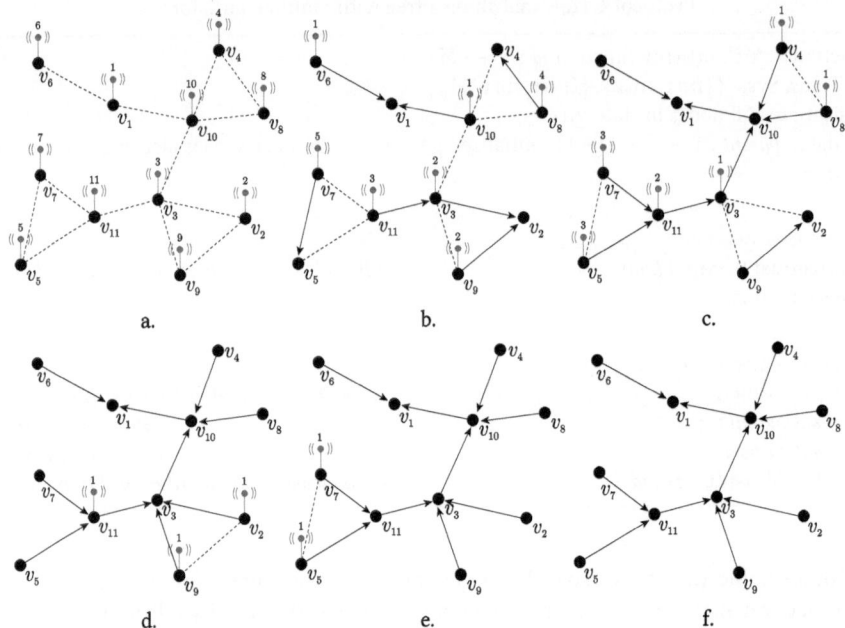

Fig. 4.8. Establishing a tree with multiple initiators (cf. Protocol 4.11). In the first round, every node broadcasts its identifier (a). In subsequent rounds, nodes only broadcast identifiers that are lower than the lowest known identifier (b – e). The result is a tree with the node with the lowest identifier as its root (f)

4.4.2 Leader Election in a Ring

The algorithm in the previous section can help to elect a leader in a distributed system regardless of the network structure. However, more efficient leader election algorithms also exist for a range of different network or overlay network structures, such as trees. To provide another example, this subsection looks at the well-studied problem of leader election in a *ring*. A ring is a connected graph where every node has exactly two neighbors.

Protocol 4.12 solves the problem of leader election in a ring using an approach known as "as far." The fact that this algorithm is suited only to ring (overlay) network structures is highlighted in the restrictions, where for all $v \in V$, $|nbr(v)| = 2$ (i.e., all nodes have exactly two neighbors). As a first step, each node sends its identifier in a `lead` message to an arbitrarily chosen neighbor. As for Protocol 4.11, `lead` messages will continue to be forwarded in the same direction in which they started around the ring, but only by nodes that have not already heard about a lower identifier in the ring (either their own identifier, or the identifier of a message they have already forwarded). In this way, all messages will sooner or later be "killed" except for the lead message originating from the node with the lowest identifier in the network. When that node receives a lead message containing its *own* identifier as a payload, it knows it has been elected leader. Note that in contrast to Protocol 4.11, the leader *does* know when the algorithm has terminated (i.e., when it receives its own message back).

Protocol 4.12. "As far": Leader election in a ring

Restrictions: $\mathcal{NB}$; for all $v \in V$, $|nbr(v)| = 2$; *injective* function $id : V \to \mathbb{N}$

State Trans. Sys.: $(\{\text{INIT}, \text{IDLE}, \text{LEAD}\}, \{(\text{INIT}, \text{IDLE}), (\text{IDLE}, \text{LEAD})\})$

Initialization: All nodes in state INIT

Local data: *parent* : $V \to \mathbb{N} \cup \{-1\}$, initialized $\overset{\circ}{parent} := -1$; smallest identifier m, initialized to $m := \overset{\circ}{id}$

INIT
  Spontaneously
    send ($\text{lead}, \overset{\circ}{id}$) to arbitrary $v \in \overset{\circ}{nbr}$ #*Broadcast identifier to arbitrarily chosen neighbor*
    become IDLE

IDLE
  Receiving (lead, i) from v
    if $i < m$ **then**   #*Check if smaller node identifier received*
      set $m := i$ #*Store new smallest id*
      send (lead, i) to unique node in singleton set $\overset{\circ}{nbr} - \{v\}$ #*Forward* lead *message*
    else
      if $i = m$ **then**   #*Check if own identifier received*
        become LEAD   #*Node receiving its own id is the leader*

Figure 4.9 shows an example of the stages in a leader election in a simple ring with nine nodes. At each "round," the figure shows the decreasing number of messages that are transmitted. Note that although Fig. 4.9 illustrates the as far process using rounds, such synchronization across the network is not required (and would violate our underlying $\mathcal{NB}$ restrictions, which only include reliable communication, not synchronization or message ordering).

Perhaps unsurprisingly given the similarities in the protocols, the "as far" leader election protocol is, in the worst case, no more efficient than Protocol 4.11. Assuming a random pattern of identifiers, on average it does achieve $O(|V| \log |V|)$ overall communication complexity, with worst case $O(|V|^2)$. However, there are more sophisticated protocols for leader election in a ring that can achieve worst case $O(|V| \log |V|)$ communication complexity [99].

4.4.3 Summary

Leader election is a fundamental and frequently encountered neighborhood-based algorithm. However, it is surprisingly inefficient, with optimal algorithms in many situations requiring worse than $O(|V|)$ communication

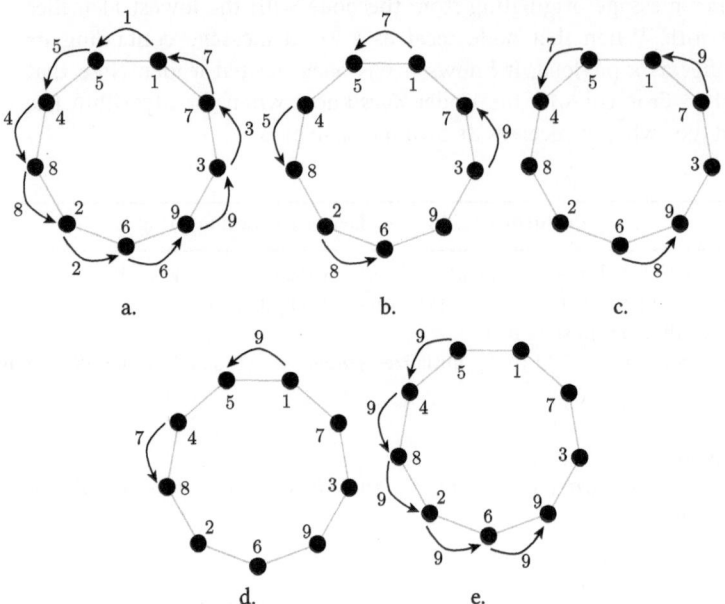

Fig. 4.9. Leader election in a ring. At the first round, every node broadcasts its identifier (a). Subsequent rounds only forward identifiers that are higher than the highest known identifier (b – d). The only node to receive a message containing its own identifier must be the leader (e)

complexity. Regardless of the specific details, all leader election protocols require some way of distinguishing between nodes. Typically this means nodes have unique identifiers, but may also involve some other element of a node's state (e.g., a node's coordinate location might in some cases substitute for a unique identifier).

4.5 Boundaries and Region Structures

The algorithms discussed previously in this section assume only minimal spatial knowledge (i.e., a node only holds information about which nodes are in its one-hop neighborhood). However, the algorithms are also only marginally spatial, in the sense that the operations performed by these algorithms are common to a wide range of applications in distributed computing. Flooding, gossiping, tree construction, and leader election are all important topics in all sorts of networks, not simply in geosensor networks.

In this, the penultimate section of the chapter on neighborhood-based algorithms, we turn at last to more explicitly spatial problems. Even with only the minimal spatial knowledge available to neighborhood-based algorithms, it is possible to construct useful spatial algorithms. More specifically, this section shows how we can design decentralized spatial algorithms to construct the boundary, interior, and exterior of a region. In turn, these important spatial structures can be used perform one of the most fundamental spatial analyses: determining the topological relations between two regions. However, first we turn to a related structure fundamental to many centralized spatial algorithms: sweeps.

4.5.1 Sweeps

Many centralized spatial algorithms, such as well-known algorithms for Delaunay triangulation construction and line segment intersection [25], rely on the concept of a *sweep line*. In a sweep line algorithm, we imagine a line sweeping across our geographic data, processing each data item (e.g., point, polyline vertex) in the order in which it encounters them (e.g., in order of increasing x-coordinate, if a vertical sweep line moves left to right). Our first truly spatial algorithm performs an analogous process in a *decentralized* way.

In our decentralized sweep algorithm, presented in Protocol 4.13, a coordinated "front" of activation sweeps across the network. It is a surprising result that this coordination can be achieved in a decentralized spatial information system, even though individual nodes at one end of the sweep line may be arbitrarily far away from nodes at the other end of the sweep line. Figure 4.10 illustrates the progress of a sweep across a network.

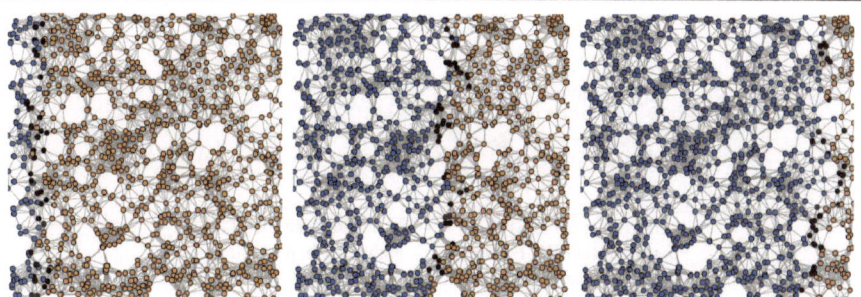

Fig. 4.10. Example of a hop-count sweep across the network, with unswept nodes in orange, swept nodes in blue, and the sweep front in black

The key to coordinating a sweep is generating a *potential function* across the network [104]. Intuitively, the potential function must form a smooth, monotonically increasing or decreasing gradient over which the sweep line can progress. In later chapters we shall see examples where environmental sensed variables can provide such a potential function. In the algorithm in this section, and in Fig. 4.10, the number of hops from the left-hand edge of the network is used as the potential function. In Fig. 4.10, the sweep has reached three hops, eight hops, and 11 hops, moving from left snapshot to right snapshot.

Protocol 4.13 first sets up the hop count required as the potential function, and then performs the sweep of the network. The sweep starts with the nodes initialized to state INIT, assumed (but not required) to be on the left-hand edge of the network. Nodes in the INIT state transition immediately to the swept SWPT state. The remaining nodes in the IDLE state must transition to SWPT via the FRNT state subject to the following constraints (summarized in Fig. 4.11):

- 'A node n may transition from IDLE to FRNT if:
  a. it is adjacent to a SWPT node; and
  b. no adjacent FRNT or IDLE nodes have a potential function value (hop count) lower than n.
- A node n may transition from FRNT to SWPT if:
  a. all adjacent FRNT nodes have the same potential function value (hop count); and
  b. all adjacent IDLE nodes have a higher potential function value (hop count) than n.

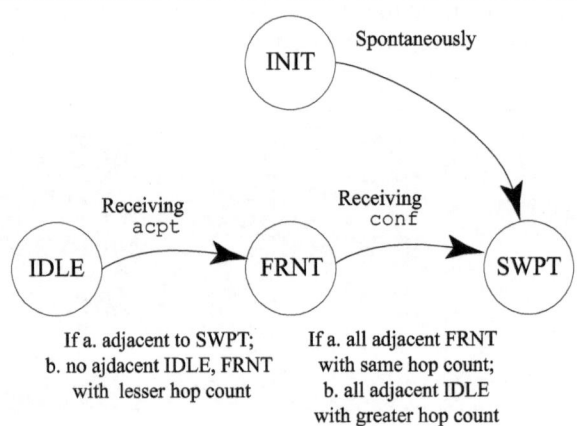

Fig. 4.11. State transition system for Protocol 4.13, highlighting conditions for transitions to FRNT and SWPT

The resulting algorithm is akin to the parallel sweep search technique sometimes used by teams of search and rescue (often seen in police dramas on television). Each individual rescuer in the team must only ensure she remains abreast with her immediate neighbors as the team advances through the search area. As long as each rescuer remains level with her immediate

Protocol 4.13. Hop count sweep

Restrictions: $\mathcal{NB}$; identifier function $id : V \to \mathbb{N}$

State Trans. Sys.: ($\{\text{INIT}, \text{IDLE}, \text{FRNT}, \text{SWPT}\}, \{(\text{INIT}, \text{SWPT}), (\text{IDLE}, \text{FRNT}), (\text{FRNT}, \text{SWPT})\}$)

Initialization: One or more nodes initiating the sweep in state INIT, all other nodes in state IDLE

Local data: Hop counter h_c, initialized to $h_c := -1$; list of neighbor identifiers of unsw message, N_u, initially empty; list of neighbor identifiers of swep message, N_s, initially empty

INIT
   Spontaneously
      set $h_c := 0$ *#Reset hop count*
      broadcast (ping, h_c)   *#Initiate hop count potential function*
      Wait for a short period
      broadcast (invt, h_c)   *#Initiate sweep algorithm inviting neighbors to join*
      become SWPT

IDLE
   Receiving (ping, h)
      if $h < h_c$ **or** $h_c < 0$ **then**
         set $h_c := h + 1$
         broadcast (ping, h_c)
   Receiving (invt, h)
      broadcast (unsw, h_c, $\overset{\circ}{id}$)
   Receiving (acpt, b, i)
      if b **then**
         set $N_u = N_{\overset{\circ}{u}} \cup \{i\}$
         if $|N_u| = |nbr|$ **then become** FRNT
   Receiving (swep, h, i)
      let $b :=$ **false**
      if $h = h_c - 1$ **then set** $b :=$ **true**
      send (conf, b, $\overset{\circ}{id}$) to node with identifier i

FRNT
   Spontaneously
      broadcast (swep, h_c, i)
   Receiving (swep, h, i)
      let $b :=$ **false**
      if $h = h_c$ **then set** $b :=$ **true**
      send (conf, b, $\overset{\circ}{id}$) to node with identifier i
   Receiving (conf, b, i)
      if b **then**
         set $N_s = N_{\overset{\circ}{s}} \cup \{i\}$
         if $|N_s| = |nbr|$ **then**
            broadcast (invt, h_c)
            become SWPT

SWPT
   Receiving (unsw, h, i)
      send (acpt, **true**, $\overset{\circ}{id}$) to node with identifier i *#Swept nodes can always respond to* acpt
   Receiving (swep, h, i)
      send (conf, **true**, $\overset{\circ}{id}$) to node with identifier i *#Swept nodes can always respond to* conf

FRNT, IDLE
   Receiving (unsw, h, i)
      let $b :=$ **true**
      if $h > h_c$ **then set** $b :=$ **false**   *#Check if invite can be accepted*
      send (acpt, b, $\overset{\circ}{id}$) to node with identifier i

neighbors, the entire search team will remain in a straight line, perpendicular to the direction of the sweep.

The potential function effectively provides a global framework to partition the network into *level sets* (i.e., subsets of nodes at one hop, two hops, three hops, ... from the INIT nodes). Assuming the subgraphs induced by these level sets are connected (by analogy, every rescuer in the sweep remains in contact with his or her immediate neighbors), the algorithm will visit each level set in strict order (e.g., all level one hop nodes will be swept before all level two hop nodes).

The efficiency of Protocol 4.13 is partly dependent on the connectivity of the communication graph. If we ignore the construction of the potential function (ping messages equivalent to the protocol for the shortest path tree with a known root; §4.3.2) and concentrate solely on the sweep algorithm itself, then we can see that in transitioning from IDLE to SWPT each node v must transmit:

- exactly one invt message,
- at most $|nbr(v)|$ unsw messages,
- at most $3|nbr(v)|$ acpt messages,
- exactly one swep message,[2]
- at most $2|nbr(v)|$ conf messages.

Thus each node must send $2 + 5|nbr(v)|$ messages, leading to a load balance of $O(|nbr(v)|)$. Every node must perform exactly the same steps (with the exception of a small number of INIT nodes that do not need to transmit the unsw or swep messages). Consequently, the total number of messages sent should be approximately $2|V| + 5|E|$. In sparse networks, where $|V|$ is linearly proportional to $|E|$, this leads to an overall communication complexity of $O(|V|)$. In the worst case (of totally connected networks) the communication complexity rises to $O(|V|^2)$.

4.5.2 Region Boundaries

After (sweep) lines, regions are amongst the most basic of extended spatial entities. A region is simply a "bounded, connected chunk of space." More precisely, in this book the term "region" is used to refer to a closed, strongly connected piece of the plane, topologically equivalent (termed "homeomorphic") to a disk. The term "(complex) areal object" is use to refer more generally to bounded chunks of space that may be disconnected or contain holes. A region and its boundary are interdependent. A boundary is defined by the region it

---

[2] In fact, as we shall see in Chapter 7, this expectation is wrong! However, I have left this analysis in place as it was a genuine error only later uncovered when empirically investigating the algorithm behavior. Can you see why this expectation is incorrect? You may also decide to skip forward to §7.3.5 to see what the error was, and how it was uncovered.

contains; a region cannot exist without its boundary. As a result, depending on the circumstances, one may be concerned with boundary of the region or with the region itself.

Regions, and their boundaries, sometimes arise from the way humans categorize the world around them. Many features of interest in the geographic environment are modeled as *fields*: unique assignments from every point in space to some scalar value. Temperature is a typical example of a field—across some space one might, in principle at least, measure the (unique) temperature at every point in that space (such as the room or the country you are in). A "hot spot" in the sea around a coral reef might be defined as a region of this field where the sea temperature is greater than 1°C above the monthly average.[3] Such regions are said to have *fiat* boundaries; they exist because the boundary condition has some meaning or significance to humans (e.g., that 1°C above the mean monthly sea temperature may indicate danger of coral reef bleaching).

Alternatively, some regions arise from boundaries that correspond to physical discontinuities in the world. For example, when a marine oil spill has occurred, it may be sensible to define regions as corresponding directly to places where there is or is not oil on the surface of the water. Regions that correspond to real discontinuities in the world are said to have *bona fide* boundaries. More information about the distinction between fiat and bona fide boundaries can be found in [105].

From the point of view of our formal model, we can represent the regions sensed by a decentralized spatial information system as a sensor function $s : V \to \{0, 1\}$, where for some node $v \in V$, $s(v) = 0$ indicates a node is *not* in the region and $s(v) = 1$ indicates a node *is* in the region. At this level, we can abstract away from whether the region corresponds to a boundary that is fiat (i.e., arose from thresholding some continuous geographic phenomenon) or bona fide (i.e., arose from physical discontinuities in the world). As a result, each node can know (based on its sensor reading) whether it is *in* or *out* of the region.

Given this information, how can we determine where the boundary of the region is? In fact, there are many answers to this question, and we shall revisit these in more detail in later chapters. However, a natural and purely neighborhood-based definition of an (inner) boundary node is as a node that is *inside* the region but has a direct, one-hop neighbor that is *outside* the region. Conversely, an alternative (but not equivalent) definition of an (outer) boundary node would be a node that is *outside* the region but has a direct, one-hop neighbor that is *inside* the region.

---

[3] Note that, in general, there is no guarantee that thresholding a field will result in a single region, and may instead yield a complex areal object (set of disconnected regions, some of which may contain holes, islands, and so on). More sophisticated algorithms for monitoring complex areal objects are tackled in later chapters.

Figure 4.12 illustrates this concept. The figure shows a network overlaid on a "true" sensed region (such as a region of high temperatures). Nodes that sense they are *outside* this region, but have a one-hop neighbor *inside* the region are labeled with an "O" (outer boundary nodes). Conversely, nodes that sense they are *inside* this region but have a one-hop neighbor *outside* the region are labeled with an "I" (inner boundary nodes). Non-boundary nodes are white (if outside the region) or gray (if inside the region).

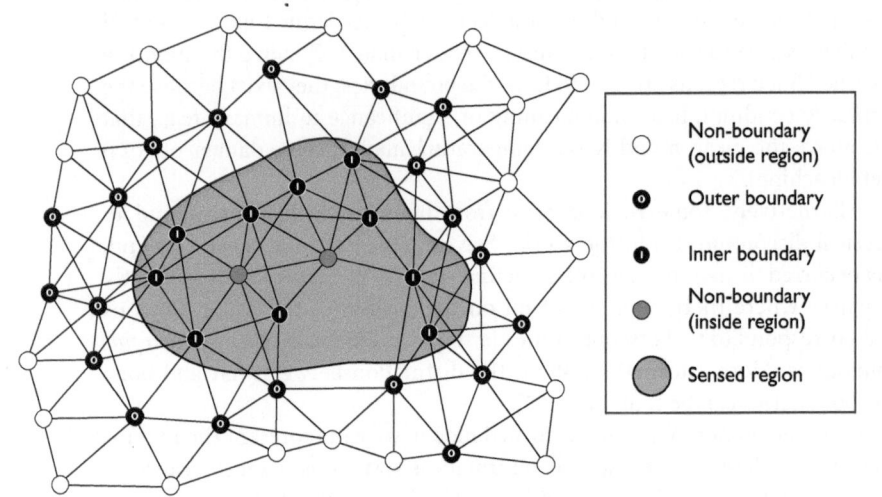

Fig. 4.12. Inner and outer boundary nodes

Using this idea, it is possible to write a simple algorithm to determine whether a node is at a boundary (inner or outer). Protocol 4.14 provides such an algorithm for nodes to locally determine whether they are at the inner boundary of a region. Each node simply broadcasts its sensed value to all its neighbors. Nodes that have sensed they are inside the region, but receive a broadcast from a neighbor that is outside the region, can deduce that they are boundary nodes.

In terms of efficiency, Protocol 4.14 requires each node to send exactly one message, leading to $\Theta(1)$ load balance and $\Theta(|V|)$ overall communication complexity. However, it is noteworthy that the boundary nodes may not directly correspond to what we might normally regard as the boundary. For example, the boundary nodes will change depending on specific connectivity in the communication graph. Thus, an adversary may note that there could be nodes that are very close to the actual region boundary, but because of their local connectivity happen not to become boundary nodes. Similarly, depend-

---

Protocol 4.14. Determining the (inner) boundary of a region

Restrictions: $\mathcal{NB}$; sensor function $s : V \to \{0, 1\}$
State Trans. Sys.: $(\{\text{INIT}, \text{IDLE}, \text{BNDY}\}, \{(\text{INIT}, \text{IDLE}), (\text{IDLE}, \text{BNDY})\})$
Initialization: All nodes in state INIT

INIT
  Spontaneously
    broadcast ($\texttt{ping}, \mathring{s}$)   *#Broadcast sensed value*
    become IDLE

IDLE
  Receiving ($\texttt{ping}, s'$)
    if $s' \neq \mathring{s}$ and $\mathring{s} = 1$ **then**   *#Check if node is in the region, but adjacent to node outside*
      become BNDY

---

ing on the maximum one-hop communication distance, it is also possible that some nodes relatively far from the boundary may become boundary nodes.

Further, although the nodes define points *on* the boundary, there is no obvious mechanism to relate nodes together to form a continuous boundary (e.g., as a cycle of nodes within the communication graph). Indeed, an adversary may correctly argue that the boundaries detected do not necessarily correspond to a single region; instead, the nodes may bound multiple disconnected region components. Later chapters will introduce some more sophisticated algorithms for detecting boundaries that begin to address these important issues. However, for now we shall stick with the simple, neighborhood-based definition of boundary node.

4.6 Protocol of the Chapter: Topological Region Relations

The last algorithm in this chapter, Protocol 4.15, integrates many of the ideas introduced in earlier algorithms into a single decentralized spatial algorithm for determining the topological relation between two regions A and B. In explaining Protocol 4.15, this subsection addresses the algorithm in three stages. First, §4.6.1 briefly introduces a famous spatial model of the topological relation between regions: the 4-intersection model. Next, §4.6.2 describes how the algorithm works. Finally, §4.6.3 presents an analysis of the algorithm using the standard tools we have already encountered.

4.6.1 The 4-Intersection Model

The 4-intersection model provides an answer to the question "how can we systematically characterize the topological relations that can exist between two spatial regions?" The model is one of the most well-known and highly cited pieces of research in the field of geographic information science [38].

The eight possible spatial relations between two regions identified by the 4-intersection model are shown in Fig. 4.13

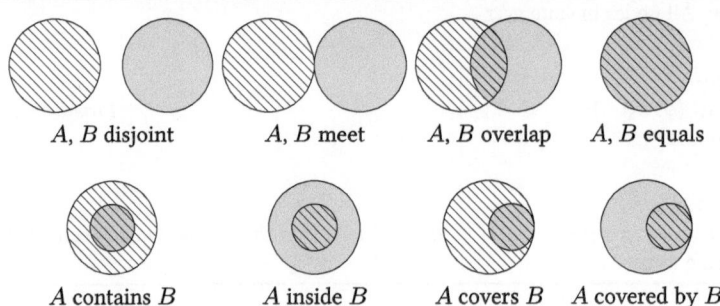

Fig. 4.13. Eight topological relations between two regions classified according to the 4-intersection model

The relations in Fig. 4.13 arise through an analysis of the interaction of two different components of each region: the boundaries of regions A and B (written ∂A and ∂B) and the interiors of regions A and B (written A° and B°). More specifically, as the name suggests, the 4-intersection model examines whether the four set-based intersections between region boundaries and interiors are empty or nonempty. Table 4.1 shows how the eight topological relations in Fig. 4.13 are arrived at. Each named relation ("disjoint," "meet," "overlap," etc.) has a different "fingerprint" in terms of whether the four intersections are empty (0) or nonempty (1). It can be shown that the other eight possible intersection "fingerprints" are not possible for two regions in the plane (although they may be possible for other shapes and spaces with other dimensions). For example, it is not possible to draw two regions in the plane with fingerprint 0111 (i.e., where the interiors of the two regions intersect, the boundary of each region intersects the interior of the other region, but the two region boundaries themselves don't intersect).

4.6.2 Algorithm Operation

The 4-intersection model is widely used in centralized spatial information systems. For example, most spatial database management systems (such as Oracle Spatial) implement operators to detect and query these eight relations. So, an obvious question is: "Can a *decentralized* spatial information system also support queries about topological relations between regions?" Protocol 4.15 provides an example of an algorithm that does exactly this. The algorithm combines the techniques for generating a rooted tree in §4.3.1 with the simple approach to region boundaries introduced in §4.5.2.

Using our structured approach to understanding algorithms, the algorithm can be decomposed into the following important components.

	$\partial A \cap \partial B$	$A^\circ \cap \partial B$	$\partial A \cap B^\circ$	$A^\circ \cap B^\circ$
A, B disjoint	0	0	0	0
A, B meet	1	0	0	0
A, B equals	1	0	0	1
A inside B	0	0	1	1
A contains B	0	1	0	1
A covered by B	1	0	1	1
A covers B	1	1	0	1
A, B overlap	1	1	1	1

Table 4.1. 4-intersections between boundary and interiors of two regions A and B (see Fig. 4.13)

Header

The algorithm assumes that nodes, in addition to neighborhood-based restrictions, have an identifier function and sensors that can detect the regions labeled A and B. This sensing capability is represented with the sensor function $s : V \to 2^{\{A,B\}}$. Consequently for a node v that detects both regions A and B, $s(v) = \{A, B\}$; for a node that detects either A or B, either $s(v) = \{A\}$ or $s(v) = \{B\}$ respectively; for a node that is outside both regions A and B, $s(v) = \varnothing$. Nodes require the ability to store locally their parent in the tree (*parent* $: V \to V \cup \{\varnothing\}$), as in Protocol 4.7. Each node also stores the set of "visited" neighbors N—the one-hop neighbors from which a node has received a tree message. Finally, each node will store a summary of the nonempty intersections it can sense or has heard about, using the function *bin* $: V \to \mathbb{B}^4$ (where $\mathbb{B}^4$ is the set of four-bit binary numbers, $\{0000, 0001, 0010, \ldots\}$).

States

Protocol 4.15 admits four states. The first two states, INIT and SINK, are reserved for the unique root node required for the algorithm. The root node begins the execution in the INIT state, and subsequently transitions directly to the SINK state. All other nodes begin in the IDLE state, and then transition to the DONE state once they have determined their own boundary status.

Events

The algorithm uses two types of messages: a tree message for setting up the rooted tree as well as determining the boundary status of each node; and a rprt message for returning sensed information to the tree root. Aside from the spontaneous event of the unique root node to initiate the algorithm when it is in the INIT state, all the other events relate to receiving tree and rprt messages.

Protocol 4.15. Determining the topological relation between two sensed regions A and B

Restrictions: $\mathcal{NB}, s : V \to 2^{\{A,B\}}, id : V \to \mathbb{N}$
State Trans. Sys.: $(\{\text{INIT}, \text{SINK}, \text{IDLE}, \text{DONE}\}, \{(\text{INIT}, \text{SINK}), (\text{IDLE}, \text{DONE})\})$
Initialization: One node in state INIT; all other nodes in state IDLE
Local data: $bin : V \to \mathbb{B}^4$, initialized $\overset{\circ}{bin} := 0000$; $parent : V \to V \cup \{\varnothing\}$, initialized $\overset{\circ}{parent} := \varnothing$;
   visited neighbors N, initialized $N := \varnothing$

INIT
   Spontaneously
      broadcast $(\texttt{tree}, \overset{\circ}{s}, \overset{\circ}{id})$ *#Sink node initiates algorithm*
      become SINK

IDLE
   Receiving $(\texttt{tree}, x, i)$ *from* v
      $N := N \cup \{v\}$ *#Update list of visited nodes*
      if $\overset{\circ}{parent} = \varnothing$ **then**   *#Check for first* tree *received*
         set $\overset{\circ}{parent} := v$ *#Store tree parent*
         broadcast $(\texttt{tree}, \overset{\circ}{s}, \overset{\circ}{id})$ *#Continue building tree*
      if $x \neq \overset{\circ}{s}$ and $\overset{\circ}{s} = \{A, B\}$ **then**   *#Check for boundary node in A and B*
         if $x = \{B\}$ **then set** $\overset{\circ}{bin} := \overset{\circ}{bin} \vee 0010$ *#Node at boundary of A only*
         if $x = \{A\}$ **then set** $\overset{\circ}{bin} := \overset{\circ}{bin} \vee 0100$ *#Node at boundary of B only*
         if $x = \varnothing$ **then set** $\overset{\circ}{bin} := \overset{\circ}{bin} \vee 1000$ *#Node at boundary of A and B*
      if $N = nbr$ **then**   *#Check if* tree *received from all neighbors*
         if $\overset{\circ}{bin} = 0000$ and $\overset{\circ}{s} = \{A, B\}$ **then set** $\overset{\circ}{bin} := 0001$ *#Check for $A° \cap B°$*
         if $\overset{\circ}{bin} \neq 0000$ **then send** $(\texttt{rprt}, \overset{\circ}{bin})$ **to** $\overset{\circ}{parent}$ *#Initiate message to sink*
         become DONE

DONE, IDLE
   Receiving $(\texttt{rprt}, b)$
      if $b \vee \overset{\circ}{bin} \neq \overset{\circ}{bin}$ **then**   *#Check for new data*
         set $\overset{\circ}{bin} := b \vee \overset{\circ}{bin}$ *#Data aggregation*
         send $(\texttt{rprt}, \overset{\circ}{bin})$ **to** $\overset{\circ}{parent}$ *#Forward aggregate data*

SINK
   Receiving $(\texttt{rprt}, b)$
      set $\overset{\circ}{bin} := b \vee \overset{\circ}{bin}$ *#Deduce topological relation between A and B from Table 4.2*

Actions

As already mentioned, the root node initiates the algorithm following a spontaneous event by broadcasting a `tree` message before transitioning to the SINK state. In the SINK state, the root node then awaits the return of `rprt` messages from across the network. The remaining IDLE nodes await the arrival of a `tree` message. When a `tree` message is received, the node first constructs the routing tree, similarly to Protocol 4.7. In addition, if it can sense both regions A and B it checks whether it is at the boundary of A and/or B. In this way, nodes suitably located can detect the first three intersections required in the 4-intersection model, $\partial A \cap \partial B$, $A° \cap \partial B$, and $\partial A \cap B°$. Nodes that do detect one or more of these intersections update their local data (the first three bits stored in *bin*) accordingly. Nodes that can sense both regions A and B, but are not at a boundary (i.e., have no neighbors not in both A and B), update their local *bin* data to 0001.

Once an IDLE node has received messages from all its neighbors, it is ready to forward a rprt message back to the sink before transitioning to the DONE state. The rprt message carries a payload of the node's observed four intersection bits. Importantly, IDLE or DONE nodes that receive a rprt message may simply combine that with their own information (using a logical disjunction operator, $\lor$). Thus, messages containing information that a node has already forwarded can be discarded. The SINK node can deduce the topological relation between the two regions based on the disjunction of all received messages (Table 4.2).

It should be immediately noticeable that Table 4.2 is not identical to Table 4.1. Although all of the possibilities in Table 4.1 appear as bit combinations in Table 4.2 (highlighted in bold), the converse is not true: there are valid bit combinations in Table 4.2 that do not correspond to possible intersections in the 4-intersection model.

The reason for this discrepancy is that nodes in a sensor network can only provide information about the regions being monitored with limited spatial granularity. As a result, nodes in both regions A and B may detect the boundaries of A and B separately (i.e., node v such that $s(v) = \{A, B\}$ with neighbors v', v'' where $s(v') = \{A\}$ and $s(v'') = \{B\}$) and/or together (i.e., neighbor v''' of v where $s(v''') = \varnothing$). Thus, a node that is at the boundary of A and B may detect the boundary of A, the boundary of B, and the combined boundary of A and B through separate interactions with neighbors. Consequently, although the states 1010, 1100, 1110, and 0110 are not possible in the 4-intersection model, in Protocol 4.15 they all relate to 1000, meet.

For similar reasons, it is possible that no node happens to be so positioned and connected as to detect directly the combined boundary between A and B. It may instead only detect separately the boundaries of A and B, leading to the state 0111 in addition to the normal overlap state 1111.

Finally, state 0001 can potentially occur in cases where *all* the nodes in the network detect both A and B (i.e., where the boundaries of both regions are beyond the spatial extents of the network). In such cases, we can only say that the regions do not meet and are not disjoint. However, determining more precisely the topological relationship is then impossible without a network with larger spatial extents.

Figure 4.14 summarizes four of the possible bit combinations that have no direct correspondence in the 4-intersection model. Together or individually these four bit combinations indicate a meet relation between the two regions (because there are no non-boundary nodes inside $A \cap B$). However, it should be immediately obvious that the meet relation in Fig. 4.14 does not look the same as the meet in Fig. 4.13. In actuality, the regions overlap slightly, rather than meet.

This effect is a result of the limited spatial granularity of the network. Discrete sensors can never detect the boundary *directly*, only indirectly through neighbors that are on the opposite side of a boundary. It is this limited granularity that makes the regions in Fig. 4.13 appear to overlap (and indeed they would overlap if we had a finer spatial granularity geosensor network).

In fact, exactly the same could be said for the conventional data capture techniques used to generate the crisp regions in today's GIS and spatial databases. However, the spatial data structures (such as polylines and polygons) used in centralized spatial information systems obscure the limited granularity of the underlying observations used to generate data. Thus, computing the four-intersection relations in a spatial database has the *appearance* of being precise, but in reality is no more precise than the process depicted in Fig. 4.14. Because data capture and data processing occur in tandem in a decentralized spatial information system, the granularity limitations of the data are clear for all to see.

$bin(v)$ for sink node $v \in V$	Topological relation
0000	A, B disjoint
0001	A and B not meet or disjoint
0010	B contains A, A has no interior nodes
0100	A contains B, B has no interior nodes
1000, 1100, **1010**, 0110, **1110**	A, B meet
0011	B contains A
0101	A contains B
1001	A, B equal
1111, 0111	A, B overlap
1011	B covers A
1101	A covers B

Table 4.2. Determining the topological relation between regions A and B (see Protocol 4.15), for $bin(v) \in \mathbb{B}^4$ of sink node $v \in V$. 1000 indicates $\partial A \cap \partial B \neq \varnothing$; 0100 indicates $A° \cap \partial B \neq \varnothing$; 0010 indicates $\partial A \cap B° \neq \varnothing$; 0001 indicates $A° \cap B° \neq \varnothing$. Bold type indicates valid bit combinations from the 4-intersection model, in Table 4.1

4.6.3 Analysis

From a computational perspective, every node must broadcast exactly one tree message, leading to $\Theta(1)$ load balance and overall $\Theta(|V|)$ tree messages sent. In the worst case, it might be expected that individual nodes close to the sink might have to forward close to $O(|V|)$ rprt messages. However, careful examination of Protocol 4.15 reveals that each node performs a small amount of processing on the received rprt messages, checking whether they

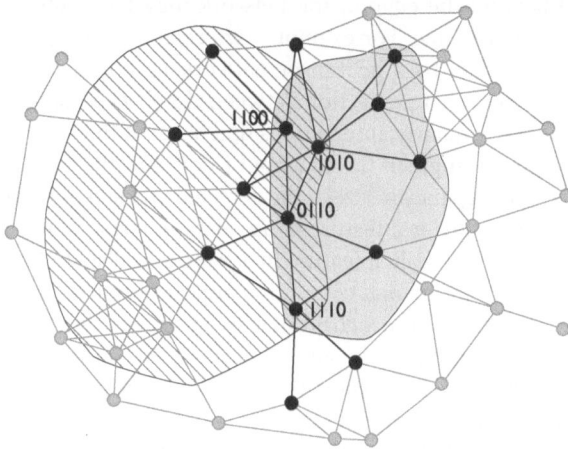

Fig. 4.14. Four instances of bit combinations that do not have direct correspondents in the 4-intersection model, related to meet relation

contain any information not already known to the node. As a result, each node can send in the *worst* case at most four `rprt` messages (1000, 0100, 0010, 0001). Thus, the algorithm results in a constant load balance ($O(1)$) and overall communication complexity of $O(|V|)$ for `rprt` messages. As a consequence, the entire algorithm requires in the worst case $5|V|$ overall messages sent—$O(|V|)$. The upshot of this result is that Protocol 4.15 is of the same order of communication complexity as computing the rooted tree and computing the boundary of a region—although in practice we must be prepared for a worst case of five times as many messages sent by Protocol 4.15 than by Protocol 4.7 or 4.14.

Adopting an adversarial position, it should be immediately obvious that the approach taken in Protocol 4.15 has an important limitation: by default it assumes *empty* intersections unless it receives information about nonempty intersections. For example, the sink node will assume the regions are disjoint unless and until it receives information about a nonempty interaction. In a very large network with regions that interact a long way from the sink, it may require arbitrarily long periods of time for the required messages to reach it. Unless the sink has already received messages that enable it to determine that the regions overlap (in which case further messages can never alter this conclusion), it can never be certain that an adversary has not delayed some new piece of information that will subsequently change the inferred topological relation. Combating this limitation would require additional (and communication-intensive) procedures to verify that all nodes have responded, as discussed in §4.4.1.

Also, as discussed above, the algorithm assumes that the spatial extent of the regions is contained within the spatial extent of the network. Regions that extend beyond the edge of the network may have interactions that cannot possibly be detected. For example, two regions that appear to be disjoint in the area covered by a geosensor network may in actuality overlap if there exists a nonempty intersection between A and B outside the area covered by the network. Similarly, interactions that occur at finer spatial granularity than can be discerned given the spacing of nodes in the network may also mean there is a discrepancy between the detected and actual spatial interactions. As argued above, this is a feature of any data in a spatial information system; it happens to be an explicit feature in decentralized spatial information systems because data capture and computing are intertwined.

Finally, because the algorithm is in part the basic rooted tree algorithm in Protocol 4.7, it is similarly subject to the criticism that the paths to the root may be substantially greater than the shortest path (see §4.3.1 for further information about rooted trees). One way to address this limitation would be to adapt Protocol 4.15 to use the shortest path tree (see §4.3.2, question 4.10).

4.6.4 Summary

Protocol 4.15 demonstrates that even with access to minimal spatial information, the neighborhoods of nodes alone, it is still possible to support fundamental spatial queries in a decentralized way. The next chapter will start to look at more powerful queries that require more sophisticated forms of spatial knowledge.

Protocol 4.15 also provides an example of how synergies between different protocols can enable their combination to operate without increasing the communication complexity. More specifically, Protocol 4.15 combines the algorithm for generating a rooted tree (Protocol 4.7) and the algorithm for detecting the boundary (Protocol 4.14), but operates with the same order of overall communication complexity, $\Theta(|V|)$.

4.7 Chapter in a Nutshell

This chapter has surveyed a wide range of decentralized algorithms that operate in neighborhood-based networks, where nodes only have minimal spatial information about their immediate one-hop neighbors in the network. The algorithms covered include basic information dissemination (e.g., flooding, gossiping, rumor routing), the generation of tree overlay networks, and the election of leaders from within a group of nodes, and some algorithms for constructing and querying the properties of sweep lines and region boundary structures.

Aside from the specific details of the algorithms themselves, some of the key design practices emphasized in the exploration of these algorithms include:

- As highlighted in Chapter 3, the design process is iterative, typically yielding a range of related algorithms that are best suited to slightly different operating environments.
- Combinations of algorithms are not only possible, but advantageous. Combining algorithms provides a way to break complex problems into smaller, more easily solvable components. One example of a technique for combining algorithms is to take the data generated by one algorithm and simply add it as a restriction to a second, derived algorithm. Later chapters introduce other techniques for algorithm modularity.
- Designing and understanding complex algorithms is helped by a systematic approach to any protocol: what are the restrictions, states, events, and actions for a protocol? The key to understanding a complex protocol is studying *interactions* before *actions*; understand the state transitions and events for communicating nodes, and the details of the actions usually follow quite naturally.

Building on this platform, the next chapter adds a new level of spatial sophistication into the mix, investigating what sorts of decentralized spatial algorithms can be supported if a node possesses more spatial information than just the identifiers of its one-hop neighbors.

Review questions

4.1 Using a literature search, discover some other types of flooding, and attempt to understand them and construct a protocol for them using our specification procedure.

4.2 The gossiping algorithm in Protocol 4.5 suffers from the problem that if the INIT node has relatively few neighbors, it is possible the message will never be rebroadcast and the gossip will prematurely die [52]. Adapt Protocol 4.5 to address this problem.

4.3 The rumor routing algorithm in Protocol 4.5 only goes so far as to route the request to the source node. Extend this protocol so that once the request is received by the source node, the sensor values from the source are periodically (e.g., every ten minutes) forwarded to the sink.

4.4 Further extend the rumor routing algorithm in Protocol 4.5 to support the detection of multiple different types of environmental events (for example, both "hot spots" and "cold spots") from multiple different sinks.

4.5 Protocol 4.8 in the worst case may require a node to send and receive $O(|V| - 1)$ messages. Prove this by example by generating a simple execution (for example, using a spatial sequence diagram) of a small network of ten nodes, where one node exhibits this worst case

4.6 Looking at Protocols 4.8 and 4.9, write a new algorithm that can generate the *bidirected* shortest path tree. What is the communication complexity of this new algorithm?

4.7 Adapt Protocol 4.10 to compute the *average* sensed value in the network.

4.8 Extend or revise Protocol 4.11 to additionally verify that the algorithm has terminated, and notify the root node of this. (Hint: One approach is to have putative roots periodically poll their tree, checking for any neighbors of nodes in that tree that are nodes of a different tree.)

4.9 Extend Protocol 4.14 to additionally determine *outer* boundary nodes.

4.10 Address the criticism that Protocol 4.15 may lead to an overlay network tree with nodes at depths that are substantially longer than those for the shortest path tree by combining Protocols 4.15 and 4.8.

4.11 In cases where the intersection of regions A and B covers most of the network, Protocol 4.15 may result in a large proportion of nodes in the network sending a message. Consider how to adapt Protocol 4.15 in order to ensure that only nodes at the boundary of the intersection of A and B generate `rprt` messages. (Hint: see [33].)

Location-Based Algorithms

5

Summary: Neighborhood-based algorithms have access to only the most basic spatial information. Location-based algorithms, however, relax this restriction and assume access to a broader range of spatial information, such as the coordinate locations of nodes. As a result, location-based algorithms can provide a much broader range of functionality, including generating planar network topologies; efficient geographic routing; computing the area and centroid of regions; and identifying the topological structure of complex areal objects.

L ocation-based algorithms extend the neighborhood-based algorithms seen in the previous chapter by using additional spatial information about the locations of neighboring nodes. This additional information is often just the coordinates of the nodes (perhaps determined using GPS). However, other types of location information may also be important, such as qualitative or quantitative direction information (e.g., cyclic ordering or bearing of neighbors), as argued in §2.3. Like the neighborhood-based algorithms in Chapter 4, the location-based algorithms explored in this chapter are restricted to purely instantaneous information at a node and its immediate one-hop neighbors, and ignore issues of change over time.

This chapter starts by revisiting some of the neighborhood structures, originally introduced in Chapter 2, which require nodes to have access to their coordinate location. Then §5.2 examines the question of routing information through the network using the nodes' spatial locations. Section 5.3 examines some examples of location-based algorithms that use region boundaries for computing a range of sophisticated properties, such as the area or centroid of regions, leading to the generation of information about the topological structure of complex areal objects (§5.4).

M. Duckham, *Decentralized Spatial Computing*, DOI 10.1007/978-3-642-30853-6_5,
© Springer-Verlag Berlin Heidelberg 2013

5.1 Planar Overlay Network Structures

Section 2.6.2 introduced a range of planar network structures. As we shall see later in this chapter, these planar network structures are useful as overlay networks for a wide range of location-based algorithms. However, in order to use these overlay network structures, they must first be decentrally computed. This section will examine briefly the computation of three important network structures previously encountered: the Gabriel graph, the relative neighborhood graph, and a localized version of the minimum spanning tree.

5.1.1 Gabriel and Relative Neighborhood Graph

The Gabriel and relative neighborhood graphs were introduced in §2.6.2. The algorithms for computing these two overlay networks (Protocols 5.1 and 5.2) are closely related, as might be expected given the close relationship between the graph structures themselves.

Turning first to the Gabriel graph in Protocol 5.1, the restrictions to the algorithm are the standard neighborhood-based restrictions plus the identifier function and the coordinate location of each node, $p : V \to \mathbb{R}^2$. All nodes are initialized in state INIT. Each node's knowledge of its neighbors' coordinate locations, as acquired, is stored locally in the set N.

Protocol 5.1. Gabriel graph protocol

Restrictions: $\mathcal{NB}$, coordinate location $p : V \to \mathbb{R}^2$, identifier function $id : V \to \mathbb{N}$
State Trans. Sys.: $(\{\text{INIT}, \text{GBRG}\}, \{(\text{INIT}, \text{GBRG})\})$
Initialization: All nodes INIT
Local data: Set N of Gabriel graph neighbors and locations, initialized $N := \varnothing$

INIT
  Spontaneously
    broadcast (ping, id, $\mathring{p}$)   *#Broadcast location and identifier*
    become GBRG

GBRG, INIT
  Receiving (ping, i, l)
    let add := **true**   *#Default flag for adding l to Gabriel graph neighbors N*
    for all $(i', l') \in N$ **do**
      if $\delta(\mathring{p}, l')^2 + \delta(l', l)^2 < \delta(\mathring{p}, l)^2$ **then**   *#Check if l violates Gabriel graph condition*
        set add := **false**   *#Flag l not to be added to Gabriel graph neighbors*
      if $\delta(\mathring{p}, l)^2 + \delta(l', l)^2 < \delta(\mathring{p}, l')^2$ **then**   *#Check Gabriel graph condition for l' if l added*
        set $N := N - \{(i', l')\}$ *#Remove l' from Gabriel graph neighbors*
    if add = **true then**   *#Check if l has satisfied Gabriel graph condition*
      set $N := N \cup \{(i, l)\}$ *#Add l to list of Gabriel graph neighbors*

Protocol 5.1 requires every node (initially in the INIT state) to broadcast its location and identifier in a ping message, before transitioning to the GBRG state. Once in the GBRG state, nodes await the receipt of ping messages from their neighbors. The algorithm operates *incrementally*; as each ping message with its new neighbor location l is received, Protocol 5.1 updates the set N of Gabriel graph neighbors according to the following two rules:

1. The new neighbor will be added to a node's list of Gabriel graph neighbors only if this addition will not violate the Gabriel graph condition with respect to any pairs of existing nodes in the set N. (See Fig. 2.10—essentially we check that the new neighbor is not in position w before adding it to N.)
2. If the new neighbor causes the Gabriel graph condition to be violated for any nodes in the set N, they will be removed from the list. (See again Fig. 2.10—essentially we now check that the new node never puts *other* neighbors in N in position w.)

These two rules give rise to four possible cases, shown in Fig. 5.1. Upon receiving a ping message, the two rules above may cause a node to: 1. do nothing (conditions 1 and 2 not met); 2. add the new neighbor to its set of Gabriel graph neighbors N (condition 1 met, condition 2 not met); 3. remove one or more existing neighbors from its set of Gabriel graph neighbors N (condition 1 not met, condition 2 met); or 4. add the new neighbor *and* remove one or more existing neighbors from the set of Gabriel graph neighbors (conditions 1 and 2 met). This is the essence of the algorithm's *incremental* approach: the Gabriel graph neighbors are recomputed as each new ping message is received.

Two further style points are worth highlighting for the protocol. First, recall that the Gabriel graph will be a spanning subgraph of the communication graph (the UDG, $G = (V, E)$). For conciseness, the algorithm does not go as far as regenerating the global overlay network structure; it only generates the set N of Gabriel graph neighbors for a node. Nevertheless, across all nodes, the union of local Gabriel graph neighbor sets N does contain all the information required to construct the global Gabriel graph (i.e., for all nodes v with Gabriel graph neighbors N_v, the set of edges $E' \subseteq E$ in the Gabriel graph is $E' = \bigcup_{v \in V} \{\{v, v'\} | (id(v'), l) \in N_v\}$).

Second, a new keyword has been introduced into the protocol syntax to assist with more sophisticated actions. The **let** keyword simply creates a new variable that can be used inside an action (i.e., **let** *add* will create a new variable *add* that can be used such as local data, but that has scope limited to the action in which it is created).

Protocol 5.2 is almost identical to Protocol 5.1. The only difference is that instead of the Gabriel graph condition, the relative neighborhood condition is applied inside the action for the *Receiving* event. Because the two protocols are essentially the same, their computational characteristics are also equivalent. Each node must send exactly one message, leading to $\Theta(|V|)$ overall communication complexity and $\Theta(1)$ load balance for messages sent. This also means that the algorithm for computing the planar Gabriel or

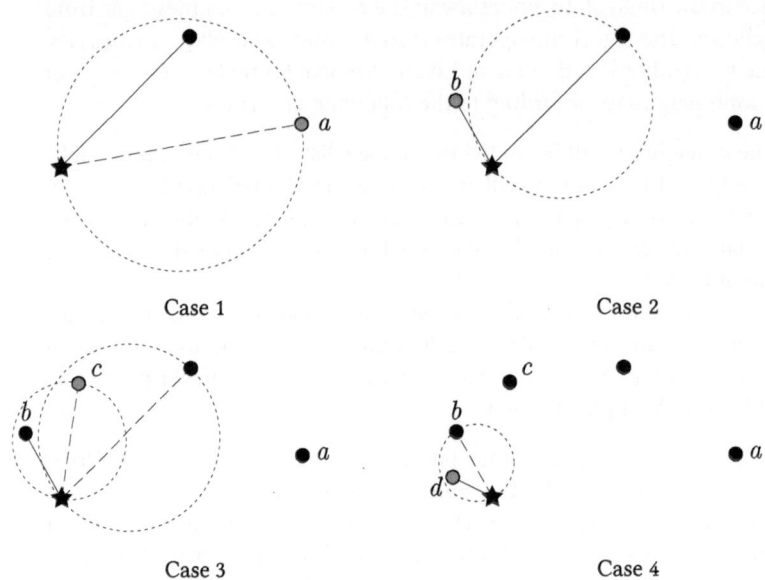

Case 1   Case 2

Case 3   Case 4

Fig. 5.1. Four possible cases for a node (star) receiving a ping messages from neighbors in Protocol 5.1: 1. receiving ping message from a, do nothing; 2. receiving ping from b, add new neighbor; 3. from c, remove existing neighbor; 4. from d, add new and remove existing neighbor

relative neighbor graph can be combined in a single step with some other fundamental algorithms, such as the boundary detection algorithm in the previous chapter, Protocol 4.14 (see question 5.1).

From an adversarial perspective, we should note that the algorithm assumes each node's coordinate location is accurate and precise. The algorithm will still terminate if nodes' knowledge of their location is in some way imperfect. However, the graph generated is not guaranteed to be (or even planar) if the coordinates are inaccurate or imprecise.

Further, the incremental nature of the algorithm does mean that at any point before the algorithm terminates the set of all neighbors may not form the Gabriel graph (although they will always be planar), and may not even be bidirected (i.e., before the algorithm has terminated, it is possible that a node v may have v' as a Gabriel graph neighbor, but that v' has not yet received the information that makes v a neighbor). The algorithms do not explicitly acknowledge termination. However, Protocols 5.1 and 5.2 could easily be extended to explicitly terminate by keeping track of which neighbor a ping message has been received from and subsequently entering a DONE state when ping messages have been received from all (communication graph) neighbors (question 5.2).

Protocol 5.2. Relative neighborhood protocol

Restrictions: $\mathcal{NB}$, coordinate location $p : V \to \mathbb{R}^2$, identifier function $id : V \to \mathbb{N}$
State Trans. Sys.: $(\{\text{INIT}, \text{RNBR}\}, \{(\text{INIT}, \text{RNBR})\})$
Initialization: All nodes INIT
Local data: Set N of Gabriel graph neighbors and locations, initialized $N := \varnothing$

INIT

   Spontaneously
      broadcast (`ping`, $\overset{\circ}{id}$, $\overset{\circ}{p}$)   *#Broadcast location and id*
      become RNBR

RNBR, INIT

   Receiving (`ping`, i, l)
      let $add :=$ **true**   *#Default flag for adding l to RNG neighbors N*
      for all $(i', l') \in N$ **do**
         if $\delta(\overset{\circ}{p}, l') < \delta(\overset{\circ}{p}, l)$ **and** $\delta(l, l') < \delta(\overset{\circ}{p}, l)$ **then**   *#Check if l violates RNG condition*
            set $add :=$ **false**   *#Flag l not to be added to RNG neighbors*
         if $\delta(\overset{\circ}{p}, l) < \delta(\overset{\circ}{p}, l')$ **and** $\delta(l, l') < \delta(\overset{\circ}{p}, l')$ **then**   *#Check RNG condition for l' if l added*
            set $N := N - \{(i', l')\}$ *#Remove l' from RNG neighbors*
      if $add =$ **true then**   *#Check if l has satisfied RNG condition*
         set $N := N \cup \{(i, l)\}$ *#Add l to list of RNG neighbors*

5.1.2 Local Minimum Spanning Tree

Section 2.6.3 introduced the minimum spanning tree (MST): the unique spanning tree with the minimum total edge length. Decentralized algorithms for computing the MST do exist (e.g., [43]). However, as one might expect from the global nature of the MST definition (i.e., that the *total* edge length is minimized across the graph), decentralized algorithms for computing the MST tend to be algorithmically and computationally relatively complex (typically $O(n \log n)$).

As a result, several alternative algorithms have been proposed that can locally compute overlay network structures based on the MST. For example, Protocol 5.3 computes a structure termed the *local minimum spanning tree* (LMST) [71]. The LMST is in fact not necessarily a tree at all, as it may contain cycles. However, the LMST can be shown to be planar [73] and connected, and have nodes with degree 6 at most [71]. Providing such guarantees of maximum degree can provide technical benefits, for example, by reducing the likelihood of network interference between broadcasting nodes.

Protocol 5.3 starts with each node broadcasting its identifier and coordinate location to its neighbors. Once a node has received information from all its neighbors, it computes its own MST for its local neighborhood. Computing the MST in this way is a well-studied problem in computational geometry, with a number of classic algorithmic solutions. In the case of Protocol 5.3, one of the most famous solutions is adopted: Prim's algorithm. Prim's algorithm begins with an arbitrarily selected node, stored as the sole element in a set of *visited* nodes. Next, the shortest edge incident with exactly *one* node in the (initially singleton) set of visited nodes is selected. This edge is stored as an edge in the MST, and the new node incident with the opposite end of this edge is added to the set of visited nodes. This step of finding and storing the shortest unvisited edge with exactly one endpoint incident with the set of visited nodes is repeated until all nodes have be added to the set of visited nodes. Figure 5.2 illustrates Prim's algorithm for the complete graph of 8 nodes, a–h, initialized with arbitrarily chosen node a.

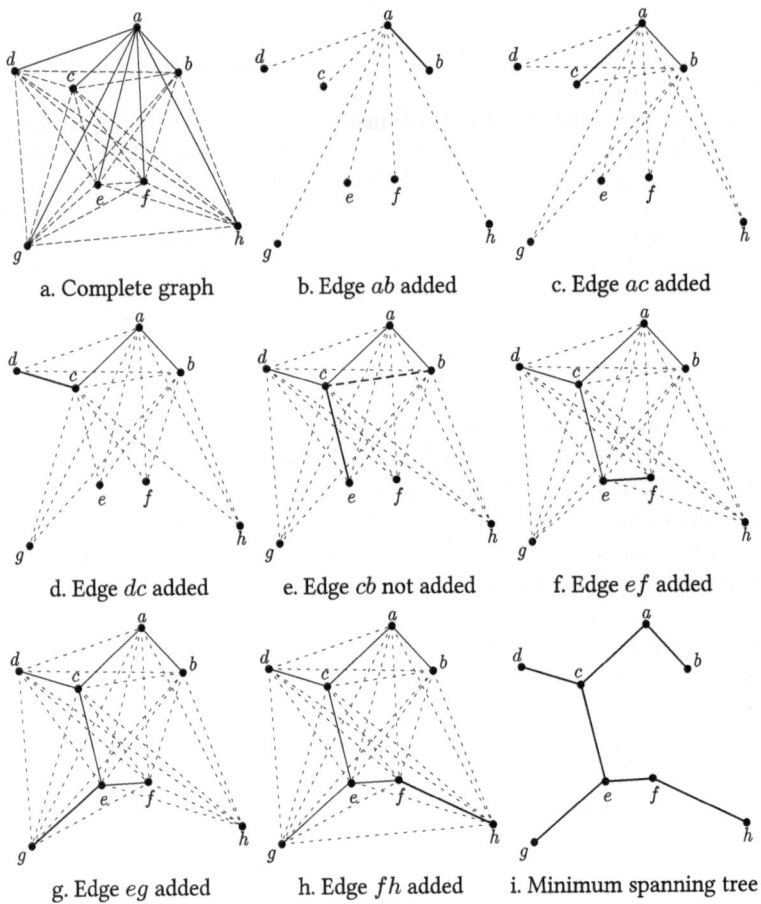

Fig. 5.2. Prim's algorithm, initialized with node a as starting node

In Protocol 5.3, each node locally computes the minimum spanning tree for its own neighborhood, as in Fig. 5.2, as a first step in the algorithm. Having computed the MST for its own neighborhood, each node then communicates with any neighbors to which it is connected in this neighborhood MST, unicasting a chck message. Where a pair of adjacent nodes are connected in both neighboring nodes' neighborhood MSTs, an edge is created in the output LMST overlay network topology. In Protocol 5.3, the local variable nbr_{lmst} is used to store this information at each node.

One new keyword has been introduced in Protocol 5.3: **defer**. It is to be expected that any pair of neighboring nodes will transition to the CHCK state at slightly different times. Consequently, nodes may often receive chck messages before they have completed their own local neighborhood MST construction. In this case, acting on the chck message receiving event needs to be delayed until after the node has transitioned to state CHCK. The **defer** statement does exactly this: it allows the node to respond to a previously received message event only after that node has transitioned into a specified new state. Deferring message receipt effectively puts a message "back in the stack," treating that message as newly received once the node is in a state to act on it.

Unlike Protocols 5.1 and 5.2, Protocol 5.3 is not incremental. Rather, it must collate coordinate information from all neighbors before proceeding to compute the LMST. In the worst case, each node in Protocol 5.3 must transmit at most seven messages. It must first broadcast its identifier and coordinate location (one message). Then it must exchange a chck message with each of its LMST neighbors. It is a well known property of the MST that each node has a maximum degree of 6. Thus, at most six further chck messages must be broadcast. Consequently, in the worst case, Protocol 5.3 will generate seven messages per node, leading to a constant load balance of $O(1)$ and overall communication complexity $O(|V|)$. This compares favorably with decentralized MST algorithms with optimal communication complexity $O(|V| \log |V|)$ [43].

The definition of the LMST can be generalized to include all k-hop (rather than only one-hop) neighbors in each node's neighborhood MST computation (see question 5.4), termed k-*localized minimum spanning trees* [73]. For $k \geq 2$, the k-LMST has the advantage of bounded (although not minimal) total edge length, which in turn can have technical advantages, for example, for topology control.

5.1.3 Unit Delaunay Triangulation

Section 2.6.2 also introduced the Delaunay triangulation and the UDT: the unit Delaunay triangulation formed from the intersection of the Delaunay triangulation and the UDG. Computing the UDT itself is not straightforward. In fact, as for the MST, a range of local algorithms have been proposed to compute various approximations to the UDT (including the restricted Delaunay triangulation, RDT; the partial Delaunay triangulation, PDT; and the local Delaunay triangulation, LDT; see [5]). We leave the further exploration of these algorithms as advanced exercises for the reader (see question 5.5).

Protocol 5.3. Local minimum spanning tree (LMST), after [71]

Restrictions: $\mathcal{NB}$, coordinate location $p : V \to \mathbb{R}^2$, identifier function $id : V \to \mathbb{N}$
State Trans. Sys.: $(\{\text{INIT}, \text{LSTN}\}, \{(\text{LSTN}, \text{CHCK})\})$
Initialization: All nodes INIT
Local data: Set N of neighbors and locations, initialized $N := \varnothing$; Set L of local candidate minimum
  spanning tree edges, initialized $L := \varnothing$;

INIT
  Spontaneously
    set $N := \{\mathring{id}, \mathring{p}\}$ *#Add this node's identifier and location to neighbor set*
    broadcast $(\texttt{ping}, \mathring{id}, \mathring{p})$ *#Broadcast location and id*
    become LSTN

LSTN
  Receiving $(\texttt{ping}, i, l)$
    set $N := N \cup (i, l)$ *#Store neighbor's identifier and location*
    if $|N| > |\mathring{nbr}|$ then *#Process stored identifier/location pairs once received from all neighbors*
      let $M := \{\mathring{id}\}$ *#Initialize Prim's algorithm with first visited node*
      while $|M| \le |\mathring{nbr}|$ do   *#Loop until all nodes in nbr plus this node have been visited*
        let $O = \{(i', l', i'', l'') \in N \times N | i_1 \in M$ and $i_2 \notin M\}$ *#Visited/unvisited node pairs*
        Find $(i_1, l_1, i_2, l_2) \in O$ such that $0 < \delta(l_1, l_2) \le \min_{\delta(l', l'')}(O)$ *#Min length edge*
        set $M := M \cup \{i_2\}$ *#Store new visited neighbor identifier*
        set $L := L \cup \{\{i_1, i_2\}\}$ *#Store new candidate MST edge*
      for all $\{\mathring{id}, j\} \in L$ do
        send $(\texttt{chck}, \mathring{id})$ to node with identifier j *#Communicate chck to candidate neighbors*
      become CHCK
  Receiving $(\texttt{chck}, i)$
    defer until CHCK

CHCK
  Receiving $(\texttt{chck}, i)$
    if $\{\mathring{id}, i\} \in L$ then LMST contains bidirected edge $\{\mathring{id}, i\}$ *#Store bidirected edges*

5.1.4 Summary

The physical network connections, typically modeled as the UDG, provide
the fundamental spatial constraints on the movement of information in a
geosensor network. Overlay networks are used to provide additional constraints on communication. Adding further constraints to the movement of
information initially sounds counterintuitive—surely additional constraints
only make effective computation and useful algorithms harder to achieve?
Nevertheless, the additional structure provided by overlay networks can provide a range of technical advantages, especially in terms of topology control,
energy budgets, and reduction in radio interference between nearby nodes.
Indeed, for these reasons there has already been substantial research in the
computer science community on the definition, properties, and computation
of different overlay network structures and their relative advantages (for
more information, the reader is referred to [106]).

However, our main objective lies in a slightly different direction: designing algorithms for spatial and spatiotemporal queries. Rather than focusing on the technical advantages, in this book we focus more on the effects of overlay network structures (and in particular planar overlay networks) on the opportunities for spatial computation. As we shall see in the following sections, planar overlay networks are fundamental to a number of basic decentralized spatial algorithms, such as those for efficient routing of information based on location.

5.2 Georouting

Geographic routing, or *georouting*, is the term used to describe the process of routing information through a (geosensor) network to a target at a known coordinate location. A fundamental type of query in (centralized) spatial databases is the *point query*, which aims to retrieve all records in the database with spatial references that intersect a particular point in space. In the case of a decentralized spatial information system, such as a geosensor network, there are likely to be occasions when it is similarly important to retrieve sensed data related to a particular coordinate location. In such cases, it is likely to be highly inefficient to query the entire network—rather, we can expect substantial efficiency gains if we can route our query to the node at, or close to, the coordinate location.

The most basic georouting protocol is *greedy* georouting (§5.2.1). However, greedy georouting has an important disadvantage that can be addressed with the use of a planar overlay network, discussed further in §5.2.2.

5.2.1 Greedy Georouting

At its most simple, georouting aims to route a message from one node (the source) through a spatial network to a known coordinate location (the sink). However, in a decentralized spatial information system it is to be expected that the source and the sink will typically not be directly connected in the communication graph. Instead, we must use multi-hop communication to route the message to its destination.

So how can we solve this problem? Clearly, one solution would be to simply flood the message through the network. Flooding will work, but consumes substantial communication resources to route the message to all nodes, for most of which the message will be irrelevant. An alternative solution is for the source to route the message only to a specific neighbor: the neighbor that is closest to the destination. Assuming that all the nodes are able to sense their own coordinate location (e.g., using a GPS), at each hop this process can be repeated, finding the next neighbor that is closest to the destination. This process is called greedy georouting.

Protocol 5.4 provides the specification of a greedy georouting protocol. The algorithm has a number of simple steps:

1. First, the source node broadcasts an init message to all its neighbors.
2. Next, each neighbor receiving an init message responds by returning a posn message containing its coordinate location.
3. Upon receiving all the neighbors' coordinates, the source node computes the Euclidean distance between the coordinates of the destination for the message and each neighbor's location.
4. The source node then unicasts the grdy message, including its destination coordinates, to the neighbor with the *smallest* distance to the destination. In the case of a tie, the source node may arbitrarily select one of the tied neighbors to forward the message to.
5. Upon receiving the grdy message, the neighbor node checks to see if it is the destination (i.e., if its coordinates match those of the message destination). If it is, the algorithm terminates; if it is not, the node now repeats the entire process as if it were the source node.

Protocol 5.4. Greedy georouting protocol

Restrictions: $\mathcal{NB}$, coordinate location $p : V \to \mathbb{R}^2$
State Trans. Sys.: $(\{\text{SEND}, \text{IDLE}, \text{LSTN}\}, \{(\text{IDLE}, \text{SEND}), (\text{SEND}, \text{LSTN}), (\text{LSTN}, \text{IDLE})\})$
Initialization: One node in state SEND, all other nodes in state IDLE
Local data: List $N \subseteq \mathring{nbr}$ of neighbor locations, initialized $N := \varnothing$, message m and destination d, initialized with actual message and destination coordinate only for unique SEND node

SEND
  Spontaneously
    become LSTN   #*Change state to listen for responses to "ping"*
    broadcast init   #*Broadcast "ping" to neighbors*

IDLE
  Receiving init
    broadcast $(\text{posn}, \mathring{p})$ #*Broadcast location in response to "ping" message*
  Receiving (grdy, m', d')
    if $\mathring{p} = d'$ **then**   #*Check if this node location is the destination for this message*
      Message m' received!   #*The message has been successfully received at the destination*
    else
      set $m := m'$ and $d := d'$ #*Store the message and the destination*
    become SEND

LSTN
  Receiving (posn, l)
    set $N := N \cup \{l\}$ #*Store received neighbor location*
    if $|N| = |\mathring{nbr}|$ **then**   #*Check if messages received from all neighbors*
      set $n \in N$ such that for all $n' \in N$, $\delta(d, n) \le \delta(d, n')$ #*Find closest neighbor*
      send (grdy, m, d) to node at location n #*Forward message to next neighbor*
      become IDLE

The greedy georouting protocol is, on average, efficient. Each node that transitions into a SEND state will need to broadcast one init message and to unicast one grdy message. Thus, the total number of init and grdy messages sent will be equal to the length of the route of the message (in terms of number of hops of the message). Additionally, each neighbor of a SEND node will need to send one posn message.

In the worst case, the grdy message might traverse a route that takes it through *every* node in the network. For example, consider an adversary that operates this algorithm in a network where every node has degree 2, except the source and sink, which have degree 1. However, it is likely in a large network that many nodes will be neither on nor neighbors of nodes on the route of the message. These nodes will play no part in the algorithm, and will communicate no information at all. Assuming nodes randomly positioned in the plane, and that the route of the message traces a Jordan (simple, not self-intersecting) curve through the network, we can expect routes to have $\sqrt{|V|}$ hops, leading to $O(\sqrt{|V|})$ hops for init and grdy messages. In this case, the load balance for init and grdy messages is constant communication $O(1)$

The situation is a little more complicated with posn messages. In the worst case, an adversary might design a network with one node that was a neighbor of *every* node on the route of grdy message. In this case, the worst case load balance could be as high as $O(|V|)$, with an overall complexity of $O(|E|) = O(|V|^2)$! A planar overlay network structure reduces the overall worst case complexity of posn to $O(|V|)$ (because $|E|$ is linearly related to $|V|$ in a planar graph), but again on average we can expect in practice a relatively small proportion of nodes, $\sqrt{|V|}$, to be involved in the computation. We return to this expectation in Part III of this book.

Leaving aside efficiency for the time being, there are a number of other important disadvantages of greedy georouting. Can you think of other ways in which an adversary might cause the algorithm to fail?

One important assumption behind Protocol 5.4 is that there *is* a node located at the destination. If this is not the case, then the algorithm will never terminate, with two (or possibly more) nodes close to the destination coordinate forever passing the grdy message in a loop between each other. This failing might be lessened (although not solved) by adapting Protocol 5.4 to terminate once the grdy reaches a node within a certain threshold distance of the destination. In many cases, using such a threshold distance may be reasonable, since spatially nearby sensors will normally sense closely related values (and so nodes *proximal* to the destination may be able to satisfy the query almost as well as sensor nodes *at* the destination).

However, a more subtle problem arises when we observe that it is possible for a node arbitrarily far from the destination to have no neighbors closer than itself to the destination, even in a connected network. Figure 5.3 illustrates just such a situation. In Fig. 5.3, node a is closer to the destination than any of its neighbors, but is by no means the closest node in the network to the destination. This problem can occur whenever there exist regions in space that are not covered by the network, termed *voids*, as labeled in Fig. 5.3.

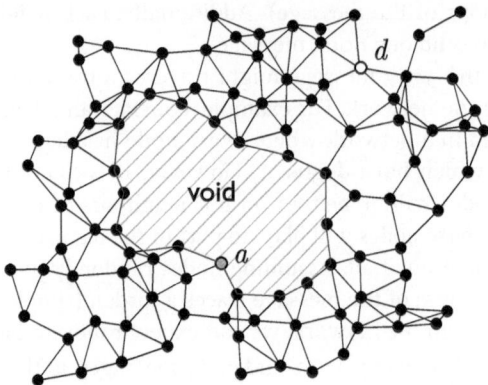

Fig. 5.3. Example void, where node a is closer to destination d than any of its neighbors

To solve this problem we shall need to use a technique called *face routing*.

5.2.2 Face Routing

So far, all our discussion of networks has focused on the graph vertices and the edges. However, in the case of plane networks, there is an alternative view. We may instead focus on the *regions* of the plane bounded by the vertices and edges in the graph, termed *faces*. In a plane graph, the number of faces, edges, and vertices are directly related to each other by the Euler-Poincaré formula, $v - e + f = 2$, where v is the number of vertices, e is the number of edges, and f is the number of faces. For example, Fig. 5.4 shows a plane graph (in fact a Gabriel graph) with 20 vertices (labeled a–t), 33 edges, and $2 - 20 + 33 = 15$ faces. If you count the faces in Fig. 5.4 for yourself and you find only 14 faces, you should note the *exterior* face (the unbounded face round the outside of the graph).

Let us now consider the cyclic ordering for this network. Recall from §2.3 that the cyclic ordering is the (counterclockwise) sequence of neighbors around a node. So, for node s in Fig. 5.4, the cyclic ordering is $rpotj$ (or equivalently $jrpot, tjrpo, \ldots$). The cyclic ordering for the entire graph can be conveniently represented as a function $cyc : E \to V$ (see Table 2.2). Thus, for example, for Fig. 5.4, $cyc(s, t) = j$, $cyc(s, j) = r$, $cyc(s, r) = p$, $cyc(s, p) = o$, and $cyc(s, o) = t$.

The next step in understanding face routing is to notice that one can always cycle around the boundary of any face by visiting edges in order, at each step finding the next edge in the cyclic order of the terminating

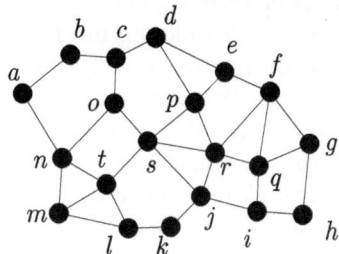

Fig. 5.4. Vertices, edges, and faces in a plane graph

node for that edge. For example, starting at edge (s, j), the next edge should be $(j, cyc(j, s)) = (j, k)$. Then comes $(k, cyc(k, j)) = (k, l)$. Next comes $(l, cyc(l, k)) = (l, t)$. And finally with $(t, cyc(t, l)) = (t, s)$, we have arrived back at the start of our cycle (since $(s, cyc(s, t)) = (s, j)$ again). In combination with the cyclic ordering, any (directed) edge can be used to find a unique face. Try this for yourself—pick any edge at random and find the unique face it bounds (and remember, some edges, such as (b, a) will bound the exterior face).

Armed with this knowledge, it is possible define a simple algorithm for routing a message around any face in a plane network. Protocol 5.5, below, provides such an algorithm. The restrictions include the new abbreviation $\mathcal{LB}$ (for "location-based" restrictions) indicating reliable communication and a connected, bidirectional *planar* communication graph G with associated neighborhood function $nbr: V \to 2^V$.

Protocol 5.5. Face routing

Restrictions: $\mathcal{LB}$; identifier function $id : V \to \mathbb{N}$; cyclic ordering $cyc : E' \to id_*$, where $E' = \{(v, id(v'))|(v, v') \in E\}$

State Trans. Sys.: $(\{\text{SEND}, \text{IDLE}\}, \varnothing)$

Initialization: One node in state SEND, all other nodes in state IDLE

SEND

   Spontaneously
      broadcast init $(\overset{\circ}{id})$ #Broadcast "ping" with node identifier to neighbors
   Receiving init (i)
      Message routed around face

IDLE

   Receiving init (i)
      send init $(\overset{\circ}{id})$ to node with identifier $\overset{\circ}{cyc}(i)$ #Route message to next neighbor in cyclic order

---

In understanding Protocol 5.5, note one important technical alteration to the definition of the cyclic ordering function *cyc*. In previous discussions, the cyclic ordering function has been defined as $cyc : E \to V$. However, the distinction between the *identifier* of a node and the node itself has already been highlighted (see §2.2). A node can only access information about neighbors that is explicitly communicated to it, such as a neighbor's identifier, but not the neighbor *itself*. Consequently, in Protocol 5.5 the cyclic ordering is represented as the function $cyc : E' \to id_*$, where $E' = \{(v, id(v'))|(v, v') \in E\}$. The notation id_* denotes the *image* of the function *id* (the set of values *mapped to* by *id*). Formally, for $id : V \to \mathbb{N}$, the image of the function $id_* = \{i \in \mathbb{N} | id(v) = i$ for some $v \in V\}$.

To help illustrate the operation of face routing, Fig. 5.5 provides a spatial sequence diagram for Protocol 5.5 based on the plane network in Fig. 5.4, using node *o* as the unique node in state SEND, with identifier function $id(a) = 1, id(b) = 2, id(c) = 3, \ldots, id(t) = 20$.

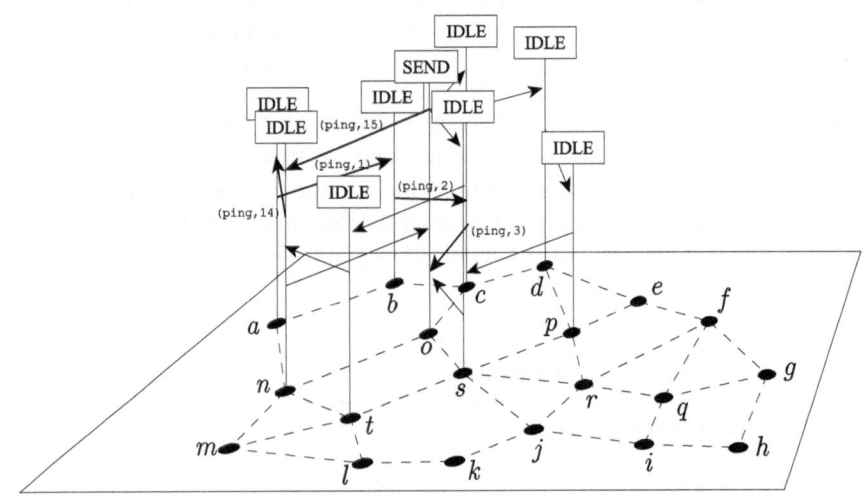

Fig. 5.5. Spatial sequence diagram for face routing, Protocol 5.5, highlighting routing around face *onabc*, and with identifier function $id(a) = 1, id(b) = 2, id(c) = 3, \ldots, id(t) = 20$

5.2.3 GPSR

The face routing algorithm described in the previous section does not on its own generate any useful new information. However, face routing is a fundamental technique in decentralized spatial computing, which as we shall see is basic to a range of more sophisticated behaviors. In the context of georouting, combining face routing and greedy georouting (introduced in §5.2.1) can help to solve the problem of voids in the communication network (Fig. 5.3), which as we have already seen can cause greedy georouting to fail.

The combination of face routing and greedy georouting is called GPSR (greedy perimeter stateless routing, [15, 61]). Like any georouting protocol, GPSR aims to route a message from a source through the network to a sink at a known coordinate location. Under normal circumstances, GPSR simply uses greedy georouting. However, if a non-sink node v detects that all its neighbors are further from the destination than itself, that node then switches to using face routing. Nodes in the network receiving the message continue to use face routing unless a receiving node v' is closer to the destination than the node v which initiated face routing. At that point, node v' switches back to using greedy georouting. The route of a message from a source to a sink may include any number of these switches between face routing and greedy georouting.

Protocol 5.6 presents a version of the GPSR algorithm. To summarize, GPSR proceeds according to the following steps.

1. A node receiving a greedy georouted `grdy` message or a face routed `face` message stores the information contained in the message. The node then transitions to an SEND state (or in the case of the unique source node, starts in a SEND state) and broadcasts an `init` message to its neighbors. Finally, the node transitions into a LSTN state to await responses from neighbors.

2. Upon receiving `posn` responses to the `init` message from all its neighbors, the LSTN node then forwards a `face` or `grdy` message to a neighbor, selected according to the following criteria:

   a) For LSTN nodes that received a greedy georouted `grdy` message (or the source node): i. if they have a neighbor closer to the destination than themselves, they use greedy georouting to forward the message; ii. otherwise they use the geometry of their neighbor locations to forward a `face` message to the next neighbor counterclockwise from the destination coordinate d (i.e., for a node v, the neighbor v' such that angle $\angle dp(v)p(v')$ is minimal). The `face` message includes a target distance t: the remaining distance to the destination.

   b) For LSTN nodes that received a face routed `face` message: i. if they have a neighbor closer to the destination than the target distance t (contained in the received `face` message), then they revert to greedy georouting to forward the message; ii. otherwise they continue to use face routing, again using the geometry of neighbors to forward the message to the next neighbor counterclockwise from the node n from which the message was received (i.e., for a node v, the neighbor v' such that angle $\angle p(n)p(v)p(v')$ is minimal).

Protocol 5.6. GPSR

Restrictions: $\mathcal{LB}$; coordinate location $p : V \to \mathbb{R}^2$
State Trans. Sys.: $(\{\text{SEND}, \text{IDLE}, \text{LSTN}\}, \{(\text{LSTN}, \text{IDLE}), (\text{SEND}, \text{LSTN}), (\text{IDLE}, \text{SEND})\})$
Initialization: One node in state SEND
Local data: List $N \subset \overset{\circ}{nbr} \times \mathbb{R}^2$ of neighbor locations, initialized $N := \varnothing$; message m and sink
 (destination) d (for single SEND node, initialized with actual message and destination coordinate);
 face routing node f; target distance t initialized $t := 0$

SEND
   Spontaneously
      broadcast init   *#Broadcast* `ping` *to neighbors*
      set $t := 0$ **and** $N := \varnothing$ *#Reset t and N*
      become LSTN   *#Change state to listen for responses to* `ping`

IDLE
   Receiving init
      broadcast $(\text{posn}, \overset{\circ}{p})$ *#Broadcast location in response to* `ping` *message*
   Receiving (grdy, m', d') *#Payload d' is destination coordinate for message m'*
      if $\overset{\circ}{p} = d'$ **then**   *#Check if this node location is the destination for this message*
         Message m' received!   *#The message has been successfully received at the destination*
      else
         set $m := m'$ **and** $d := d'$ *#Store the message and the destination*
         become SEND
   Receiving $(\text{face}, m', d', f', t')$ *#Destination d'; message m'; previous node f; target distance t*
      if $\overset{\circ}{p} = d'$ **then**
         Message m' received!   *#The message has been successfully received at the destination*
      else
         set $m := m', d := d', f := f', t := t'$ *#Store* `face` *message payload*
         become SEND

LSTN
   Receiving (posn, l)
      set $N := N \cup \{l\}$ *#Store received neighbor location*
      if $|N| = |\overset{\circ}{nbr}|$ **then**   *#Check if messages received from all neighbors*
         let n such that $n \in N$ and for all $n' \in N, \delta(d, n) \le \delta(d, n')$ *#Find closest neighbor*
         if $t \le 0$ **then**   *#Check if received message through face routing*
            if $\delta(n, d) < \delta(\overset{\circ}{p}, d)$ **then**
               send (grdy, m, d) to node at location n *#Forward message to next neighbor*
            else
               let $n := n \in N$ such that for all $n' \in N, \angle d\overset{\circ}{p}n \le \angle d\overset{\circ}{p}n'$ *#Find ccw neighbor from d*
               send $(\text{face}, m, d, \overset{\circ}{p}, \delta(\overset{\circ}{p}, d))$ to node at location n
         else
            if $\delta(n, d) < t$ **then**
               send (grdy, m, d) to node at location n *#Forward message to next neighbor*
            else
               let $n := n \in N$ such that for all $n' \in N, \angle f\overset{\circ}{p}n \le \angle f\overset{\circ}{p}n'$ *#Find ccw neighbor from f*
               send $(\text{face}, m, d, \overset{\circ}{p}, t)$ to node at location n
      become IDLE

Although a full spatial sequence diagram of GPSR in operation quickly becomes too complex to draw or understand, Fig. 5.6 shows an example sequence of msge and face messages based on the planar (Gabriel) graph derived from Fig. 5.3. In Fig. 5.6, a switch from greedy to face routing occurs at the node at coordinate $(169, 171)$, at distance 122.4 from the destination node $(130, 55)$. Face routing continues until the node at coordinate $(82, 153)$ is reached at distance 109.1 from the destination, less than the target distance.

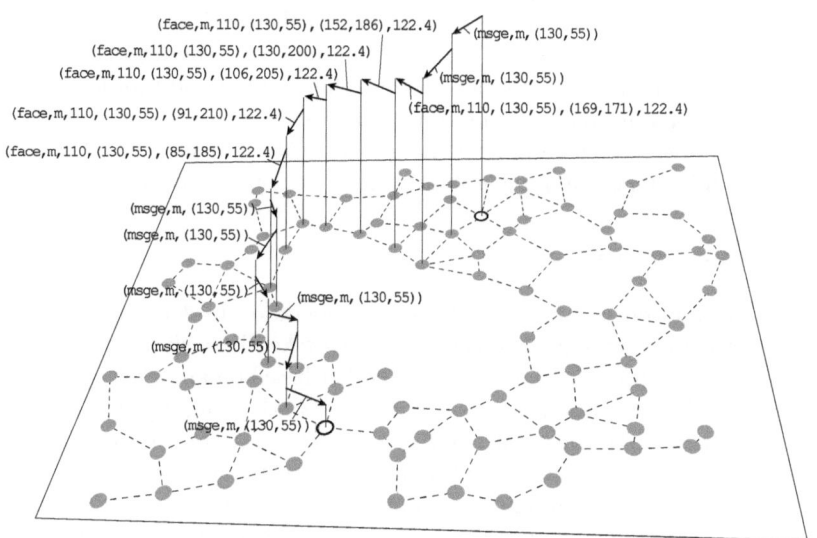

Fig. 5.6. Example sequence of msge and face messages generated using GPSR (Protocol 5.6) from source at coordinate $(205, 215)$ to destination $(130, 55)$

The computational characteristics of GPSR are the same as those of greedy routing, with each message requiring $O(n)$ messages, where n is the number of hops from source to destination. As already seen, we expect $n \approx \sqrt{|V|}$, leading on average to $O(\sqrt{|V|})$ overall communication complexity for each message, and $O(1)$ load balance. In addition, it may be necessary to consider the cost of constructing the planar network necessary for GPSR, for example, by generating the Gabriel or relative neighborhood graph ($\Theta(|V|)$; see §5.1.1).

Using GPSR does ensure that the message can traverse the boundary of any voids in the network, and so eliminates the possibility of delivery failure inherent in greedy routing, with messages stuck in local minima (see Fig. 5.3). However, it should also be noted that when we repeatedly use GPSR to route multiple messages, nodes at the boundary of a large void are likely to fall on a greater proportion of routes. This will result in load imbalance, where nodes at the boundary of a void bear a greater proportion of the communication load, and may have their energy reserves depleted more quickly.

Other problems can be created for GPSR by an adversary. In particular, the arbitrary choice of whether clockwise or counterclockwise cyclic ordering is used can dramatically change the route around the void. For example, trace out for yourself the route of the message in Fig. 5.6 assuming *clockwise* (as opposed to counterclockwise) cyclic ordering. What is different about it? In degenerate cases, a message that may only require two or three hops to the destination with one ordering direction may end up being routed around almost every edge in the exterior face of the network if the other ordering direction happens to be used (see question 5.7). These sorts of issues have led to intense research activity on refining the basic georouting strategies presented in this section (see [42]). However, at this point we leave further exploration of georouting to the reader, and turn to our final set of algorithms in this chapter: routing around geographic region boundaries.

5.2.4 Summary

Georouting is a fundamental spatial operation in decentralized spatial information systems. The basic approach taken here shows how information can be routed from one node to another based on knowledge of each node's coordinate location. Georouting continues to be a highly active research area, with an ever widening range of ingenious variations on these approaches appearing in the literature. Similar techniques can also be used as the basis of *geocasting*: routing information in a spatial network to nodes in a target *region* with known coordinate or polygonal boundaries (see question 5.8). If georouting is analogous to point queries in a spatial database, then geocasting can be seen as the companion to *range queries* in spatial databases (retrieving all records in the database with spatial references that intersect a defined spatial *region*).

5.3 Region Boundaries

In the previous section on georouting, face routing was used simply to solve a problem with greedy routing: to circumscribe voids in the network that would otherwise cause greedy routing to fail. However, the ability to route *around* a face can also provide a mechanism to route information around the boundary of a *geographic* region being monitored by the network, such as a hot spot. As we shall see in this section, this ability to route around the boundary of a geographic region is invaluable, as it provides a way of structuring more complex computations at the boundaries of geographic regions of interest. In general, restricting computation solely to the boundary of a region (rather than considering the entire region) can also substantially reduce the computational overheads of an algorithm.

5.3.1 Boundary Cycles

The idea for routing a message around the boundary of a sensed, geographic region combines two techniques already encountered: region boundaries (§4.5.2) and face routing (§5.2.2). The objective is to find a closed path in the graph (called a *cycle*) that visits all the boundary nodes. We call this the *boundary cycle*. Figure 5.7 illustrates the boundary cycle for a planar network and sensed geographic region similar to that in Fig. 4.12.

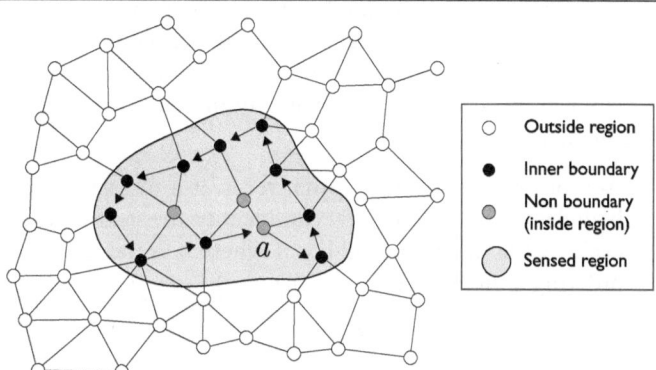

Fig. 5.7. Inner boundary cycle, highlighting boundary condition

More precisely, imagine the *subgraph* induced by the set of nodes inside (i.e., that sense) a particular connected geographic region. Formally, for a graph $G = (V, E)$ and a set of nodes $V' \subseteq V$ each of which sense the same (simple and connected) geographic region, $v \in V'$ $s(v) = 1$, subgraph $G' = (V', E')$ induced by V' has edges $E' = \{(v', v'') \in E | v', v'' \in V'\}$. The inner boundary cycle is formed by the edges of the *exterior* face of this subgraph. The ring of arrows in Fig. 5.7 shows the region's inner boundary cycle.

The question, then, is how can nodes find the inner boundary cycle? The answer to this question requires two distinct conditions, which relate to the two different types of nodes which may occur in the boundary cycle: boundary nodes and non-boundary nodes. A requirement of the boundary cycle is that it visit every boundary node in the region. However, the converse may not be true; not every node in the boundary *cycle* is necessarily a boundary *node* (except in the special case where G is a maximal plane graph). For example, node a in Fig. 5.7 is in the boundary cycle, but is not itself a boundary node (it is inside the region, but has no one-hop neighbors that are outside the region).

The two conditions needed are illustrated in Fig. 5.8. In short, each boundary node can identify the next node in the boundary cycle by finding a pair of neighbors that are adjacent in the cyclic ordering, and where the first node of the pair is *outside* the region but the second node of the pair is *inside* the region (see Fig. 5.8, left-hand side). More formally:

1. Any boundary node b can identify the next node in the boundary cycle as the (unique) neighbor n where neighbor m is *outside* the region (i.e., $s(m) \neq s(b)$) and neighbor n is *inside* the region (i.e., $s(n) = s(b)$), where n is the next node in the cyclic ordering around b from m (i.e., $cyc(b, m) = n$).

The second condition simply requires that to bridge the gaps between boundary nodes in the boundary cycle, any non-boundary nodes that receive a message to be cycled around the region boundary from a neighbor must forward this message to its next neighbor in their cyclic ordering (see Fig. 5.8, right-hand side):

2. Any non-boundary node a that receives a message to cycle around the boundary from neighbor m can identify the next node in the boundary cycle as the (unique) neighbor n where n is the next node in the cyclic ordering around a (i.e., $cyc(a, m) = n$).

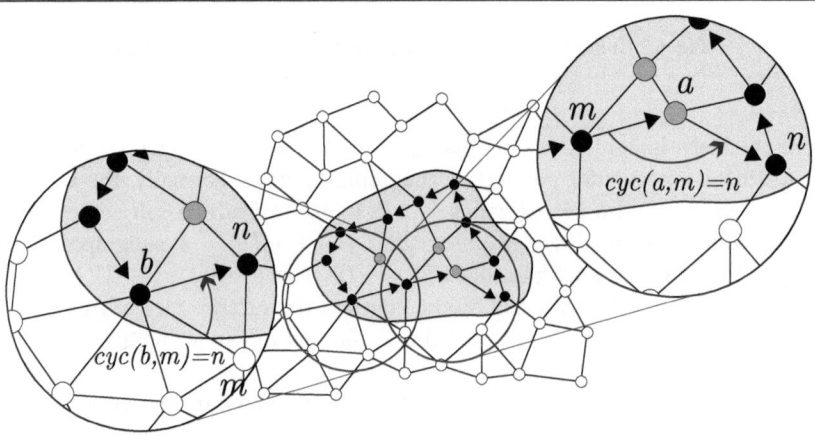

Fig. 5.8. Two conditions for cycling around the boundary (condition 1 left-hand side; condition 2 right-hand side)

The alert reader may have noticed that in the formal conditions above, the word "unique" has been sneakily inserted! For condition 2, uniqueness is unproblematic, as there must by definition always exist a unique neighbor in the cyclic ordering (see §2.3). However, can we guarantee there will always be a *unique* neighbor that satisfies condition 1? Unfortunately, in general we cannot. There may exist cases where a boundary node b has *more* than one neighbor v such that the sensed value $s(v) \neq s(b)$, but $s(cyc(b,v)) = s(b)$. There are a number of ways of dealing with this issue, one of which is left as an exercise (see question 5.9) and another of which we will visit in the next chapter. However, to keep things simple for now, we shall add the constraint that the subgraph G' formed by the nodes inside the sensed region must be *2-connected*, meaning there must exist at least two distinct paths between any pair of nodes in the subgraph (see Appendix A). Subject to this constraint, it can be proved that each boundary node must have a *unique* neighbor that satisfies condition 1 (see [35]).

Two other points are worth highlighting here. First, we have required that the network is plane. The approach taken here can fail for graphs that are not plane. Figure 5.9 shows an example of such a problem, where a boundary cycle message routed from a to b will get stuck in a loop, being sent back and forth between b and c. Second, here we have only defined the process for finding the *inner* boundary cycle. It should, however, be clear from the discussion how one could adapt the approach slightly to also route around *outer* boundary cycles (see question 5.10).

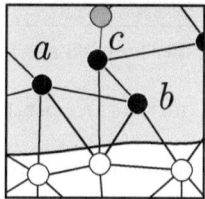

Fig. 5.9. Failure of condition 1 in non-planar graphs. A boundary cycle message routed from a to b will get stuck in a loop between b and c

5.3.2 Protocols for Boundary Cycles

To simplify the explanation of the protocol for routing a message around a boundary cycle, we shall build up the algorithm in two stages. In the first stage, Protocol 5.7 presents the algorithm for finding the inner boundary nodes (similarly to Protocol 4.14), additionally finding for those boundary nodes the next neighbor in the boundary cycle. This is achieved with the *wind* function, which applies condition 1 from the previous section. The algorithm proceeds according the following steps:

1. The algorithm restrictions consist of a communication network structured as a plane graph; sensor and identifier functions; and the cyclic ordering (as with face routing in Protocol 5.5). In addition, the local variable that will be populated with data by the algorithm is the *wind* function.

2. As in Protocol 4.14, each node begins in the INIT state by spontaneously broadcasting an `init` message containing its sensed value in addition to its identifier to its neighbors, before transitioning to an IDLE state.

3. Upon receiving a `ping` message from a neighbor, a node stores this information in the local relation D. Once messages have been received from all neighbors, the relation D is restructured into the local function $data : I \to \{0, 1\}$, where $I = \{i'|(i', d') \in D\}$ and $data(i') \mapsto d'$ such that $(i', d') \in D$. This step is only to assist in readability and conciseness of the algorithm (i.e., we may write "$data(i) = d$" rather than "for some i, the d such that $(i, d) \in D$"). No data is changed or created by this step, the relation D must necessarily contain a unique $d \in D$ for each i.

4. For any nodes that are inside the region (i.e., $\mathring{s} = 1$) and have neighbors that are outside the region (i.e., $0 \in data_*$; recall $data_*$ is the *image* of the function $data$), the local function $\mathring{wind}$ is set to the next node in the boundary cycle (condition 1 in §5.3.1, left-hand side of Fig. 5.8). Finally, these nodes also transition to the BNDY state.

Protocol 5.7. Determining the (inner) boundary nodes and cycle for a region (cf. Protocol 4.14)

Restrictions: $\mathcal{LB}$; $s : V \to \{0, 1\}$; identifier function $id : V \to \mathbb{N}$; cyclic ordering $cyc : E' \to id_*$, where $E' = \{(v, id(v'))|(v, v') \in E\}$
State Trans. Sys.: $(\{\text{INIT}, \text{IDLE}, \text{BNDY}\}, \{(\text{INIT}, \text{IDLE}), (\text{IDLE}, \text{BNDY})\})$
Initialization: All nodes in state INIT
Local data: $\mathring{wind} : V \to V \cup \{\varnothing\}$, initialized $\mathring{wind} := \varnothing$; relation $D \subset \mathbb{N} \times \{0, 1\}$

INIT
  Spontaneously
    broadcast $(\texttt{ping}, \mathring{id}, \mathring{s})$ #*Broadcast sensed value and identifier*
    become IDLE
  Receiving $(\texttt{ping}, i, d)$
    defer until IDLE   #*Only respond to* `ping` *message receipt in* IDLE *state*

IDLE
  Receiving $(\texttt{ping}, i, d)$
    set $D := D \cup (i, d)$ #*Store neighbor identifier and sensed value*
    if $|D| = |\mathring{nbr}|$ **then**   #*Check whether* `ping` *received from all neighbors*
      Create function $data : I \to \{0, 1\}$, where $I = \{i'|(i', d') \in D\}$ and $data : i' \mapsto d'$
      if $\mathring{s} = 1$ and $0 \in data_*$ **then**   #*Check for node inside region with neighbor outside*
        set $\mathring{wind} := \mathring{cyc}(i'')$, where $data(\mathring{cyc}(i'')) = \mathring{s}$ and $data(\mathring{cyc}(i'')) \neq data(i'')$
      become BNDY

As for Protocol 4.14, the algorithm is clearly of $\Theta(|V|)$ overall communication complexity, with optimal, $\Theta(1)$, load balance.

The outcome of operating Protocol 5.7 is that nodes at the boundary of a sensed region transition to the BNDY state, and have local information about which of their neighbors is the next node in the boundary cycle, through the local *wind* variable. Protocol 5.8 takes the next step of routing a single message from each boundary node around the entire boundary cycle. However, instead of rewriting all of Protocol 5.7 in Protocol 5.8, we introduce a new notation for *extending* Protocol 5.7 with the additional capabilities required by Protocol 5.8. Up to now, all protocols have been written out in full. As our algorithms become more sophisticated, it becomes more important to allow *modularity*, where complex protocols can be constructed from simpler components. Modularity can be achieved by specifying protocol *fragments* which either add *new* restrictions, states and/or transitions, local data, events, and actions to existing protocols (indicated with the restriction in the header "Fragment extends"), or *override* (subject to certain constraints) existing actions (indicated with the *Updates* keyword). A more detailed discussion on the structure of protocol fragments can be found in Chapter 9 in the section on modularity.

For example, Protocol 5.8 is a protocol fragment which *extends* Protocol 5.7 by using exactly the same restrictions, states and transitions, local data, and INIT events and actions as Protocol 5.7; adding new *Receiving* events for the IDLE and BNDY states; and updating the action associated with the *Receiving* ping message event for the BNDY state. The algorithm operates much as Protocol 5.7, except that when a BNDY node receives a ping message, the updated action of Protocol 5.8 sends a msge containing that node's identifier to its next neighbor in the boundary cycle (*wind*). The additional *Receiving* events ensure that IDLE and BNDY nodes are able to respond to the receipt of msge messages, forwarding it to the next neighbor in the boundary cycle (condition 1 in Fig. 5.8), or the next neighbor in the cyclic ordering (condition 2 in Fig. 5.8) respectively.

Taking this approach a step further, it is possible to extend Protocol 5.8 with the additional capability to elect a *leader* for the boundary cycle. This extension, given in Protocol 5.9, only requires the addition of the approach already used for leader election in a ring (Protocol 4.12). Protocol 5.9 extends the header of Protocol 5.8 with a new state, LEAD, and a new transition (BNDY, LEAD) in the state transition system; with new local data to store the smallest encountered identifier; and with an update to the action associated with a BNDY node receiving a msge message. The update applies the "as far" leader election test, with a BNDY node only forwarding received messages as long as the source node identifier i contained in the payload of the message is strictly smaller than any other source node identifier previously encountered by that node (including its own identifier). A node that receives back its own msge message must have the smallest identifier of all BNDY nodes in the boundary cycle, and so may transition to the LEAD (leader) state.

Computationally, Protocols 5.8 and 5.9 again require $\Theta(1)$ ping messages to be sent by each node, leading to an overall communication complexity of $\Theta(|V|)$. But how much additional communication is associated with msge messages? Each boundary node will generate a msge message, subsequently passed around boundary cycle nodes. Therefore, deducing the communi-

---

Protocol 5.8. Routing around a boundary cycle (see Protocol 5.7)

Fragment extends: Protocol 5.7

IDLE

  Receiving (ping, i, d) *Updates* Protocol 5.7

    set $D := D \cup (i, d)$ *#Store neighbor identifier and sensed value*

    if $|D| = |n\mathring{b}r|$ then   *#Check whether* ping *received from all neighbors*

      Create function $data : I \rightarrow \{0, 1\}$, where $I = \{i' | (i', d') \in D\}$ and $data : i' \mapsto d'$

      if $\mathring{s} = 1$ and $0 \in data_*$ then   *#Check for node inside region with neighbor outside*

        set $w\mathring{i}nd := c\mathring{y}c(i'')$, where $data(c\mathring{y}c(i'')) = \mathring{s}$ and $data(c\mathring{y}c(i'')) \neq data(i'')$

        send (msge, $\mathring{id}$, $\mathring{id}$) to $w\mathring{i}nd$ *#Initiate message to first boundary neighbor*

        become BNDY

  Receiving (msge, i, i')

    defer until $|D| = |n\mathring{b}r|$

    send (msge, i, $\mathring{id}$) to node with identifier $c\mathring{y}c(i')$ *#Forward message to bndy cycle neighbor*

BNDY

  Receiving (msge, i, i')

    if $i = \mathring{id}$ then   *#Check whether message returned to sender*

    Message traversed entire region boundary!

    else

      send (msge, i, $\mathring{id}$) to node with identifier $w\mathring{i}nd$ *#Forward message to next bndy neighbor*

Protocol 5.9. Electing a leader for the boundary cycle (see Protocol 5.8 and Protocol 4.12)

Fragment extends: Protocol 5.8

State Trans. Sys.: $(\{\text{INIT}, \text{IDLE}, \text{BNDY}, \text{LEAD}\}, \{(\text{INIT}, \text{IDLE}), (\text{IDLE}, \text{BNDY}), (\text{BNDY}, \text{LEAD})\})$

Local data: Smallest identifier m, initialized to $m := \mathring{id}$

BNDY

  Receiving (msge, i, i') *Updates* Protocol 5.8

    if $i = \mathring{id}$ then   *#Check whether message returned to sender*

      become LEAD

    else

      if $i < m$ then

        set $m := i$

      send (msge, i, $\mathring{id}$) to node with identifier $w\mathring{i}nd$ *#Forward to next bndy cycle neighbor*

---

cation complexity of msge messages relies on knowledge of how many boundary and boundary cycle nodes we can expect in the network.

In the worst case, it appears that $O(|V| - 1)$ nodes are boundary (cycle) nodes. An adversary may set up a network with a cycle of $|V| - 1$ nodes inside a monitored region, and one node outside the monitored region (see, for example, Fig. 5.10, with six nodes, five of which are boundary nodes). In such a case, we can expect Protocol 5.8 to generate $O(|V|^2)$ msge messages, with optimal $O(|V|)$ load balance: each boundary node must unicast one message for itself and for every other boundary node. Similarly, the "as far" algorithm for leader election in a ring, which forms the basis of Protocol 5.9, will ensure that in the worst case Protocol 5.9 will have communication complexity $O(|V| \log |V|)$ (see §4.4.2). In short, it appears Protocols 5.8 and Protocols 5.9 can be expensive!

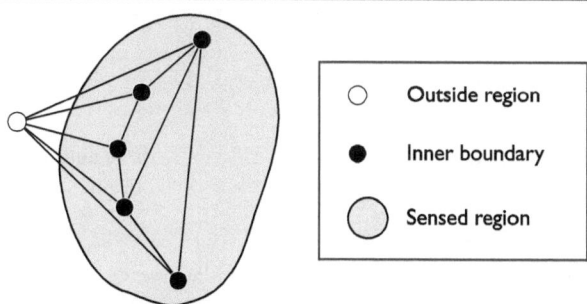

○	Outside region
●	Inner boundary
◉	Sensed region

Fig. 5.10. In the worst cases, it is possible to have (close to) $O(|V|)$ boundary cycle

However, on average things are not as bad as they might at first seem. In most cases, we expect relatively few boundary and boundary cycle nodes as a proportion of the total number of nodes in the network. For example, Fig. 5.11 gives an example of the same geographic region monitored by randomized plane networks at increasing node density. For each network, the *total* number of nodes in the network doubles, from 15 in Fig. 5.11a to 30, to 60, to 120 nodes in Fig. 5.11d. It is noticeable that the number of *boundary* nodes does not increase so rapidly, from six in Fig. 5.11a to eight, to 13, to 20 in Fig. 5.11d. The total number of boundary *cycle* nodes (which includes all the boundary nodes plus a small number of black, non-boundary nodes in the boundary cycle) increases more rapidly than the boundary nodes, but comparably (to a maximum of 24 boundary cycle nodes in Fig. 5.11d).

In general, the precise relationship between the number of nodes in the network and in the boundary (cycle) will depend on the geometric character-istics of the network and region being monitored. Luckily, this dependency is already an established topic of study within the area of *fractal geometry*.

Fractal geometry is concerned with the scaling characteristics of shapes. For many shapes such as Euclidean squares, circles, or polygons, the length of the boundary is independent of the scale at which the shape is viewed. However, for other shapes, called *fractals*, looking more closely at the shape reveals more detail: "the more you look, the more you see."

A mathematician called Lewis Fry Richardson noted that adjacent coun-tries often obtained different official measurements for the length of their shared borders. In one famous example, official measurements by Spain of its shared border with Portugal put the length at 987 km. By contrast, on the other side of the boundary, Portugal measured its shared border with Spain

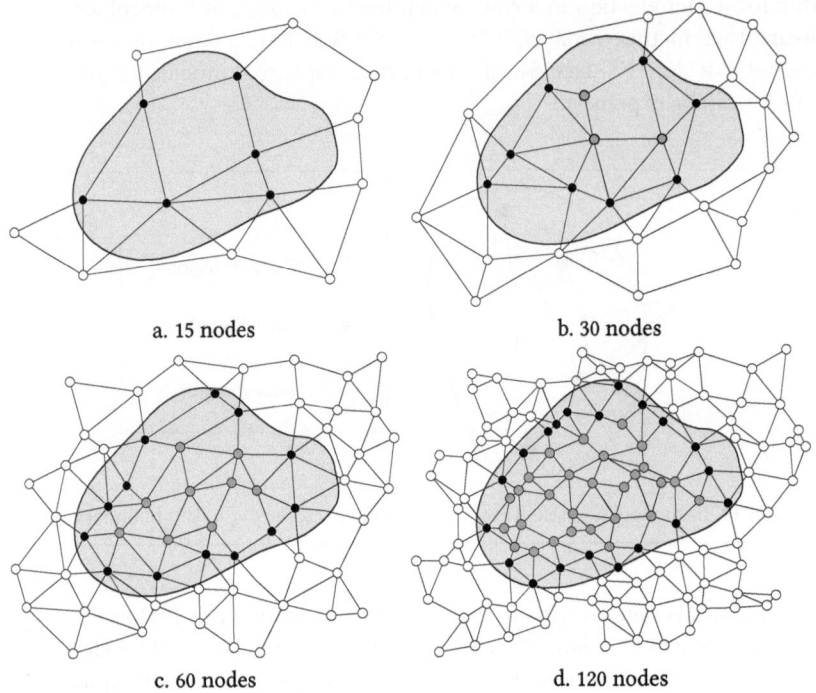

a. 15 nodes   b. 30 nodes

c. 60 nodes   d. 120 nodes

Fig. 5.11. Scaling of boundary nodes (gray fill) with increasing network size

as 1,214 km [92]. The discrepancy arises because of the level of detail at which the measurements are made. Smaller Portugal, bordering only one other country, measured the shared border at higher levels of detail than its larger, more connected neighbor. Like many geographic shapes, the border between Spain and Portugal has the scale-dependent characteristics of a fractal.

One of the surprising results of fractal geometry is that the relationship between length and scale of observation can be encapsulated with a fractional ("fractal") dimension [80], which lies somewhere 1 (a Euclidean curve) and 2 (a piece of the Euclidean plane) for fractal curves in the plane. More specifically, the number of steps $n(\epsilon)$ required to traverse the boundary of a shape increases in inverse proportion to the length of the step ϵ used to measure the boundary, such that $n(\epsilon) = \epsilon^{-D}$, where D is the (fractal) dimension of the shape.

This discussion has direct relevance to the networks and geographic regions in Fig. 5.11. The cycle of nodes that approximates the region's boundary is analogous to a measurement of the length of boundary—the more nodes in the boundary, the longer the observed length. Increasing the network density is analogous to "zooming in" on the region, measuring it more closely. Doubling the number of nodes in the network (i.e., in the plane) will on average decrease the spacing between the nodes by a factor of $2^{-1/2}$. (An an analogy, imagine doubling the *area* of a square; this only increases the length of the boundary of the square $2^{1/2}$ times.) Thus, the relationship between the number of nodes in the network $|V|$ and the number of boundary nodes for a fixed region $|B|$ is expected to vary in proportion to $|V|^{-D} * |V|^{-1/2} = |V|^{D/2}$.

The direct consequence of this result is that, on average, we expect the number of nodes in the boundary cycle for a region to increase in proportion to $|V|^{D/2}$. For Euclidean shapes, $D = 1$. A wide range of natural geographic shapes (such as "hot spots") routinely exhibit fractal properties [80]. As a rule of thumb, the measured fractal dimension of these shapes is often in the range 1.2–1.3 [79] (although the measurement of fractal dimension of natural features is problematic, with other work reporting higher values, e.g., 1.6–1.9 [20]).

For Protocol 5.8, this results in $|V|^{D/2}$ msge messages passed around $|V|^{D/2}$ boundary (cycle) nodes, leading to $O(|V|^{D})$ overall communication complexity. However, Protocol 5.8 is a stepping stone to demonstrating cycling around the boundary. Electing a leader for the boundary, using Protocol 5.9 above, requires an overall communication complexity of $O(|V|^{D/2} \log |V|^{D/2})$ (recall that the "as far" leader election protocol has overall complexity of $O(n \log n)$). For any $D < 2$ (which is safe to assume for any geographic shape), in effect this reduces to fewer than $O(|V|)$ msge messages sent in total. For this scenario, the elected leader will have to pass on the most msge messages, $|V|^{D/2}$, leading at worst to a load balance of $O(|V|^{D/2})$. This expectation is explored empirically in later chapters (see §7.3.2).

5.3.3 Area and Centroid

The discussion above shows that it is possible to route a message around the boundary of a sensed geographic region, and even efficiently elect a leader for that region. To illustrate the usefulness of this approach, we can now use these abilities to perform some basic spatial computations with these regions, such as computing their area and centroid.

The area *area* of a polygon P, made up of points (x_1, y_1), (x_2, y_2), $\dots (x_n, y_n)$ where $(x_1, y_1) = (x_n, y_n)$, can be found with the formula

$$area(P) = 1/2 \sum_{i=1}^{n-1} x_i y_{i+1} - x_{i+1} y_i.$$

There is a direct analogy between a polygon and the boundary cycle we have defined above for monitoring geographic regions. The nodes in the network can be seen as vertices of the polygon; and the communication links connecting adjacent nodes in the boundary cycle can be regarded as edges connecting adjacent vertices in the polygon. The key observation in designing a decentralized spatial algorithm for computing the area of a geographic region is that the formula for area only ever requires the summation of *adjacent* vertices in the polygon, (x_i, y_i) and (x_{i+1}, y_{i+1}). Thus, each node requires only one of its boundary cycle neighbors' coordinate locations in order to compute one iteration of the area sum.

The formulas for the x and y coordinates of the centroid for the shape are closely related to the formula for area:

$$x\text{-}centroid(P) = \frac{1}{6} * area(P) \sum_{i=1}^{n-1} (x_i + x_{i+1})(x_i y_{i+1} - x_{i+1} y_i),$$

$$y\text{-}centroid(P) = \frac{1}{6} * area(P) \sum_{i=1}^{n-1} (y_i + y_{i+1})(x_i y_{i+1} - x_{i+1} y_i).$$

Consequently, it is possible to compute the centroid in the same way as computing the area (see [98]). Protocol 5.10 provides a protocol fragment based on Protocol 5.9 which computes both area and centroid together. Starting with the elected leader for the boundary cycle, each node passes to its neighbor an `area` message containing its coordinate location along with the partial sum of the area, and x and y centroid coordinates (all initialized to 0). Upon receiving an `area` message, a boundary cycle node computes one further iteration of the area and centroid sums, before forwarding these to the next boundary cycle node. Upon receiving its own returned `area` message, the boundary cycle leader can finalize the computations to complete the area and centroid sum.

Figure 5.12, a spatial sequence diagram for Protocol 5.10, shows the first five `area` messages (see question 5.11).

After computing the boundary nodes ($\Theta(|V|)$ `ping` messages) and electing a leader for the boundary cycle ($O(|V|)$ `msge` messages), a single cycle of the boundary requires only $O(|V|^{D/2})$ messages, where $1 \leq D < 2$ is assumed for all geographic shapes. Thus, the overall communication complexity is still only $O(|V|)$, and after leader election further cycles of the boundary are relatively inexpensive in terms of overall communication costs.

5.3.4 Summary

Aside from the specific details of the algorithm, this section has addressed two important topics. First, we have again seen how efficient decentralized spatial algorithms can be achieved through computing across *spatial structures* such as boundary nodes and boundary cycles. Although finding the boundary nodes and electing a leader require $O(|V|)$ overall communication, some of the messages (e.g., `ping` "handshake" messages) would be arguably required

---

Protocol 5.10. Computing the area and centroid for a boundary cycle (see Protocol 5.9)

Fragment extends: Protocol 5.9

Restrictions: Coordinate location $p : V \to \mathbb{R}^2$

IDLE

   Receiving $(\texttt{area}, i, (x, y), a, c_x, c_y)$
     let $(x', y') := \mathring{p}$
     let $u := (x * y' - x' * y)$
     send $(\texttt{area}, \mathring{id}, \mathring{p}, a + u, c_x + (x + x') * u, c_y + (y + y') * u)$ to node with identifier $\mathring{cyc}(i)$

BNDY

   Receiving $(\texttt{msge}, i, i')$ *Updates* Protocol 5.9
     if $i = \mathring{id}$ **then**   #*Check whether message returned to sender*
       send $(\texttt{area}, \mathring{id}, \mathring{p}, 0, 0, 0)$
       become LEAD
     else
       if $i < m$ **then**
         set $m := i$
         send $(\texttt{msge}, i, \mathring{id})$ to $\mathring{wind}$ #*Forward message to next boundary cycle neighbor*
   Receiving $(\texttt{area}, i, (x, y), a, c_x, c_y)$
     let $(x', y') := \mathring{p}$
     let $u := (x * y' - x' * y)$
     send $(\texttt{area}, \mathring{id}, \mathring{p}, a + u, c_x + (x + x') * u, c_y + (y + y') * u)$ to node with identifier $\mathring{wind}$

LEAD

   Receiving $(\texttt{area}, i, (x, y), a, c_x, c_y)$
     let $(x', y') := \mathring{p}$
     let $u := (x * y' - x' * y)$
     let $area := \frac{a+u}{2}$ #*Final area calculation*
     let $cenx := \frac{c_x + (x+x') * u}{6 * area}$ #*Final centroid x calculation*
     let $ceny := \frac{c_y + (y+y') * u}{6 * area}$ #*Final centroid y calculation*

---

for initialization of any algorithm. Further, once a leader is elected, it can coordinate efficient communication and computation around the boundary of the region, instead of over the entire region. It is worth noting that the area algorithm in Protocol 5.10 is truly decentralized: it can compute the area enclosed by the boundary cycle of any size without any node ever requiring any information other than its own coordinates and sensed value and those of its immediate one-hop neighbors.

Second, this section has introduced tools to support *modularity*, critically important to any algorithm design process. Functional decomposition, the process of breaking down large problems into smaller manageable units, is essential to any complex algorithm design. Using protocol fragments, which allow the definition of new protocols built on existing protocols, has allowed us to take our first steps into functional decomposition.

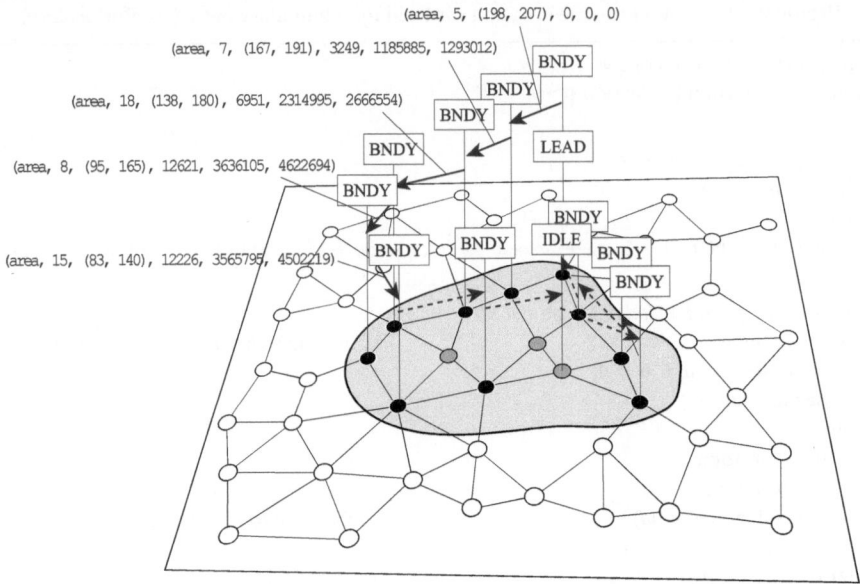

Fig. 5.12. Spatial sequence diagram for Protocol (fragment) 5.10, showing first five area messages (comprising: message type, node identifier, (x, y) coordinate, partial area sum, partial x centroid sum, and partial y centroid sum)

5.4 Protocol of the Chapter: Topology of Complex Areal Objects

To show the potential of location-based algorithms, this section showcases a sophisticated algorithm for determining the topological structure of complex areal objects, first introduced in [35]. We begin by exploring what is meant by "complex areal objects," and what topological structure we are interested in (§5.4.1), before exploring the algorithm itself in §5.4.3.

5.4.1 Complex Areal Objects

In defining and computing with "regions," we have has to be careful so far to ensure that our regions are "closed, strongly connected piece of the plane ...homeomorphic ...to a disk" (p. 116). However, in many cases the objects we are interested in will not be homeomorphic to a disk; they will instead have holes, islands, and disconnected parts. Figure 5.13a shows just such a complex areal object, containing five disconnected *islands*, a, b, c, g, h, and four *holes* in those islands, d, e, f, i. By abstracting away from the specific geometry, the topological structure of this kind of complex areal object is often summarized using a tree, with edges in the tree representing spatial containment [117]. Figure 5.13b shows the containment tree for the complex areal object in Fig. 5.13a. It shows, for example, that hole i is contained within island g. In turn, island g is contained within hole e, which is inside island b. Ultimately island c as well as islands a and b and their inclusions, are contained within the exterior component x.

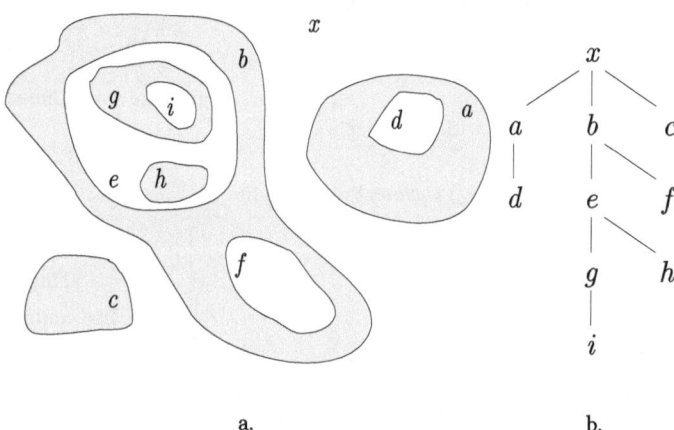

a.   b.

Fig. 5.13. A complex areal object (a), and its topological structure represented as a containment tree (b) after [117]

Determining containment is a basic function in conventional GIS and spatial databases. With this in mind, the algorithm presented below assumes we have a sensor network monitoring a complex areal object, as in Fig. 5.13a. The objective of the algorithm is then to generate the containment tree, as in Fig. 5.13b. The algorithm is presented in two parts. First, §5.4.2 presents an algorithm for enabling boundary cycle nodes to distinguish between holes and islands. Then, this information is used in §5.4.3 to determine the topological containment structure of the entire complex areal object.

5.4.2 Holes and Areal Components

Protocol fragment 5.11 extends Protocol 5.10, the algorithm for generating the area enclosed by a boundary cycle. Having found the area, each boundary cycle leader in Protocol fragment 5.11 sends a `ring` message once around the boundary cycle. This `ring` message contains a vital piece of information: not the area itself, but whether the boundary cycle bounds a *hole* or an *island*.

The information about whether a boundary cycle is a hole or an island can be deduced from the *sign* of the computed area. The formula for computing the area of a polygon introduced in §5.3.3 generates a *positive* area if the

---

Protocol 5.11. Informing boundary cycle nodes about leader identifier and boundary type (hole or island, see Protocol 5.10)

Fragment extends: Protocol 5.10
Local data: Sink node location (s_x, s_y); *hole*, initialized *hole* $:= -1$; ring identifier *ring*, initialized *ring* $:= -1$

IDLE
    Receiving $(\texttt{ring}, i, o, r)$
      send $(\texttt{ring}, \mathring{id}, o, r)$ to node with identifier $\mathring{cyc}(i)$

BNDY
    Receiving $(\texttt{ring}, i, o, r)$
      set *hole* := and *ring* := r #*Update stored hole flag and ring id*
      send $(\texttt{ring}, \mathring{id}, o, r)$ to node with identifier $\mathring{wind}$

LEAD
    Receiving $(\texttt{area}, i, (x, y), a, c_x, c_y)$ *Updates* Protocol 5.10
      let $(x', y') := \mathring{p}$
      let $u := (x * y' - x' * y)$
      let *area* $:= \frac{a+u}{2}$ #*Final area calculation*
      let *cenx* $:= \frac{c_x + (x+x') * u}{6 * area}$ #*Final centroid x calculation*
      let *ceny* $:= \frac{c_y + (y+y') * u}{6 * area}$ #*Final centroid y calculation*
      if *area* > 0 **then send** $(\texttt{ring}, \mathring{id}, 0, \mathring{id})$ to node with id $\mathring{wind}$ #*Positive area bounds island*
      if *area* < 0 **then send** $(\texttt{ring}, \mathring{id}, 1, \mathring{id})$ to node with id $\mathring{wind}$ #*Negative area bounds hole*

---

vertices are ordered *counterclockwise* around the polygon. Conversely, the formula generates a *negative* area if the vertices are ordered *clockwise* around the polygon. Assuming that the cyclic ordering of neighbors around each node is counterclockwise, Protocol 5.10 will generate a *negative* area if the boundary cycle bounds a hole; and a *positive* area if the boundary cycle bounds an island (see Fig. 5.14).

Protocol 5.11 uses the `ring` message to circulate information about the identifier of the elected leader for that cycle, as well as information about whether the boundary cycle bounds a hole or an island. This information is stored at each boundary cycle node as local data *ring* and *hole* respectively.

5.4.3 Containment Relationships Between Holes and Areal Components

Finally, Protocol 5.12 extends Protocol 5.11 to report the containment relationships between all holes and islands to a central, designated sink node contained in the exterior of the complex areal object. The sink node in state ROOT spontaneously initiates the construction of the shortest path tree, using the technique already encountered in Protocol 4.8. The shortest path tree will be used to report direct containment between boundaries to the central sink, contained in a `rprt` message. It will also be possible to use the known coordinates of nodes to georoute the `rprt` message back to a sink at a known location (see question 5.12).

Protocol 5.12. Determining the topological structure of a complex areal object

---

Fragment extends: Protocol 5.11

State Trans. Sys.: ($\{$ROOT, INIT, IDLE, BNDY, LEAD$\}$, $\{($ROOT, INIT$)$, $($INIT, IDLE$)$, $($IDLE, BNDY$)$, $($BNDY, DONE$)\})$

Initialization: All nodes in state INIT, except one node in state ROOT

Local data: *parent* : $V \to \mathbb{N} \cup \{-1\}$, initialized $par\mathring{e}nt := -1$; hop count to root r, initialized $r := -1$

ROOT

  Spontaneously

    set $r := 0$

    broadcast $(\texttt{tree}, i\mathring{d}, r)$ #*Broadcast root identifier and hop count of zero*

    become INIT

INIT

  Receiving $(\texttt{rprt}, r_a, r_b, s_p, c)$

    defer until IDLE, BNDY, or LEAD   #*Ensure* rprt *messages only processed after initialization*

IDLE

  Receiving $(\texttt{rprt}, r_a, r_b, s_p, c)$

    if $par\mathring{e}nt \neq -1$ **then**   #*Check if this is the root node*

      send $(\texttt{rprt}, r_a, r_b, \mathring{s}, c)$ to node with id $par\mathring{e}nt$

    else

      Message r_a contained in r_b received!

INIT, IDLE, BNDY, LEAD

  Receiving $(\texttt{tree}, i, h)$

    if $r < 0$ **or** $h + 1 < r$ **then**

      set $par\mathring{e}nt := i$ #*Store parent id*

      set $r := h + 1$ #*Store hop count to root*

      broadcast $(\texttt{tree}, i\mathring{d}, r)$ #*Broadcast node identifier incrementing hop count*

BNDY, LEAD

  Receiving $(\texttt{rprt}, r_a, r_b, s_p, c)$ #r_a/r_b *contained/containing ring id;* s_p *prev. sensed val.;* c *score*

    if $hole < 0$ **then defer until** $hole >= 0$ #*Defer* rprt *messages until after* ring *message*

    if $r_b < 0$ **then**   #*Check for* rprt *without containing region*

      if $s_p \neq \mathring{s}$ **then set** $c := c + 1 - 2 * hole$ #*Increment score based on incoming neighbor*

      if $(par\mathring{e}nt, \mathring{s}) \notin D$ **then set** $c := c - 1 + 2 * hole$ #*Increment score based on outgoing nbr*

      if $c = 0$ **then set** $r_b := ring$ #*Update containing region if required*

    send $(\texttt{rprt}, r_a, r_b, \mathring{s}, c)$ to node with id $par\mathring{e}nt$

LEAD

  Receiving $(\texttt{ring}, i, o, r)$

    set $hole := o$, $ring := r$,

    if $\big((par\mathring{e}nt, \mathring{s}) \notin D$ **and** $hole = 0\big)$ **or** $\big((par\mathring{e}nt, \mathring{s}) \in D$ **and** $hole = 1\big)$ **then**

      send $(\texttt{rprt}, i\mathring{d}, \text{-1}, \mathring{s}, 1)$ #*Initiate report with score 1 (cases I1/I4)*

    else

      send $(\texttt{rprt}, i\mathring{d}, \text{-1}, \mathring{s}, 2)$ #*Initiate report with score 2 (cases I2/I3)*

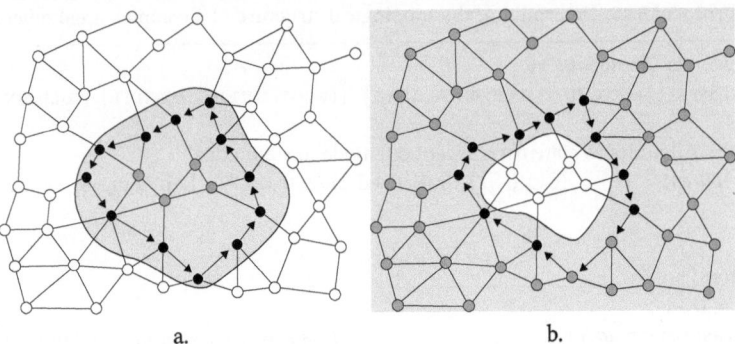

Fig. 5.14. Boundary cycles bounding islands (a) and holes (b) have counterclockwise and clockwise orientations, and so positive and negative areas, respectively

The `rprt` messages are initialized with a "score," either 1 or 2. As a `rprt` message proceeds back to the sink, this score is either incremented or decremented by 1 each time the message crosses a boundary cycle. The precise rules for the initializing, incrementing, and decrementing the score depend on whether the `rprt` message is crossing into/out of the boundary for a hole/island.

Figure 5.15 summarizes the 12 different cases that Protocol 5.12 deals with: eight (H1–8) for crossing a boundary cycle, and four (I1–4) for initializing the score. With reference to Protocol 5.12 and to Fig. 5.15, the four possible cases for initialization are:

I1 From an island boundary leave the island: $(parent, \mathring{s}) \notin D$ and $hole = 0$ (score initialized to 1).

I2 From an island boundary remain in the island: $(parent, \mathring{s}) \in D$ and $hole = 0$ (score initialized to 2).

I3 From a hole boundary leave the island: $(parent, \mathring{s}) \notin D$ and $hole = 1$ (score initialized to 2).

I4 From a hole boundary remain in the island: $(parent, \mathring{s}) \in D$ and $hole = 1$ (score initialized to 1).

The further eight possible cases for incrementing/decrementing the score are:

H1 Touch an island boundary from outside the island (no change in score).

H2 Enter via an island boundary, $(parent, \mathring{s}) \in D$ and $hole = 0$ (score incremented by 1).

H3 Touch an island boundary from inside the island (no change in score).

H4 Leave via an island boundary, $(parent, \mathring{s}) \notin D$ and $hole = 0$ (score decremented by 1).

H5 Touch a hole boundary from outside the island (no change in score).

H6 Enter via a hole boundary, $(\mathring{parent}, \mathring{s}) \in D$ and $hole = 1$ (score decremented by 1).

H7 Touch the hole boundary from inside the island (no change in score).

H8 Leave via a hole boundary, $(\mathring{parent}, \mathring{s}) \notin D$ and $hole = 1$ (score incremented by 1).

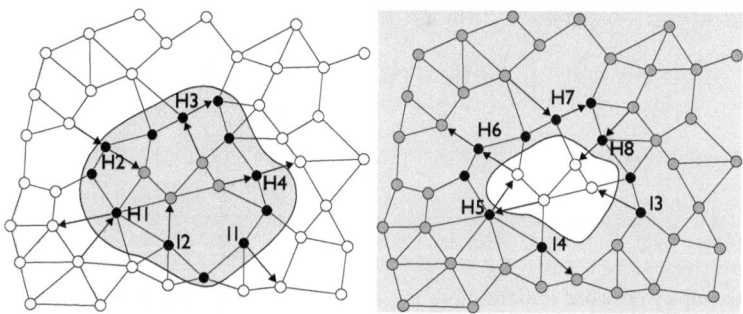

Fig. 5.15. Twelve cases of score initialization to 1 (I1, I3) or 2 (I2, I4), incrementing score by 1 (H2, H8), decrementing score by 1 (H4, H6), and no change in score (H1, H3, H5, and H7) covered by Protocol 5.12

When a boundary node decrements a rprt message score to 0, it knows that message must have come from a region that is directly contained by its own boundary cycle! The boundary node can then update the rprt message to include the identifier of the (leader of) the boundary cycle. This information can then proceed unaltered to the sink node along the shortest path tree. The sink node can then combine all the containment relations received from each boundary into a single containment tree (Fig. 5.16).

Despite this admittedly bewildering range of cases, the intuition behind the scoring process will be familiar to any student of GIS. Protocol 5.12, its scoring process depicted in Fig. 5.16, is essentially a decentralized version of the famous point-in-polygon algorithm (e.g., see [119]). The point-in-polygon algorithm solves the question of whether a point is inside a polygon by computing the number of intersections of the polygon boundary with a line from the point to infinity (called a half-line). Zero or an even number of intersections means the point is outside the polygon; an odd number of intersections indicates the point is inside the polygon. The algorithm must also deal with a number of special cases where the half-line just touches the boundary of the polygon (similarly to cases H1, H3, H5, and H7 above). By counting the number of intersections that the rprt message makes with boundaries, and considering whether it crosses into or out of an island or hole, we can similarly keep track of whether the leader for a boundary cycle

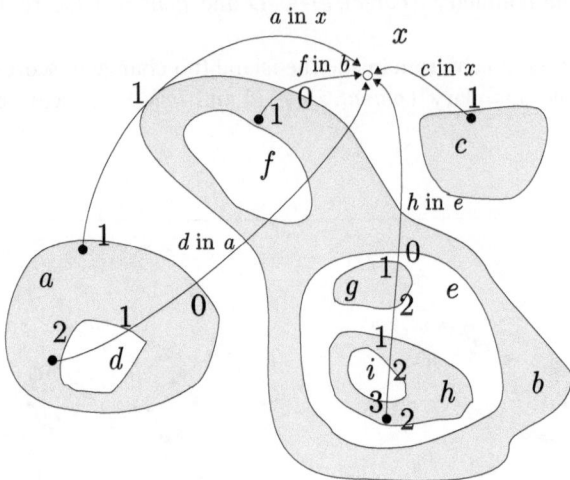

Fig. 5.16. Examples of score initialization, increment, decrement, and containment detection for Protocol 5.12

(and so the island or hole enclosed by that boundary cycle) is contained by another boundary cycle it meets.

5.5 Chapter in a Nutshell

This chapter has explored location-based algorithms, where nodes in the network have knowledge not only of their neighborhoods, but also of their locations. Location will typically mean coordinate position, but can also mean other forms of location information, such as cyclic ordering of neighbors around a node (e.g., see Protocol 5.5).

The most basic techniques in location-based algorithms concern the generation of planar overlay networks (including the Gabriel graph, relative neighborhood graph, local minimum spanning tree, and unit Delaunay triangulation). Two further fundamental techniques were introduced in this chapter: greedy georouting (routing information from a source to a sink at a known coordinate location, at each step using the geometry of neighbors relative to the known destination); and face routing (routing a message around a face of a planar overlay network).

All the other algorithms introduced in this chapter rely on these three basic location-based algorithms, as do other algorithms encountered in previous chapters. For example, GPSR (greedy perimeter stateless routing) is fundamentally a combination of a planar overlay network, greedy georouting, and face routing. Routing a message around the boundary of a geographic region detected by the network combines a planar overlay network and a version of face routing with the boundary node detection technique first encountered in §4.5.2. Electing a leader for a geographic region adds our approaches to leader election in a ring (§4.4). Computing the area of a region follows in a straightforward way once a leader is elected for a boundary cycle. Computing the topology of a complex area object, with holes and islands, combines the area computation with shortest path trees (§4.3.2), but could also be adapted to work with greedy georouting or GPSR.

Finally, this chapter has demonstrated the use of modular decentralized spatial algorithm design: reusing existing algorithms to build more sophisticated ones. Top-down problem solving, where hard problems are broken down into successively smaller, more easily solved problems, is a fundamental technique in designing algorithms to solve complex problems. Modularity supports top-down problem solving by providing a mechanism to help combine smaller steps (such as routing around a boundary, leader election, and shortest path trees) into bigger solutions (such as our program of the chapter, determining the topological structure of complex areal objects).

Review questions

5.1 Adapt Protocol 5.2 to additionally detect inner *boundary* nodes, using the approach already set out in Protocol 4.14. Only those nodes that are inside a region and have a one hop neighbor *in the relative neighborhood graph* overlay network (not the underlying communication graph) should be considered at the (inner) boundary. Your algorithm should send exactly $\Theta(|V|)$ messages (i.e., combine the algorithms such that exactly the same number of messages is sent).

5.2 Extend Protocols 5.1 and 5.2 to explicitly detect the termination of the algorithms, i.e., when a node has received a `ping` from every neighbor. After termination, generate the full overlay network by repopulating a global neighborhood function $nbr' : V \to 2^V$ (e.g., for a node v with neighbor set N, $nbr'(v) \mapsto \{v'|\{v,v'\} \in E$ and $(id(v'),l) \in N\})$.

5.3 Nodes in the LMST generated by Protocol 5.3 are connected only if the nodes are also connected in *both* of the nodes' neighborhood MSTs. In [71], this structure is denoted by G_0^-. However, [71] also defines an alternative structure, G_0^+, where an edge is created in the output LMST if *either* node's neighborhood MST contains a connection to the other node. Adapt Protocol 5.3 to generate the G_0^+ variant of the LMST.

5.4 With reference to [73], adapt Protocol 5.3 to construct the $LMST_2$ graph— the LMST where each node uses its *two*-hop neighborhood to compute its neighborhood MST.

5.5 With reference to [72], construct the protocol required to compute the partial Delaunay triangulation (PDT).

5.6 The greedy georouting algorithm presented in Protocol 5.4 minimizes at each hop the *distance* to the destination. However, greedy georouting also works just as well if we minimize the *angular deviation* from the bearing of the destination at each hop (i.e., a node forwards the message to the neighbor whose *bearing* is closest to that of the destination node), sometimes called "compass routing." Adapt Protocol 5.4 to perform compass routing, and check for yourself the routes generated through a network such as that in Fig. 5.3. (See also [65].)

5.7 In certain cases, GPSR can follow a route that traverses almost the entire *exterior* face of the planar network (§5.2.3). Looking at the underlying network in Fig. 5.6, find a source and destination node that result in a near-complete traversal of the exterior face of the network. (Hint: if you look in the top right-hand corner of Fig. 5.6, assuming counterclockwise cyclic ordering, it is possible to find a source and a destination node that results in a GPSR route that traverses 48 of the 51 edges in the exterior face, with some edges even being traversed twice!)

5.8 Adapt the greedy georouting or GPSR protocols to perform geocasting: routing a message from a single source to all nodes in a specified geographic *region*. For simplicity, assume that the region is an axis-parallel rectangle.

5.9 Protocol 5.7 was used to route information around a region boundary, but only under the condition that the subgraph formed by the region be 2-connected (see p. 148). Adapt Protocol 5.7 to relax this condition so that information can be successfully routed around the boundary even if the subgraph formed by the region is simply 1-connected. (Hint: try drawing some examples of sensed regions that have narrow concavities, where one boundary node may have more than two boundary node neighbors, and examine carefully what path a sensible boundary cycle should take in such cases.)

5.10 Protocol 5.7 and subsequent protocols that build on it only find the *inner* boundary. Adapt Protocol 5.7 to also find the *outer* boundary cycle. What do you notice about the orientation of the outer boundary cycle, when compared with the inner boundary cycle?

5.11 Figure 5.12 shows the first five `area` messages in a spatial sequence diagram for Protocol 5.10. Complete the figure by writing down the remaining six `area` messages in sequence, and calculate the actual area and centroid for the region in Fig. 5.12. In order, the six remaining boundary (cycle) node coordinates are: $(106, 105)$, $(158, 123)$, $(202, 131)$, $(247, 109)$, $(239, 139)$, $(208, 173)$.

5.12 Adapt Protocol 5.12 to use georouting instead of the shortest path tree to return containment information to a known sink. You may use greedy georouting (simpler but subject to failure in cases where there are voids in the network) or GPSR (reliable, but more complex to specify) to achieve this.

5.13 Adapt Protocol 5.12 to report on the *connectivity* of the monitored areal object (i.e., whether the region has disconnected parts), identifying how many disconnected parts it has.

5.14 The neighborhood-based sweep algorithm presented in §4.5.1 used hop-count as its potential function. Where nodes know their coordinate positions, this algorithm can be considerably simplified, using instead x- or y-coordinates for the potential function. Verify this, by adapting Protocol 4.13 to use coordinate position instead of hop-count.

Monitoring Spatial Change over Time

6

Summary: Monitoring change is essential for many practical applications of decentralized spatial computing, because the world around us is highly dynamic. The neighborhood- and location-based algorithms discussed in the previous two chapters do not consider change in the monitored environment. Nor do they account for change in the decentralized spatial information system monitoring that environment. Relaxing these assumptions, this chapter takes the final step in the foundations of decentralized spatial computing, exploring the design of decentralized spatiotemporal algorithms that can monitor spatial change over time.

ALL the algorithms discussed in the previous chapters are *atemporal*: they do not involve change over time. Yet, in practice we are almost always interested in changes to monitored environments. One can argue, for example, that geosensor networks are *only* useful in cases where we want to monitor change over time. If we simply want a single *snapshot* of the world at a particular time, conventional data capture methods, such as ground survey or aerial photography, are typically simpler and more appropriate.

This chapter takes the final step in our exploration of the design of decentralized spatial algorithms by incorporating change. Change may occur both in the environment being monitored, and in the monitoring system itself. For example, if we are interested in monitoring the health of a coral reef, the sensed environmental qualities, such as sea temperature, turbidity, and pH, will surely change over time. But it is also very likely that the monitoring system itself will change, perhaps with sensor nodes moving with sea currents, network connectivity changing, and even new sensor nodes being introduced to the network as older nodes become depleted or fail.

We begin with a discussion of exactly what we mean by change, and of the sorts of changes we are interested in. Building on these definitions we introduce some increasingly sophisticated spatiotemporal algorithms,

M. Duckham, *Decentralized Spatial Computing*, DOI 10.1007/978-3-642-30853-6_6,
© Springer-Verlag Berlin Heidelberg 2013

which illustrate the key decentralized algorithm design techniques needed for monitoring spatial change over time.

6.1 Change over Time

A basic philosophical distinction is made usually between things that *endure* through time, and things that *happen* in time [116]. Examples of things that endure or continue through time, often called *endurants* or *continuants*, include you and me, this book, the University of Melbourne, and Australia. Things that occur in time, often called *perdurants* or *occurrents*, include your life, reading this book, the founding of the University of Melbourne, and the Federation of Australia. Boundaries and regions are typical examples of geo-spatial endurants that we have already encountered in earlier chapters; the appearance, splitting, merging, and disappearance of regions are all examples of geospatial perdurants that we shall encounter later in this chapter.

Making a clear distinction between endurants and perdurants is a pre-requisite to good decentralized spatiotemporal algorithm design. Because the same underlying environmental changes can often be viewed from different perspectives, depending on the specific needs of the application, confusion can easily occur without this distinction. For example, imagine a geosensor network tasked with monitoring the spread of an oil spill. We might wish the system generate an alert when parts of the oil spill *appear* or *break up*. Alternatively, we might (also) wish the system report on the *connectivity* of the oil spill every ten minutes over the course of a day. Although based on the same underlying scenario, the former is an example of an application that monitors meaningful events that *occur* to the oil spill (perdurants); the latter is an example of monitoring the state of the oil spill over time (endurants).

6.1.1 Histories and Chronicles

The distinction between endurants and perdurants maps directly to two fundamentally different types of information that may be generated by a decentralized spatiotemporal algorithm: *histories* and *chronicles* [45]. A history provides a spatiotemporal record of the states of monitored endurants (e.g., point locations, regions, boundaries, moving objects) through time. A chronicle provides a record of the occurrences (perdurants) that happened through time. For example, Fig. 6.1 contrasts the two perspectives through the changes to the Deepwater Horizon disaster oil spill over six days shortly after the disaster occurred. The left-hand side of Fig. 6.1 shows a history for the period, where for each day the observed extent of the oil slick is pictured. For example, on days 17 and 18, the slick forms one connected component; on days 19 and 20 there are two connected components; and on days 21 and 22 there are seven connected components. In contrast, the right-hand side of Fig. 6.1 shows a chronicle for the same period and changes. Some time on day

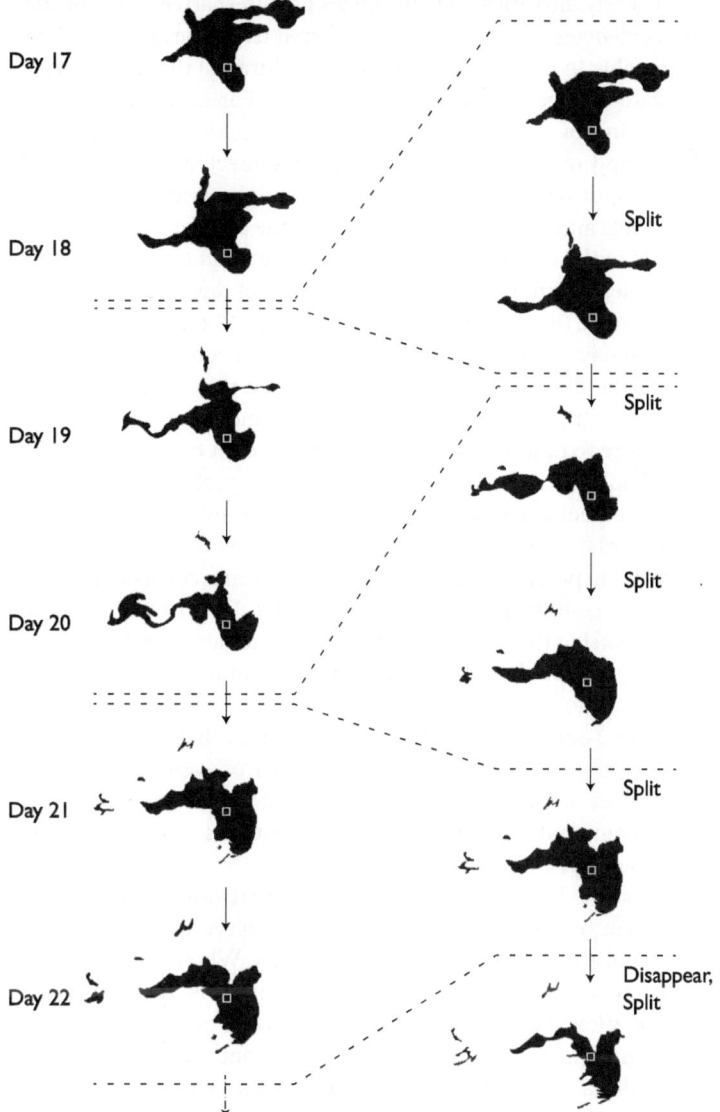

Fig. 6.1. Histories and chronicles: Two views of the changes in the extent of the Deepwater Horizon oil spill, Gulf of Mexico, from day 17, 7 May 2010. (Source: *Times-Picayune*)

18 the slick splits; on day 20 the region splits three more times (i.e., the slick "breaks up"); and on day 22 further splits occur and some region components disappear (i.e., parts of the slick start to disperse).

Clearly, histories and chronicles are alternative views of the same underlying changes, and can often be inferred from one another. Given a history, it is possible to infer the occurrences (perdurants) that may have led to changes in state between consecutive time steps. For example, if I know that at the beginning of day 18 there was one connected component, and that at the beginning of day 19 there were two connected components, I might infer that a split occurred between these two observations.[1] Conversely, given a chronicle and a starting state, it is possible to infer the history by tracking the effects of the occurrences in the chronicle upon the endurants of interest. For example, if I know that at the beginning of day 18 there was one connected component (initial state), and that on day 18 one split occurred, I can infer that on day 19 there must be two connected components.

In the discussion above we have been very careful not to use the term "event." Although this term is widely and entirely correctly used in the literature to describe what are called above "occurrences" (perdurants, things that happen), we reserve "event" to refer specifically to a *system* event (i.e., a spontaneous, trigger, or message receipt event for a node) in this book. In contrast, the terms "occurrences" or "perdurants" (and occasionally in earlier chapters "environmental events") are consistently used to prevent possible confusion with (system) events. Occurrences themselves may be *instantaneous* (happening at a particular moment in time, such as the splitting of a region) or *durative* (happening over an extended time interval, such as the expansion of a region) [44].

Thus, our first important observation is that different applications may require monitoring of histories, chronicles, or even both. Neither histories nor chronicles alone will suffice. We can expect to need to design decentralized spatiotemporal algorithms to monitor histories in some cases, and to monitor chronicles in other cases.

A second important observation is that raw sensed data is normally in the form of *histories*. Sensors usually observe the states of the world over time, but rarely sense occurrences *directly*. When we deploy sensors through space to monitor, for example, pollution levels, these sensors may capture information about the presence or absence of pollutants in different locations over time (histories). Any information about the occurrences, such as about the pollutant spreading or dispersing, must be inferred from these changing states of the world. Only occasionally can it be argued that occurrences are being monitored by sensors. For example, a card reader on a building security door monitors the close proximity or absence of an authorized swipe card over time (state of the world). However, in this case, the state is so tightly

---

[1] Note that while this inference might seem natural and in some senses "minimal," it is not necessarily correct. There are innumerable other more complex sequences of occurrences that might account for the changes (e.g., if the region disappeared and two completely new components appeared in between the history records for days 18 and 19).

linked with the associated occurrence—of a person *entering* a building—that it could be argued that the card reader also monitors this occurrence itself.

6.1.2 Characteristics of Mobile Objects

In addition to the clear distinction between histories and chronicles as two complementary views of change, there are a number of other important distinguishing characteristics when considering moving objects (such as people or cars) monitored by a geosensor network, including:

- whether the movement takes place within a network, or whether objects are free to roam in the plane;
- if mobile objects are tracked, whether the movement of objects is monitored in terms of position over time (termed a *trajectory*) or as the times an object is detected at specific locations in space (called cordons or checkpoints); and
- whether the sensor nodes themselves are mobile, or whether static nodes monitor the movements of mobile objects.

Figure 6.2 summarizes these three characteristics, discussed in more detail below.

First, any object's movement is in part influenced by the constraints on movement imposed by the environment. When we conceptualize movement, we often think of unconstrained movement on the plane, on the surface of the Earth, or even in three dimensions—in the air or the sea. However, unconstrained movement is in actuality rare in geographic environments. In most situations, the world around presents substantial constraints on movement. Buildings, roads, railways, fences, bridges, rivers, national borders, and topography all present constraints on the movement of people, vehicles, livestock, and other animals. As often as not, these constraints are well modeled by a transportation network in which objects move.

Second, although the movement of an object through space must always be continuous, in computing it is common to discretize it into a *trajectory*: a sequence of locations of an object tracked through time. For example, the movement of a GPS-enabled vehicle would typically be monitored as a sequence of latitude and longitude coordinates every few seconds or minutes. However, there is an alternative, complementary view of movement: as a sequence of timestamps for an object tracked through locations in space. For example, a vehicle monitored by an electronic road tolling system will be tracked at fixed points in space (toll gates) when it passes those points. The fixed points in space are often referred to as *checkpoints* or *cordons*. Checkpoints can exist in planar space as well as in networks (consider, for example, cellular mobile phone systems, where the location of a device may be known in terms of its most proximal cell phone tower).

Third, the sensors in the network themselves may be mobile; or static sensors may be tracking mobile objects. In the case of mobile sensors, each

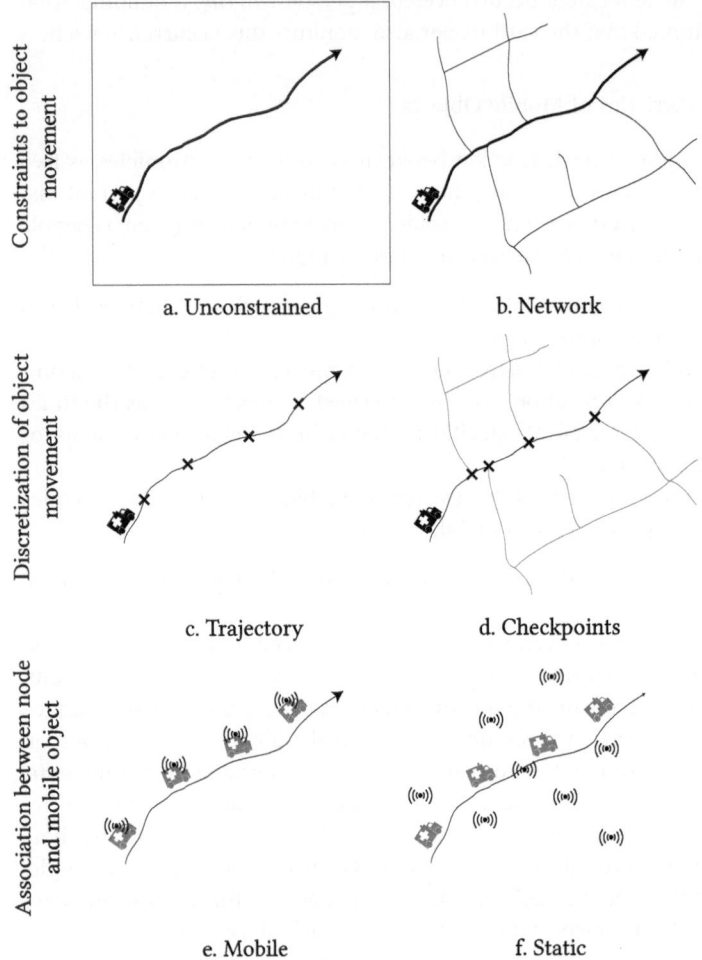

Fig. 6.2. Three distinguishing characteristics of mobile objects monitored by a geosensor network: constraints to movement (e.g., a. unconstrained, b. network constrained); discretization (e.g., c. trajectory, d. cordon-structured); and node mobility (e.g., e. mobile nodes; f. mobile object, but static nodes)

sensor will often be directly associated with a mobile object. For example, a person's mobile smartphone can be thought of as a sensor node associated with that individual. In such cases, information about the one-to-one link between sensor node and its associated mobile object may be central to the application. Indeed, there is a subtle but important ontological distinction to make between situations where the locations of (static) nodes do not have any particular significance (as for most of our algorithms up to this point) and situations where the nodes are known to be co-located with (mobile) objects which are relevant to the application. The distinction is akin to that often made between data-centric and address-centric approaches in wireless sensor network routing [66].

6.2 Histories of Dynamic Environments

We have seen that the raw material for decentralized spatiotemporal algorithms—the sensed data—is typically in the form of histories. As a result, algorithms for monitoring histories are a simpler starting point for our exploration of monitoring of change. We can further simplify the domain of interest by restricting our focus to decentralized spatial information systems where the network itself is static and immobile, and only the environment changes. This models the situation, for example, where a set of fixed sensors monitors environmental changes to temperature, humidity, or CO_2 concentrations.

This section extends the basic boundary detection algorithms introduced in §4.5.2 to track boundary movements over time. Although such an extension is not especially challenging, it is a good departure point for our voyage into spatiotemporal algorithms, because it illustrates two of the most basic techniques for monitoring histories. First, when monitoring change over time it is always possible to repeatedly trigger an existing decentralized (static) spatial algorithm. Clearly, if an existing algorithm generates a snapshot of the state of the world at a particular time, repeated triggering of this same algorithm will generate a sequence of snapshots, i.e., a history. Second, information generated in one snapshot may often be useful in subsequent snapshots. As a result, our second technique aims to retain relevant information about previous states in order to improve the efficiency of repeated triggers.

6.2.1 Boundary Tracking: Repeated Trigger

Protocol 4.14 provided a simple algorithm for detecting the inner or outer boundary of a region. Protocol 6.1 directly extends this earlier algorithm by repeatedly triggering the reinitialization of the static algorithm. One slight change in Protocol 6.1 is that the co-domain of the sensor function s is the set of real numbers $\mathbb{R}$ (rather than a simple Boolean value $\{0, 1\}$; cf. Protocol 4.14). This change models a little more naturally most sensors that capture continuous data about some scalar value, such as temperature or light sensors. The header to Protocol 6.1 defines a threshold r below which sensors are *outside* the region. Thus, this change makes no substantial difference to the operation of the algorithm, as there is a direct mapping from sensed value of less than/greater than or equal to r (in Protocol 6.1) to the Boolean values 0 or 1, respectively (in Protocol 4.14).

There are however three more substantial technical extensions to Protocol 6.1 which relate to spatiotemporal, rather than just spatial, capabilities:

- As described in §2.4, changes in sensed values are modeled using a time-varying sensor function $s : V \times T \to \mathbb{R}$. Thus, at time $t \in T$, $s(v,t)$ denotes the sensed value at node v. Similarly, "my s" ($\mathring{s}(t)$) can be used where the algorithm must refer to an individual node's partial knowledge of its own sensed value (see §2.5).
- The shorthand *now* is borrowed from the literature on temporal databases to refer to "the present time $t \in T$." For example, $s(v, now)$ denotes the sensed value of node v at the present time.
- In order to enable the repeated triggering at of an action, the event "*When* time step elapsed" is used. It would be more precise to define some specific trigger time interval d. However, the specific time interval is not directly relevant to the operation of the algorithm, and will surely vary from application to application.

Protocol 6.1. Tracking the (inner) boundary of a region, defined as sensed value above threshold r

---

Restrictions: $\mathcal{NB}$; $s : V \times T \to \mathbb{R}$; region threshold r
State Trans. Sys.: $(\{\text{INIT}, \text{IDLE}, \text{BNDY}\}, \{(\text{INIT}, \text{IDLE}), (\text{IDLE}, \text{BNDY}), (\text{IDLE}, \text{INIT}), (\text{BNDY}, \text{INIT})\})$
Initialization: All nodes in state INIT

INIT
  Spontaneously
    broadcast (ping, $\mathring{s}(now)$)   #*Broadcast sensed value*
    become IDLE

IDLE
  Receiving (ping, s')
    if $s' < r \le \mathring{s}$ **then**   #*Check if node is in the region, but adjacent to node outside*
      become BNDY

IDLE, BNDY
  When time step elapsed
    become INIT

In terms of complexity, the algorithm is essentially unchanged, only repeatedly triggering over time. As a result, the overall communication complexity (messages sent) can be summarized as $\Theta(|V| * |T|)$ (assuming the set of times T contains only the times over which the algorithm executes). Similarly, the load balance for the algorithm is optimal, $O(|T|)$.

6.2.2 Boundary Tracking: Maintain State

Of course, it is to be expected for relatively small time steps and relatively slow-changing environmental phenomena that the boundary at one time step will not change substantially in the next time step. Protocol 6.1 takes no account of this temporal autocorrelation, where the state of the world is more similar at more proximal times. Reinitializing the entire network at every time step is clearly a highly inefficient approach in the presence of such correlation.

Thus, our second basic technique for moving from decentralized spatial to spatiotemporal algorithms is to maintain some information about the state of the nodes in previous time steps. Protocol 6.2, for example, only reinitializes a node when the sensed value has changed enough to cross the threshold between the inside and the outside of the region. By storing the sensor reading the time of a state change, s_l, a node v can tell if at some later time t, the current sensed value has crossed the threshold r, i.e., whether $s_l < r \leq s(v,t)$ or $s(v,t) < r \leq s_l$.

Protocol 6.2. Tracking the (inner) boundary of a region, with state maintenance

---

Restrictions: $\mathcal{NB}$; $s : V \times T \to \mathbb{R}$; $id : V \to \mathbb{N}$; region threshold r
State Trans. Sys.: $(\{\text{INIT}, \text{IDLE}, \text{BNDY}\}, \{(\text{INIT}, \text{IDLE}), (\text{IDLE}, \text{BNDY})\})$
Initialization: All nodes in state INIT
Local data: sensor reading at time of state change s_l; neighbor data $d : \mathring{nbr} \to \{-1, 0, 1\}$, initialized to $d(v) := -1$

INIT
    Spontaneously
        set $s_l := \mathring{s}(now)$ *#Store last sensed value*
        broadcast $(\texttt{ping}, \mathring{s}(now), \mathring{id})$ *#Broadcast sensed value and identifier*
        become IDLE

IDLE, BNDY
    Spontaneously
        if $\mathring{s}(now) = 1$ and $0 \in d_*$ **then**   *#Check if inner boundary condition satisfied*
            become BNDY
        else
            become IDLE
    Receiving $(\texttt{ping}, s', i)$
        set $d(i) := s'$ *#Store neighbor's sensed values*
    When $\mathring{s}(now) < r \leq s_l$ or $s_l < r \leq \mathring{s}(now)$
        become INIT

The effect of this change on the communication complexity is to require that $|V| + |T| * |V'|$ messages be sent, where $V' \subseteq V$ is the set of nodes whose sensed values have crossed the threshold r (i.e., the nodes that have changed from being in/out of the region to out of/in the region). In the worst case, this is as efficient as Protocol 6.1. However, in the best case, where, for example, only some constant number of nodes change over any time step, the overall complexity becomes $\Omega(|V|)$. If we further ignore the initialization step, as a one-off setup cost, the best case ongoing complexity of the algorithm is essentially constant communication, $\Omega(1)$! Although this is unlikely—in reality we can expect the number of nodes that change to be proportional in some way to the size of the network—this idea illustrates a dramatic, important, and counterintuitive principle: that a decentralized spatiotemporal algorithm has the potential to equal or even *improve* on the computational efficiency of a static spatial alternative.

6.2.3 Summary

Just as moving from decentralized non-spatial to spatial algorithms presents an opportunity to take advantage of the inherent autocorrelation in the geographic world (§1.3.2), so does moving from decentralized spatial to spatio-temporal algorithms. For most geographic phenomena, we can expect near-contemporaneous sensor observations to be highly correlated. As a result, algorithms that maintain the state of past snapshots hold the promise of sub-stantially improving on naïve repeated triggering of static spatial algorithms.

6.3 Histories of Mobile Objects

The previous section was concerned with basic decentralized spatiotemporal algorithms for monitoring dynamic environments. In this section, we look for the first time at mobile objects, such as vehicles or people. The common theme is that in both cases we shall be approaching change from the perspective of a history—sequences of states of the world, rather than events that occur.

6.3.1 Tracking Mobile Objects through Checkpoints

We begin with a very simple algorithm for tracking mobile objects using static sensor nodes (checkpoints or cordons). The algorithm could also be used for unconstrained movement in planar space (see question 6.1), but is primarily motivated by constrained movement through a network. Imagine a transportation network, with mobile objects (e.g., vehicles) moving on it. Sensors capable of detecting moving objects passing by are embedded at a number of locations in that network, for example, at the intersections, or perhaps at traffic lights and so forth in a road network.

For example, Fig. 6.3a shows a transportation network with several check-points (black squares) at certain intersections and network edges. The mobile objects and their identities (e.g., vehicles) are assumed to be detectable by the checkpoints. The combination of the transportation network and the checkpoints induces a *connectivity graph*, showing which checkpoints are directly accessible from which other checkpoints. Figure 6.3b shows the connectivity graph induced from Fig. 6.3a; an edge in the connectivity graph between two checkpoints in Fig. 6.3b indicates that it is possible to travel

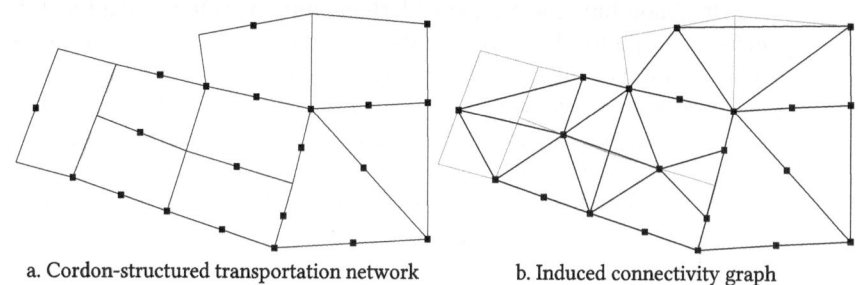

a. Cordon-structured transportation network   b. Induced connectivity graph

Fig. 6.3. A cordon-structured transportation network (black squares indicate checkpoint locations) and the connectivity graph induced by the checkpoints

through the transportation network between them without passing through any other checkpoints.

In this scenario, the question that Protocol 6.3 answers is "Which mobile objects are on which connectivity graph edges at which times?" In Protocol 6.3, each checkpoint is a sensor node, which maintains as local data a database table m listing: mobile object identifiers (*oid*) at traversed edges of the connectivity graph incident with that checkpoint; the time they entered the edge (*enter*); the time they exited the edge (*exit*); the identifier of the checkpoint they entered the edge from (*in*); and the identifier of the checkpoint they exited the edge from (*out*). When a mobile object is detected passing a node, the following steps are taken by Protocol 6.3:

1. When a checkpoint detects a mobile object passing it, that checkpoint updates its movement table m with the mobile object identifier, the time of entry to the edge, and its own identifier as the *in*-checkpoint, broadcasting the same information to its neighbors in the communication graph as an exit message.
2. Upon receiving an exit message, a node first checks if its movement table m contains any "open" records for the specified mobile object identifier. An open record is one where the mobile object identifier, entry time, and *in*-checkpoint are known, but the exit time and *out*-checkpoint are unknown (NULL values in m). A node with such an open record then updates this record with the exit time and *out*-checkpoint contained in the exit message, "closing" the record. Finally, it returns an entr message, containing the complete record for that movement, to the node that originally sent the exit message.
3. A checkpoint receiving an entr message inserts a new record in its movement table m with the information about the object movement contained in the exit message.

One important assumption underlying this algorithm is that the communication graph for the sensor network contains all the edges in the connectivity graph (i.e., the connectivity graph induced by the checkpoints on the transportation network is a subgraph of the geosensor network's connectivity graph). If this assumption holds (specified as a restriction in Protocol 6.3), then any checkpoint will always be able to directly communicate with those neighboring checkpoints which are directly accessible in the transportation network.

Protocol 6.3. Tracking histories of mobile objects in cordon-structured networks

---

Restrictions: $\mathcal{NB}$, where connectivity graph induced by sensor nodes V on transportation network is a subgraph of the communication graph G; sensor function $s : V \times T \to O \cup \{\varnothing\}$ where O is a set of unique mobile object identifiers; identifier function $id : V \to \mathbb{N}$

State Trans. Sys.: $(\{\text{IDLE}\}, \varnothing)$

Initialization: All nodes in state IDLE

Local data: Table $m = \langle oid : O, enter : T, exit : T, in : \mathbb{N}, out : \mathbb{N} \rangle$, initialized with zero records.

IDLE

    When $\mathring{s}(now) \neq \varnothing$

        let $o = \mathring{s}_c(now)$ *#o is mobile object id*

        INSERT INTO m VALUES $(o, now, \text{NULL}, \mathring{id}, \text{NULL})$ *#Create new open record*

        broadcast $(\texttt{exit}, o, now, \mathring{id})$

    Receiving $(\texttt{exit}, o, t_x, i)$

        if SELECT $count(*)$ FROM m WHERE $exit = $ NULL AND $oid = o$ AND $in = \mathring{id} > 0$ **then**

            let $t_n :=$ SELECT $enter$ INTO t_n FROM m WHERE $exit = $ NULL AND $oid = o$

            UPDATE m SET $exit = t_x, out = i$ WHERE $exit = $ NULL AND $oid = o$ *#Close record*

            send $(\texttt{entr}, o, t_n, t_x, \mathring{id})$ to node with identifier i

    Receiving $(\texttt{entr}, o, t_n, t_x, i)$

        INSERT INTO m VALUES $(o, t_n, t_x, i, \mathring{id})$

In order to maintain the movement table m, Protocol 6.3 uses SQL (structured query language) statements. SQL is the standard language for manipulating data in relational databases. Readers unfamiliar with SQL can find out more in a host of books on the topic, including many in the geographic information science domain (e.g., [93, 119]), as well as in the short primer in Appendix B of this book. As a brief summary and refresher:

- Tables in a relational database are structured as a number of attributes (columns), each of which can contain data from a specified domain (e.g., numbers, times, dates, string literals, etc.). Thus, $m = \langle oid : O, enter : T, exit : T, in : \mathbb{N}, out : \mathbb{N} \rangle$ is a table called m, with five attributes: an oid (drawn from the set of mobile object identifiers, O); $enter$ and $exit$ times (drawn from the set of times T); and in and out checkpoints (drawn from the set of node identifiers $\mathbb{N}$).

- Specific rows and columns from the table can be retrieved using a "select" statement of the form: SELECT *attributes* FROM *table* WHERE *conditions*. The selected attributes specify the subset of attributes required from the table; the conditions specify the subset of rows required from the table. Where it is necessary to store the retrieved data in a local variable v, we use the syntax: SELECT *attribute* INTO v FROM *table* WHERE *conditions matching exactly one row*.
- New records (rows) can be inserted into the table using the "insert" statement: INSERT INTO *table* VALUES $(r_1, r_2, \ldots)$. In the basic insert statement, there must be as many values specified as attributes in the table, and in the same order.
- Existing records (rows) can be updated using the "update" statement: UPDATE *table* SET *attribute* $= r_1, \ldots$ WHERE *conditions*. Updating a record in a specified table requires one or more attributes and updated values to be set, and may additionally include conditions that restrict the rows that will be updated, as in a select statement.

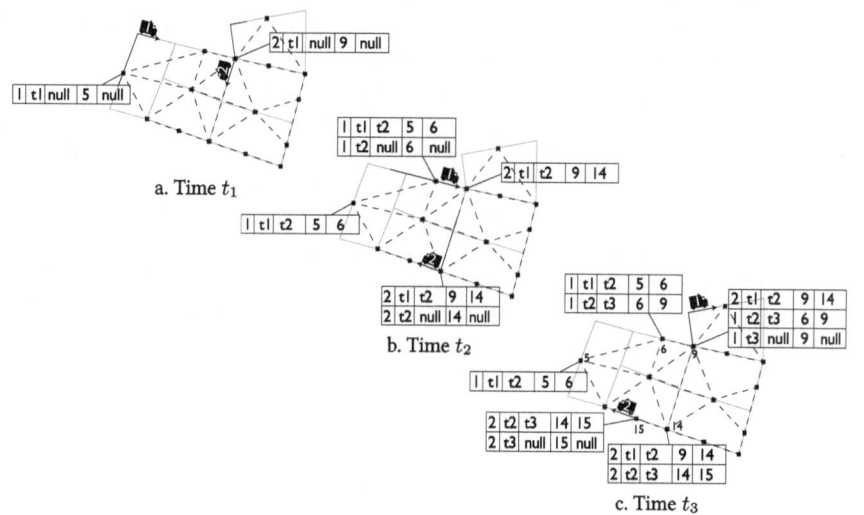

Fig. 6.4. Example states of tables tracking two objects, *oids* 1 and 2, over three time steps (cf. Protocol 6.3). The transportation network is shown with solid lines; the connectivity network (and so subgraph of communication network) is shown with dashed lines

Figure 6.4 shows a specific instance of the nodes' local databases over three time steps of two vehicles moving around a transportation network. Studying the changes to the individual node's table will help us understand Protocol 6.3, in particular the role of open records (which include NULL values) and closed records (which do not). Two messages will be generated, one exit and one entr message, every time a vehicle passes a checkpoint. Thus, the communication complexity of the algorithm is entirely dependent on the number of times vehicles pass checkpoints. In turn, we can reasonably assume the number of vehicle passing occurrences is linearly dependent on $|T| * |O|$ (the length of time for which the algorithm runs times the number of objects).

The algorithm is surely simple, and not too difficult for an adversary to undermine. For example, Protocol 6.3 assumes that vehicles never double-back, exiting an edge by the same checkpoint they passed to enter it. Developing a more sophisticated algorithm to deal with this issue is left as an exercise for the reader (see question 6.2). The algorithm also requires that at most one vehicle passes a checkpoint at any moment. While this is a reasonable simplifying assumption, it is also not especially challenging to extend the algorithm to deal with such issues.

A more subtle issue is that the algorithm only *maintains* distributed information about the movements of vehicles. Specifically, the algorithm stores at each checkpoint the movement events that occur on any connectivity graph edges incident with that checkpoint. Together, all the checkpoints will store all the movement events; however, this information is spatially distributed across the network, so checkpoints only hold information directly relevant to movement events that occurred close to that checkpoint. The algorithm does not go as far as actually *querying* this information. There is a wide variety of different queries that might be formulated based on this distributed information, from queries about the moments of individual vehicles to those about the movements of groups of vehicles. Answering these queries would require further decentralized spatial algorithms to be developed (see questions 6.3 and 6.4).

6.3.2 Mobile Information Dissemination: Flooding and Epidemics

At the beginning of Chapter 4, we explored flooding as a basic technique for moving information around a network. Those techniques can be relatively easily translated into a dynamic setting, where sensor nodes are mobile and/or the environment is dynamic.

Protocol 6.4 provides an example of hop-count flooding (cf. Protocol 4.3) for mobile nodes exploring an environment. The sensor function in Protocol 6.4 monitors a time-varying Boolean value—monitoring the presence or absence of some environmental phenomenon, for example, a pollutant. When the sensor detects the presence of the environmental phenomenon at a node v and time t (i.e., $s(v,t) = 1$), the node broadcasts a new message with hop count 0, before transitioning from IDLE to DONE state. The message then continues to be forwarded by nodes, up to a maximum hop count (ten hops in the case of Protocol 6.4). The node can only be reset, transitioning from DONE back to an IDLE state at some later time t' once $s(v, t') = 0$, when the absence of the sensed environmental phenomenon is detected. In this way, only one message will be generated per detection per node.

Protocol 6.4. Hop-count flooding for dynamic environments, with sensor value trigger

---

Restrictions: Reliable communication; time-varying connected communication graph $G(t) = (V, E(t))$; $s : V \times T \to \{0, 1\}$
State Trans. Sys.: $(\{\text{IDLE}, \text{DONE}\}, \{(\text{IDLE}, \text{DONE}), (\text{DONE}, \text{IDLE})\}\})$
Initialization: All nodes in state IDLE

IDLE
    When $\mathring{s}(now) = 1$ *#Check for presence of sensed variable*
        broadcast (msge, 0)   *#Broadcast msge to neighbors*
        become DONE

DONE
    When $\mathring{s}(now) = 0$ *#Check for absence of sensed variable*
        become IDLE

IDLE, DONE
    Receiving (msge, h)
        if $h < 10$ then   *#Maximum hop count of 10*
                broadcast (msge, $h + 1$)   *#Broadcast msge to neighbors*

Protocol 6.4 is designed for mobile nodes monitoring dynamic environments. It assumes a time-varying graph $G(t) = (V, E(t))$, where the connections between nodes, $E(t)$, may change over times $t \in T$ (assuming the graph remains connected). Indeed, the algorithm would operate correctly even if the restrictions were relaxed further, allowing volatile networks where mobile nodes can be added or removed, $G(t) = (V(t), E(t))$ (see §2.4). Of course, the algorithm will also operate correctly when the nodes are immobile. Indeed, this is in general a feature of algorithms for mobile nodes—immobility can be thought of as a special case of mobility when designing decentralized spatio-temporal algorithms. The protocol is oblivious to whether node movement is constrained by a network or not. This feature is clear because the header for Protocol 6.4 contains no reference to the space in which the nodes are moving (and hence it cannot be a restriction in the algorithm). The protocol is also oblivious to whether or not nodes are associated with meaningful moving objects—it does not matter if the nodes are attached to vehicles, say, or are being carried by water or air currents.

As with Protocol 4.3, the efficiency of Protocol 6.4 is strongly dependent on the hop count threshold h. The lower the hop count, the fewer the messages generated. However, message delivery can only be guaranteed through higher hop counts (see §4.1). The efficiency of Protocol 6.4 also depends strongly on the number of environmental occurrences detected over time. If we assume that occurrences of detection are linearly related to the length of time the algorithm has been executing, then the overall communication complexity becomes $O(h * |V| * |T|)$ messages sent (where, again, $|T|$ is the length of time for which the algorithm executes).

Protocol 6.5 adopts a more explicitly spatiotemporal approach than Protocol 4.3, based on the idea of an epidemic (see §4.2.1 and [27]). Rather than counting hops, in an epidemic approach we count the number of neighbors we "infect" with a piece of information. Protocol 6.5 defines an epidemic infection count, e. When a node senses an environmental event, it informs the next e neighbors it encounters.

One important change in designing Protocol 6.5 is the removal of the restriction that the network is connected. The communication graph, $G(t) = (V, E(t))$, is time varying, with a fixed set of nodes making and breaking one-hop connections with other nodes over time. However, at any given time it is entirely possible that some or even all nodes are disconnected from the rest of the network. In such cases, the node simply waits until it moves to a new location where it can communicate with other nodes. The approach of allowing mobile nodes to *physically* move information around in the absence of network connectivity is fundamental to and widely used with mobile nodes. The approach is often termed *mobility diffusion* [51].

In Protocol 6.5, nodes make full use of mobility diffusion, only "infecting" one-hop neighbors with information about sensed environmental occurrences. In terms of communication complexity, the epidemic approach adopted in Protocol 6.5 is highly efficient, requiring only infection threshold e messages per environmental event that is detected. Assuming the best case, that the number of detected environmental occurrences is linearly related to the length of time the algorithm operates, the overall communication complexity becomes $\Omega(e * |T|)$ (i.e., independent of the size of the network).

However, one cost of using mobility diffusion is a lack of delivery guarantees. In most applications, it is highly unlikely that a mobile node will become a one-hop neighbor of every other node in the network in a reasonable amount of time. Hence, once discovered, information will typically only infect a small proportion of nodes in the network, irrespective of the epidemic infection threshold. To address this, the approach might also easily be combined with hop-count flooding, where received information is rebroadcast up to a certain hop-count threshold (question 6.5), but only by incurring some efficiency costs.

Further, Protocol 6.5 assumes that each node only encounters one environmental event at a time. While a node is in the SEND state and infecting neighbors with information about a sensed environmental event, it is not receptive to trigger events detecting new environmental events. One solution to this issue would be to combine the epidemic approach in Protocol 6.5 with the surprise flooding approach of Protocol 4.4 (question 6.6).

Protocol 6.5. Epidemic information diffusion for mobile nodes

---

Restrictions: Reliable communication; (possibly disconnected) communication graph $G(t) = (V, E(t))$; identifier function $id : V \to \mathbb{N}$; neighborhood function $nbr : V \times T \to \mathbb{N}$, where $nbr(v, t) \mapsto \{id(v')|\{v, v'\} \in E(t)\}$; sensor function $s : V \times T \to \{0, 1\}$

State Trans. Sys.: $(\{\text{IDLE}, \text{SEND}\}, \{(\text{IDLE}, \text{SEND}), (\text{SEND}, \text{IDLE})\}\})$

Initialization: All nodes in state IDLE

Local data: Infection count e, initialized $e := 10$; set N of visited neighbors, initialized $N := \varnothing$

IDLE
> When $\overset{\circ}{s} = 1$ #*Check for presence of sensed variable*
>> become SEND

SEND
> When $|\overset{\circ}{nbr}(now)| > 0$ #*Check for presence of sensed variable*
> if $e \geq |\overset{\circ}{nbr}(now) - N|$ then
>> send (msge) to $\overset{\circ}{nbr}(now) - N$ #*Send msge to unvisited neighbors*
>> set $e := e - |\overset{\circ}{nbr}(now)| - N$
> else
>> send (msge) to e arbitrarily chosen neighbors in $\overset{\circ}{nbr}(now) - N$
>> set $e := 0$
> if $e = 0$ then
>> set $e := 10$ #*Infection count set to 10 neighbors*
>> set $N := \varnothing$ #*Reset visited neighbors*
>> become IDLE

6.3.3 Summary

The most important new situation encountered in this section is that at any particular moment in time the communication graph may be *disconnected*. Thus far, the algorithms we have examined have as a basic restriction that the network be connected (recall the $\mathcal{NB}$ restrictions, p. 86). Without this connectivity, nodes in disconnected components of the static networks in the previous two chapters have no hope of interacting, exchanging information, or performing coordinated computing tasks.

However, with mobility comes the potential for mobility diffusion: deferring communication until nodes come into close proximity. As in §6.2.3, while change makes some aspects of decentralized algorithm design harder, the surprising fact is that in other ways change can be a boon for efficiency and practicality.

Finally, Protocol 6.3 generated information about the movements of monitored objects in the network itself, and stored it at transportation checkpoints. But querying this information was left as a separate issue, requiring further decentralized algorithms (questions 6.3 and 6.4). This separation of *generation* and *querying* of information is common in decentralized spatiotemporal algorithms, much more so than in decentralized spatial algorithms. Because change is occurring continually, the network will need to record changes as they happen. However, the information about the changes may not be required until some later date, and it is often convenient to deal with queries of this stored information using separate algorithms.

6.4 Chronicles of Mobile Nodes

As we have seen, in most cases raw sensor data is in the form of histories (states of the world over time). Consequently, the switch from monitoring histories to monitoring chronicles is primarily about inferring meaningful occurrences from sequences of states. The identification of movement patterns is a typical example of this inference process: high-level knowledge about movement patterns is constructed from low-level data about trajectories [46].

Flocking is amongst the most studied and the most challenging of movement patterns. Intuitively, a group of mobile objects are "flocking" when they move together for some period of time. Flocks are of interest in monitoring both animal and human movement. For example, identifying pedestrian flocks can be important in crowd management, helping us avoid congestion during an emergency evacuation [47]. For animals, knowledge of flocking behavior can be important for protecting pastures, where urine patches caused by dense groups of livestock can adversely affect soil nutrients [7].

Actually computing with flocks can be tricky. One of the most common definitions of flocking is the nkr-flock: a group of n mobile objects which remain within a disk of radius r for a period of time k [11]. Figure 6.5 illustrates the idea with an example flock, with $n = 4$, $k = 4$, and radius r represented in the figure. In Fig. 6.5 this flock is not necessarily composed of the *same* mobile objects during its existence; in some cases it may be important to distinguish flocks where the same n entities comprise the flock over its lifetime.

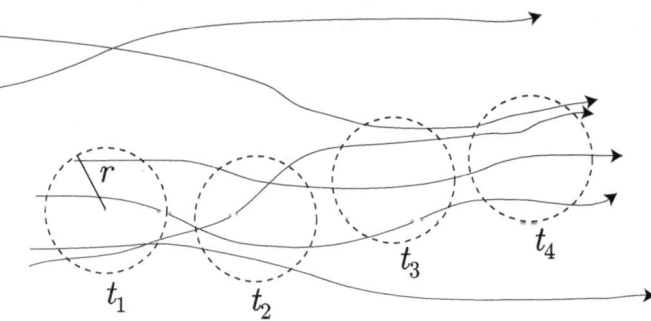

Fig. 6.5. An example nkr-flock, with flock size $n = 4$, flocking duration $k = 4$ time steps, and flock radius r illustrated on the first disk

Unfortunately, this apparently simple nkr-flock definition is computationally intractable, and so even centralized spatial information systems require heuristics in order to compute with nkr-flocks. Protocol 6.6 provides a concise algorithm for approximating flocks, using a simple heuristic. Informally, the algorithm operates in the following way:

- Each node periodically broadcasts a posn message containing its identifier, coordinate position, and a time stamp, in addition to storing a record for its own location in its movement table m.
- Nodes receiving a posn message first check whether:
  a. the distance between their own position and that indicated in the message is greater than the flocking radius r; and
  b. they have already received and stored a similar message.
  If neither of these conditions is met, the node may store the information from the posn message in its own movement table m.

Protocol 6.6. Basic flocking algorithm

Restrictions: Reliable communication; $G(t) = (V, E(t))$; identifier function $id : V \rightarrow \mathbb{N}$; coordinate location $p : V \times T \rightarrow \mathbb{R}^2$; flock parameters n (minimum number of objects in a flock), k (minimum duration of flock), r (radius of flock)

State Trans. Sys.: $(\{\text{IDLE}\}, \varnothing)$

Initialization: All nodes in state IDLE

Local data: Epoch e, initialized $e := t_0$; movement table $m = \langle nid : \mathbb{N}, epoch : T \rangle$, initialized with one record: $\langle id, t_0 \rangle$

IDLE

    When time trigger elapsed

        set $e := now$ *#Store current time as latest epoch*

        INSERT INTO m VALUES $(\overset{\circ}{id}, e)$ *#Store new data with relevant epoch*

        broadcast (posn, id, $\overset{\circ}{p}(now)$, e)   *#Broadcast identifier, position, and current epoch*

    Receiving (posn, i, p', e')

        if $\delta(\overset{\circ}{p}(e'), p') < r$ **then**   *#Process message if it originated within flock radius r*

            let $g :=$ SELECT $max(epoch)$ INTO g FROM m WHERE $epoch \leq e'$ *#Find relevant epoch*

            let d be SELECT COUNT (*) INTO d FROM m WHERE $nid = i$ AND $epoch = g$

            if $d = 0$ **then**   *#Check if data is new*

                INSERT INTO m VALUES (i, g) *#Store new data with relevant epoch*

                broadcast (posn, i, p', e')   *#Rebroadcast message*

Importantly, nodes cannot be assumed to be synchronized in triggering the posn broadcasts. posn messages generated by different nodes at the same time may have slightly different timestamps. Those messages may in turn experience different delays in traversing the communication network. As a result of this temporal granularity, when a node stores a record from a posn message in its movement table m, the node references the record to a local "epoch"—the last time it triggered a posn message. Figure 6.6 provides an example sequence diagram to illustrate the operation of Protocol 6.6. In Fig. 6.6, the changes to the movement table m highlight how all the posn messages stored at v_3 reference the epochs for v_3.

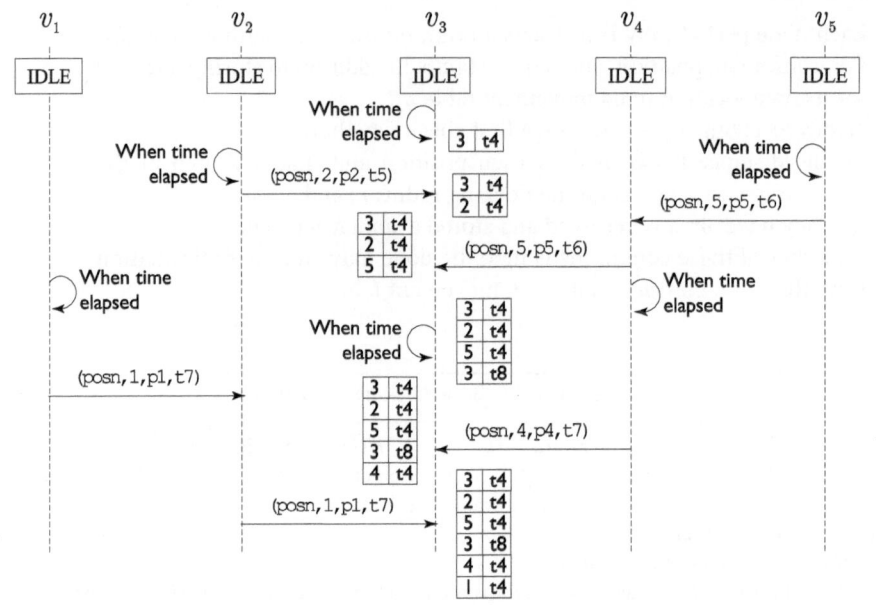

Fig. 6.6. Sequence diagram for five nodes, $v_1, \ldots v_3$, highlighting changes to movement table m of v_3 resulting from received posn messages

In cases where more than n records are to be found in consecutive epochs spanning time period k, it follows that the node has detected the existence of a flock. Nodes could check for this condition programmatically, periodically iterating through the records in the movement table m. However, the same information can also be found using an (admittedly rather complex) SQL query, using *views* (SQL queries that are realized as "virtual" tables, which automatically reflect any changes to the underlying tables):

CREATE VIEW f AS SELECT MIN(nid) $\geq n$ FROM

        (SELECT $count(nid)$ AS cnt, $epoch$ FROM m GROUP BY $epoch$)

        WHERE $epoch \leq t$ AND $epoch \geq t - k$

The view f identifies cases where there are at least n records in every epoch of a time period k, ending at some user-defined time point t. For example, after setting $t = now$, the single data item in the view will be TRUE if a flock is currently known to the node, and FALSE otherwise. In summary, the view identifies for a node the collection of records where nodes:

- have stored epochs that span flock duration k; and
- have a count of at least n flock neighbors for all epochs in the flock duration.

As already stated, the intractability of computing flocking means that Protocol 6.6 approximates the flock. In fact, it will provide a consistent underestimate of the flocks—it will generate no errors of commission, but may be subject to errors of omission, for the following reasons:

- Protocol 6.6 only searches for flocks where the disk of radius r is centered at a node in the network. It is possible, especially for flocks with smaller n or more autocorrelated node positions, that some flocks may be undetectable using this heuristic (see Fig. 6.7a).
- The posn messages can never travel further than distance r from their origin (because of the rebroadcasting condition in Protocol 6.6 in the action for the *Receiving* posn event). Consequently, in cases where the communication subgraph induced from the set of nodes within a disk is disconnected, the size of the flock may be underestimated. Disconnection may occur either simply because the unit distance c in comparison to r is too small to achieve connections, or because the only paths between nodes in the disk pass through nodes outside the disk (Fig. 6.7b).
- The clocks for the nodes need not be synchronized. However, if in addition to a lack of synchronization, some nodes' clocks have substantial drift (i.e., run faster or slower than other nodes' clocks) it is possible that this drift will cause a node that is part of a flock to skip an epoch from time to time. In cases where the flock is exactly or very close to n nodes, it may happen that the records for a skipped epoch could cause the count of nodes in the movement table at a key node to drop below the minimum number of objects for a flock.

It would take us too far from our path to begin investigating further ways to address these problems here. But for interested readers, some more sophisticated heuristics can be found in [69, 70].

Monitoring flocks is surely computationally expensive. In the worst case, where the flock radius r is comparable to the diameter of the entire network, every node may need to rebroadcast posn messages from every other node, in every epoch. This leads to a worst case of $O(|T| * |V|^2)$ overall communication complexity and $O(|T| * |V|)$ load balance for messages sent.

Such high communication complexity might call into question the usefulness of this algorithm. For example, an alternative approach would be to periodically build a routing tree with multiple initiators (Protocol 4.11, with overall complexity $O(|T| * |V|^2)$). The root node could then query this tree

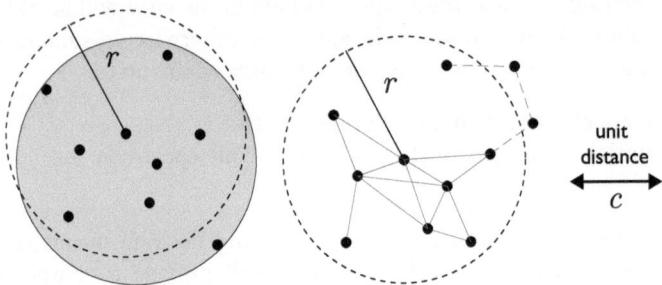

a. Disk not centered on node   b. Disk induces disconnected subgraph

Fig. 6.7. Two examples of node configurations leading to undetectable flocks of $n = 9$ nodes: a. 9 nodes contained in gray disk of radius r are contained in no disk centered on nodes; b. disconnected subgraph induced by nodes contained in disk

using data aggregation for the total count of nodes within a certain distance (for example, adapting the TAG algorithm in Protocol 4.10, overall complexity $O(|T| * |V|)$), perhaps then forwarding this data to a centralized server for monitoring over time.

However, in many applications, such as livestock monitoring, a lack of network connectivity might make such an approach impractical. Further, in some cases it is to be expected that the flocking radius r may be much smaller than the extent of the network, restricting messages to only a small proportion of nodes. Thus, on average the communication complexity may be much closer to the best case $O(|T| * |V|)$ overall communication complexity.

6.4.1 Summary

The key algorithm design technique used in this section is to infer meaningful occurrences (the emergence of flocks) from sequences of states (in this case, the trajectories of groups of mobile objects). While flocking is one of the most researched of all movement patterns, similar techniques would potentially be useful in any of a wide range of other important movement patterns, such as those of following, leadership, convergence, and divergence (cf. [29]).

6.5 Chronicles of Dynamic Environments

The fourth and final category in our classification of decentralized spatio-temporal algorithms concerns monitoring chronicles of dynamic environments. As in the previous section, the design of an algorithm for chronicling dynamic environments hinges on in-network mechanisms for inferring occurrences from changes in state.

Specifically, this section presents an algorithm for monitoring occurrences in dynamic fields. Previous algorithms in this book have tended to focus on qualitative representation of fields as thresholded regions (e.g., "hot spots"). While we will return to this representation in the course of this chapter (in the following section), there is an alternative qualitative representation of fields: as passes (saddle points), peaks, and pits (Fig. 6.8a), and the ridges and valleys that connect them.

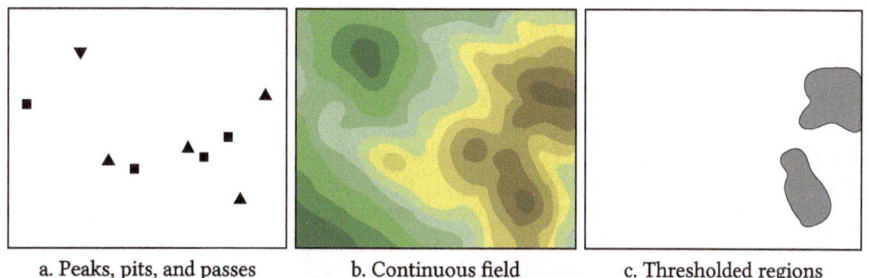

a. Peaks, pits, and passes   b. Continuous field   c. Thresholded regions

Fig. 6.8. Two qualitative representations of b. a continuous field: c. as thresholded regions (e.g., "hot spots"); and a. as peaks (triangle), pits (inverted triangle), and passes (square)

In line with the central aims of this book—to explore the process of decentralized spatiotemporal algorithm design, rather than to provide a compendium of sophisticated decentralized spatiotemporal algorithms—this section limits itself to investigating the identification of peaks and pits, and the things that can happen to peaks and pits: appearance, disappearance, and movement.

Identifying peaks and pits in a monitored field (such as a temperature surface) using a decentralized spatial algorithm is straightforward. A peak is simply any node which senses a higher value than all its neighbors; similarly, a pit senses a lower value than any of its neighbors. Protocol 6.7 performs the initialization stage for our dynamic algorithm. To start, all nodes in the INIT state broadcast their sensed value, and then listen for neighbors' broadcasts. Nodes that detect they have a lower sensed value than all their neighbors transition to a PITX state; nodes that have a higher sensed value than their neighbors become PEAK; all other nodes transition to IDLE.

The initialization in Protocol 6.7 is structurally very similar to those in algorithms we have seen before, such as the region boundary detection algorithms. It clearly has overall communication complexity $O(|V|)$ and optimal load balance $O(1)$.

Things get more interesting when, building on this initial step, we extend Protocol 6.7 to identify the changes that occur to peaks and pits in Proto-

---

Protocol 6.7. Detecting peaks and pits in a sensed field

Restrictions: $\mathcal{NB}$; identifier function $id : V \to \mathbb{N}$; sensor function $s : V \times T \to \mathbb{R}$

State Trans. Sys.: $(\{\text{INIT}, \text{IDLE}, \text{LSTN}, \text{PEAK}, \text{PITX}\}, \{(\text{INIT}, \text{LSTN}), (\text{LSTN}, \text{IDLE}), (\text{LSTN}, \text{PEAK}),$
  $(\text{LSTN}, \text{PITX})\})$

Initialization: All nodes in state IDLE

Local data: Peak id pk, pit id pt; neighbor sensor readings $r : N \to \mathbb{R}$, where $N = \{id(v) | v \in \mathring{nbr}\}$

INIT
  Spontaneously
    broadcast (ping, $\mathring{s}(now)$, $\mathring{id}$)   *#Broadcast sensed data and identifier*
    become LSTN   *#Transition to LSTN state*

LSTN
  Receiving (ping, s, i)
    set $r(i) := s$ *#Store received neighbor sensed value and identifier*
    if ping received from all neighbors **then**
      become IDLE   *#Nodes that are not peaks or pits are by default IDLE*
      if $\max(r_*) < \mathring{s}(now)$ **then**   *#Peak sensed value greater than max of image of r*
        set $pk :=$ new non-negative peak id   *#Duplicate peak ids need to be avoided*
        become PEAK   *#Transition to PEAK state*
      if $\min(r_*) > \mathring{s}(now)$ **then**   *#Pit sensed value less than min of image of r*
        set $pt :=$ new non-negative pit id   *#Duplicate pit ids need to be avoided*
        become PITX   *#Transition to PITX state*

---

col 6.8. Although it is a long algorithm, the intuition behind Protocol 6.8 is reasonably simple to explain.

1. When a node detects a change in its sensed value (such as a rise in temperature) it broadcasts an upd8 message with details of that change. A node that detects it is no longer a peak, either because its own sensed value drops below its neighbors' or a neighbor's sensed value rises above it, transitions to an IDLE state and unicasts a peak message to its maximum ascent neighbor.

2. The peak message continues to be forwarded to maximum ascent neighbors until it reaches another PEAK node. If we assume there are no plateaus in the field (neighbors with identical sensed values), the peak message must, sooner or later, arrive at another unique peak.

3. A node that newly transitions to an PEAK state must then deal with three cases:
   a) If the new PEAK node receives exactly one peak message, this indicates that the peak has *moved*.
   b) If the new PEAK node receives no peak messages, then that peak has *appeared*.
   c) If the new PEAK node receives more than one peak message, then this indicates that the additional peak messages have come from peaks that have *disappeared*.

In cases where multiple `peak` messages are received, it is in reality indeterminate which one of the originating peaks has moved and which have disappeared. One might equally regard all peaks as having *merged*, or, equivalently, all originating peaks as having disappeared and new peaks as having appeared. For example, a natural interpretation of moving from Fig. 6.9a to b is that the lower right peak has moved a little, while the left-hand peak has disappeared. However, a number of other interpretations are possible, including that the lower right peak has disappeared, while the left-hand peak has moved a lot; or indeed that all the peaks have disappeared, and two new peaks have appeared.

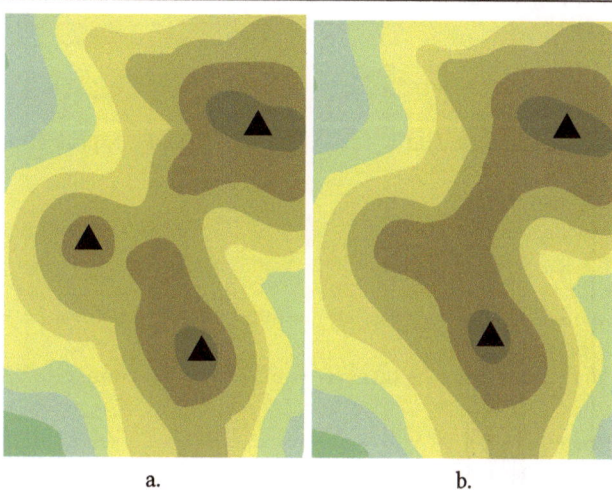

a.   b.

Fig. 6.9. Example of peak disappearance and movement over two time steps (a. and b.)

Protocol 6.8 arbitrarily treats the first received `peak` message as indicating a movement (on the assumption that the first `peak` message to be received will be from the closest ex-peak), with subsequent received `peak` messages treated as peak disappearances. For conciseness, the events and actions for the PITX state have been omitted from Protocol 6.8; however, these correspond directly to the PEAK events and actions (see question 6.8).

Figure 6.10 contains the full state transition diagram for Protocol 6.8.

As already mentioned, Protocols 6.7 and 6.8 may fail if any neighboring sensors located at peaks or pits have identical sensed values. In practice, this assumption of "no plateaus" is not particularly limiting. Real data from geosensor networks rarely contains *identical* sensed values. Identical values tend to occur only when sensors are malfunctioning (e.g., sensors continually record some system default value), have exceeded their sensitivity range (e.g., where indoor light sensors with a maximum sensitivity of 10,000 lux are

---

Protocol 6.8. Identifying peak/pit appearance, disappearance, and movement events

---

Fragment extends: Protocol 6.7

State Trans. Sys.: $(\{\text{INIT}, \text{IDLE}, \text{LSTN}, \text{PEAK}, \text{PITX}\}, \{(\text{INIT}, \text{LSTN}), (\text{LSTN}, \text{IDLE}), (\text{LSTN}, \text{PEAK}),$
$(\text{LSTN}, \text{PITX}), (\text{PEAK}, \text{IDLE}), (\text{PITX}, \text{IDLE}), (\text{IDLE}, \text{PEAK}), (\text{IDLE}, \text{PITX})\})$

IDLE

    When $\mathring{s}(now)$ *changes*

      broadcast $(\text{upd8}, \mathring{s}(now), \mathring{id})$

      if $\max(r_*) < \mathring{s}(now)$ **then**   *#Check for peak*

        set $pk :=$ new negative peak id   *#Duplicate peak ids need to be avoided*

        become PEAK

      if $\min(r_*) > \mathring{s}(now)$ **then**   *#Check for pit*

        set $pt :=$ new negative pit id   *#Duplicate pit ids need to be avoided*

        become PITX

    Receiving $(\text{upd8}, s, i)$

      set $r(i) := s$

      if $\max(r_*) < \mathring{s}(now)$ **then**   *#Check for peak*

        set $pk :=$ new negative peak id   *#Duplicate peak ids need to be avoided*

        become PEAK

      if $\min(r_*) > \mathring{s}(now)$ **then**   *#Check for pit*

        set $pt :=$ new negative pit id   *#Duplicate pit ids need to be avoided*

        become PITX

    Receiving (peak, p)

      send (peak, p) to neighbor i such that $r(i) = \max(r_*)$ *#Forward peak up max ascent*

    Receiving (pitx, p)

      send (pitx, p) to neighbor i such that $r(i) = \min(r_*)$ *#Forward peak down max descent*

PEAK

    When $\mathring{s}(now)$ *changes*

      broadcast $(\text{upd8}, \mathring{s}(now), \mathring{id})$

      if $\max(r_*) > \mathring{s}(now)$ **then**

        if $pk < 0$ **then set** $pk := -pk$ *#Peak appearance*

        send (peak, pk) to neighbor i such that $r(i) = \max(r_*)$ *#Forward peak up max ascent*

        become IDLE

    Receiving $(\text{upd8}, s, i)$

      set $r(i) := s$

      if $\max(r_*) > \mathring{s}(now)$ **then**

        if $pk < 0$ **then set** $pk := -pk$ *#Peak appearance*

        send (peak, pk) to neighbor i such that $r(i) = \max(r_*)$ *#Forward peak up max ascent*

        become IDLE

    Receiving (peak, p)

      if $pk < 0$ **then**

        set $pk := p$ *#Peak p movement detected*

      else

        No specified operation   *#Peak p disappearance detected*

PITX

    Events and actions for PITX correspond directly to PEAK state

---

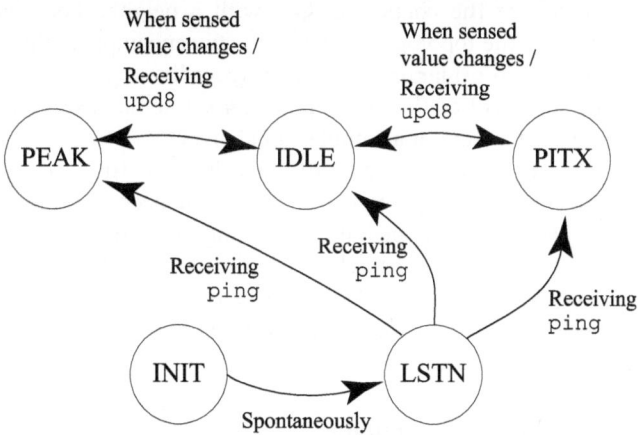

Fig. 6.10. State transition diagram for Protocol 6.8

in bright direct sunshine, 50,000+ lux), or are measuring an environmental parameter with a maximum and/or a minimum value (e.g., relative humidity, from 0 to 100%).

In terms of communication complexity, Protocol 6.8 is expected to be quite efficient. Following the $O(|V|)$ of overall messages sent and the $O(1)$ (load balance) initialization in Protocol 6.7, all subsequent upd8 messages are broadcast strictly one-hop, and all peak/pitx messages are unicast, following direct paths up spurs/down valleys to their unique destination peak/pit. In many cases, such paths can be expected to be of length $\sqrt{|V|}$. However, the actual communication complexity will depend strongly on the correlation and rate of change of the field. Random fields are expected to incur high computational costs; slowly changing or highly correlated fields, often encountered in monitoring geographic environments, will incur dramatically lower computational overheads.

When an algorithm such as Protocol 6.8 is strongly dependent on the characteristics of the monitored environment, there is sometimes no alternative other than to use empirical investigation to further understand its computational characteristics. As we near the end of Part II of this book, it is worth noting that such empirical investigations are the primary focus of the third and final part of this book.

6.6 Protocol of the Chapter: Topological Changes to Regions

In this penultimate section of Part II, we investigate a second algorithm for chronicles of dynamic fields, this time returning to the region-based, thresholded representation of the sensed environment. We started our sequence of protocols of the chapter in §4.6 with a neighborhood-based algorithm for detecting the topological relations between simple regions. Adding more sophisticated information about space (coordinate position and cyclic ordering), the protocol of the chapter in §5.4 set about discovering the internal topological structure of a complex areal object.

Completing the sequence, the algorithm in this section adds change, and chronicles the topological changes that occur to monitored regions—their appearance, disappearance, merging, and splitting. For example, our algorithm might be used to chronicle the changes that occur to an oil slick, as already seen in Fig. 6.1. However, before presenting the algorithm itself, it is necessary to first discuss some important assumptions that the final algorithm of Part II uses.

6.6.1 Simplifying Assumptions

The protocol of this chapter, Protocol 6.9, relies on a number of strong simplifying assumptions about the network, regions, and changes being monitored. However, most—all except one—of these assumptions are purely for the purpose of shortening and simplifying, and, as we shall see, would be relatively easy to relax, often using techniques we have encountered earlier in this book.

Simple region components

In Protocol 6.9, it was assumed that each connected component of the areal object forms a simple region, homeomorphic to a disk (i.e., with no holes; see §4.5.2). This is in most cases an unrealistic assumption. In general, we may expect regions over time to evolve with the full complexity of the areal structure explored in §5.4, with holes, islands, holes in islands, etc. In the context of Protocol 6.9, the restriction to simple region components helps by restricting the topological changes that can occur to region appearance, disappearance, merging, and splitting. However, [58] has shown that even without this restriction, all topological changes to a complex areal object can be decomposed into a combination of only six atomic topological changes to regions: the four already mentioned, plus the so-called "self-merge" and "self-split," where a region captures or releases a hole (see Fig. 6.11).

Happily, it is not especially difficult to define a more sophisticated algorithm to capture these additional self-merge and self-split events, and indeed such algorithms already exist (e.g., [97]). The key is to detect and maintain information at each region boundary about what, if any, regions contain it, much as has already been investigated in §5.4.

Initialization with no regions

The initialization conditions for Protocol 6.9 assume that *no* regions exist to begin with. In other words, the algorithm starts with a blank slate—all nodes

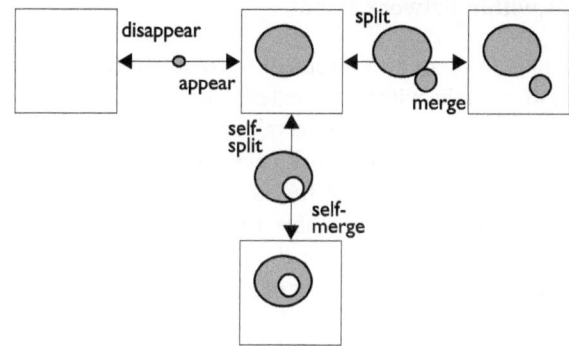

Fig. 6.11. Topological atoms: Six fundamental topological changes to regions (after [58]), only four of which (appear, disappear, merge, split) can occur to simple regions, homeomorphic to a disk

$v \in V$ sensing $s(v) = 0$. This simplifies the algorithm exposition, as any regions must first be detected to appear, before they can subsequently split, merge, or disappear. In this way, the initial state of the algorithm becomes trivial, ensuring that nodes do not have to discover boundaries or create new region identities at algorithm initialization. However, to relax this assumption (whether or not we assume simple regions) only requires a single pass of an algorithm for determining the static topological structure, exactly as already achieved using Protocol 5.12.

Meeting and splitting in pairs of regions

Protocol 6.9 also relies on the assumption that at most two regions are involved in a meeting or splitting. Of course, in reality it is possible that multiple regions meet at the same time and place, or that a region splits into three or more regions at the same time (Fig. 6.12). Ignoring these possibilities simplifies Protocol 6.9 somewhat. But the same techniques can be relatively easily extended to deal with a split or a merge that involves more than two regions. Indeed, again, this has already been successfully investigated in [97].

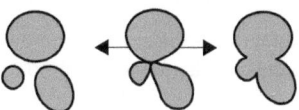

Fig. 6.12. Protocol 6.9 disallows "three-way" splits and merges, although it can be extended to deal with such cases

Regions within network extent

Rather like Protocol 5.12, Protocol 6.9 requires that the regions being monitored fall entirely within the extent of the sensing network. In effect, this assumption prevents "edge effects," where, for example, a region appears to split at the edge of the network coverage, but is in actuality still connected (albeit beyond the extent of the network). Changes that occur outside the network extent are unknowable to the network. So we may regard this restriction as reasonable under the circumstances. However, with additional knowledge about which nodes are located at the edge of the network extent, a more robust protocol could additionally filter out events that might have resulted from edge effects.

Maximal plane communication graph

The communication graph is maximal plane (i.e., a triangulation). At first sight this may appear to be a puzzling restriction, given that it was noted back in §2.6.2 that the UDG for real geosensor networks hardly ever contains a maximal plane subgraph. However, this restriction is purely to make the algorithm more concise when routing around a region boundary. As noted in §5.3.1, using a maximal plane communication graph ensures that each node in the boundary cycle enclosing a region has two boundary node neighbors (Fig. 6.13).

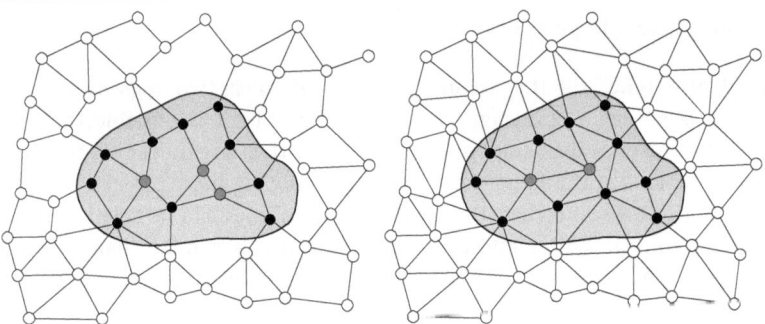

a. General plane communication graph   b. Maximal plane communication graph

Fig. 6.13. In b., a maximal plane graph, the boundary cycle contains only boundary (black) nodes; in a., a general plane graph, the boundary cycle may contain non-boundary interior (gray) nodes (cf. Fig. 5.7)

For Protocol 6.9, relaxing this assumption would only require a more sophisticated technique for handling non-boundary nodes in the boundary cycle, such as those already discussed in §5.3.

Incremental change

After our having addressed most of the assumptions behind Protocol 6.9, there is one final assumption that presents a more fundamental restriction to the algorithm. Protocol 6.9 assumes that the changes monitored are *incremental*, in the specific sense that only one sensor in any neighborhood can change its sensed value *at a time*. Putting it another way, the algorithm assumes that the temporal granularity of the sensor network is fine enough to capture all the details of the changes occurring. Figure 6.14 illustrates this assumption of incremental change with the example of two merging regions.

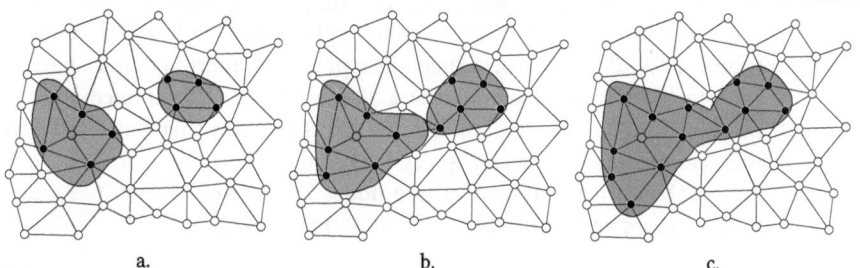

a.   b.   c.

Fig. 6.14. Example of incremental change assumption, where only one node in any neighborhood may change its sense value in a single time step. The sequence of figures a., b., c., is incremental, but the direct sequence a., c. is not

Conceptually, this assumption may be sound—we are monitoring dynamic fields undergoing continuous change, so, in theory at least, for all but the most rapidly changing fields it should be possible to increase the sampling frequency of the nodes in the network to capture all the details of the changes.

However, in practice, this assumption may often not hold. A number of extensions to Protocol 6.9 could help relax, if not remove, the assumption of incremental change. Some research is already aiming to allow increments in contiguous groups of neighboring nodes (e.g., [101]). Further, as we shall see, in many cases the algorithm may be able to detect when the assumption has been violated. It will also be possible to "reset" the algorithm, by periodically executing the static Protocol 5.12 to determine the actual topological structure of the regions, either in response to detected violations, or as a regular sanity check.

6.6.2 Protocol

The assumptions listed in the previous subsections form a part of our restrictions to Protocol 6.9. Even though most would be straightforward to relax, they are still important to document. In summary, in addition to the restrictions listed in the header, Protocol 6.9 assumes the further restrictions:

- each region component is simply connected (no holes);
- the set of regions is initially empty ($s(v) = 0$ for all $v \in V$);
- at most two regions are involved in any topological change at any moment;
- the monitored regions remain within the spatial extent of the network;
- the graph G is maximal plane; and
- changes in the dynamic field are incremental, in the sense that only one sensor in any neighborhood changes its sensed value at any point in time.

Having thoroughly addressed the simplifying assumptions inherent in the algorithm, it is relatively easy to explain the two key intuitions behind Protocol 6.9. Region appearance, disappearance, merging, and splitting can be detected *locally* by examining a single node and its immediate one-hop neighbors. Of particular importance are the nodes that have just changed their sensed value (i.e., from $s(v, t-1) = 0$ to $s(v, t) = 1$, or $s(v, t-1) = 1$ to $s(v, t) = 0$). We refer to these nodes as "active" at time t. Assuming a binary sensor function $s : V \times T \rightarrow \{0, 1\}$:

1. A region appears if a single active node v where $s(v, t) = 1$ has neighbors all of which are outside the region $s(v', t) = 0$, where $v' \in nbr(v)$. Conversely, a region disappears if a single active node v where $s(v, t) = 0$ has neighbors that are outside the region $s(v', t) = 0$, where $v' \in nbr(v)$. Figure 6.15 illustrates the appearance and disappearance of a sensed region.

2. A region merges and splits at a "pinch point": a node that if removed would disconnect two distinct components of the monitored region. Pinch points can be found using the cyclic ordering of sensed values around a node, as illustrated in Fig. 6.16. More formally, if for a node v, $|\{n \in nbr(v)| d(n) = 0 \text{ and } d(c\tilde{y}c(n)) = 1\}| = 2$ then this node is a "pinch point." Note that the assumptions that each region component has no holes (and so disconnection at a pinch points entails disconnection of the regions) and that at most two regions are involved in any topological change are required to ensure that this rule is valid.

With all these preliminaries out of the way, the algorithm can now proceed in a reasonably straightforward way:

- Any nodes that detect a change in their sensed value (i.e., active nodes) inform neighbors with an upd8 message, causing one-hop neighbors to store this change in their local data store d.
- If an active node v has no one-hop neighbors that sense part of the region, then they have detected the appearance or disappearance of a region (i.e., $s(v, now) = 1$ or $s(v, now) = 0$ respectively).

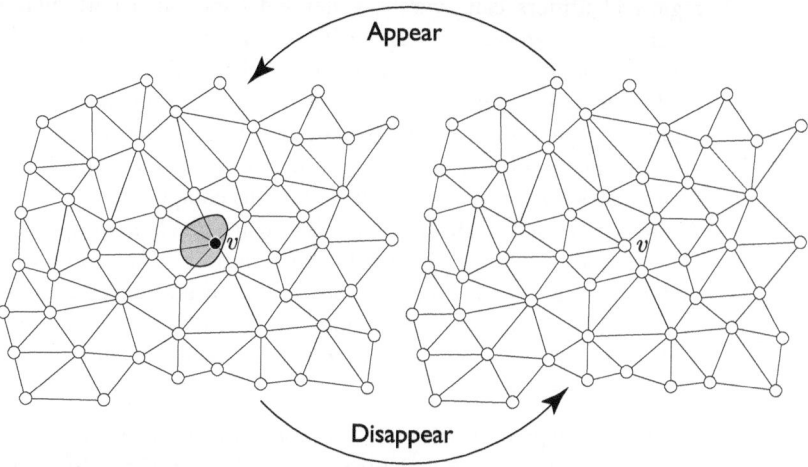

Fig. 6.15. An active node v (changing its sensed value) surrounded by neighbors outside the region can detect appearance or disappearance

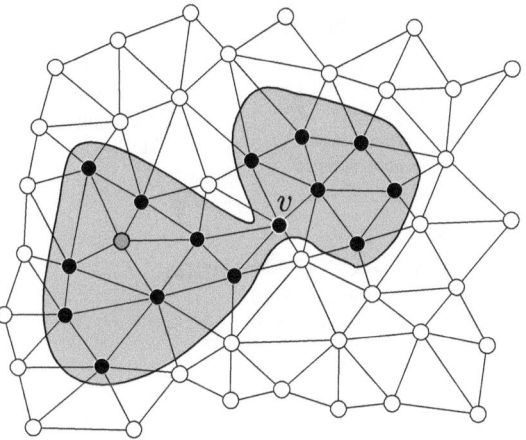

Fig. 6.16. A "pinch point" node (v) if removed would disconnect two distinct components of the monitored region. Pinch points can be found by examining the cyclic ordering of sensed values around a node

- An active node that detects that it is at a pinch point has detected the merging of two regions. This pinch point generates a new unique merged region identifier. The function $u : \mathbb{N} \times T \rightarrow \mathbb{N}$ is responsible for generating unique new region identifiers. Although not explicitly specified, this function is assumed to combine the (unique) node identifier with the current timestamp *now* in a way that means that no duplicate region identifiers can ever be generated (since at a particular time, a specific node can only be responsible for one new region, e.g., $u(n, t) \mapsto n * |T| + t$). The pinch point node then updates all other boundary nodes in the merged region with this new region component identifier via a regn message, routed around the boundary using the winding procedure already encountered in Protocol 5.8 (albeit in a version much simplified by its restriction to maximally plane regions components).

- An inactive node that detects that it is also at a pinch point upon receiving an upd8 message will transition to state PNCH. If this PNCH node v subsequently becomes active (i.e., $s(v, now)$), then it has detected the splitting of a region component. It then updates its two new region components with new unique region identifiers, again using the mechanism of an regn message.

Figure 6.17 depicts graphically the state transition diagram for Protocol 6.9. And (at last!) Protocol 6.9 presents the algorithm for monitoring topological changes to regions itself.

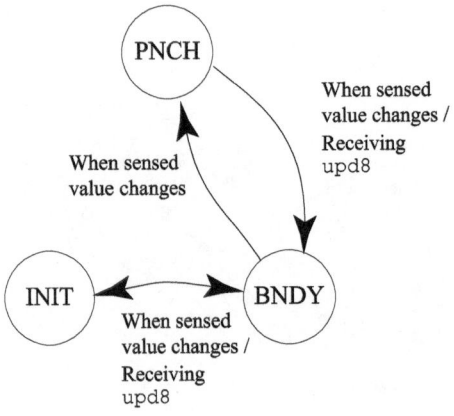

Fig. 6.17. State diagram for Protocol 6.9

Like Protocol 6.8, Protocol 6.9 is expected to be relatively efficient as long as the rate of change of the field is relatively slow. Individual node updates result in a single upd8 message broadcast. Routing regn messages around the boundary for merge and split events is expected to require $O(|V|^{\frac{D}{2}})$ unicast messages (see §5.3.2). Protocol 6.9 does not go as far as actually reporting the detected topological changes to a sink node (see question 6.10), but again this would be as efficient as the routing procedure used for this purpose, many examples of which we have already seen. In cases where the rate of change of the field *is* expected to be high, and so Protocol 6.9 is inefficient, a viable alternative would be instead to opt for periodic execution of a static

---

Protocol 6.9. Monitoring topological changes in dynamic regions

---

Restrictions: Connected plane graph $G = (V, E)$; sensor function $s : V \times T \to \{0, 1\}$; identifier function $id : V \to \mathbb{N}$; cyclic ordering $cyc : E' \to id_*$, where $E' = \{(v, id(v')) | (v, v') \in E\}$; neighborhood function $nbr : V \to \mathbb{N}$, where $nbr(v) \mapsto \{id(v') | \{v, v'\} \in E\}$; unique new region identifier function $u : \mathbb{N} \times T \to \mathbb{N}$, plus six further restrictions set out in §6.6.1

State Trans. Sys.: $(\{\text{IDLE, BNDY, PNCH}\}, \{(\text{IDLE, BNDY}), (\text{BNDY, IDLE}), (\text{BNDY, PNCH}), (\text{PNCH, BNDY})\})$

Initialization: All nodes in state IDLE

Local data: Data $d : id(\mathring{nbr}) \to \{0, 1\}$, init. to $d(i) := 0$; region id $rid \in \mathbb{N}$ init. to $rid := 0$

IDLE

  When $\mathring{s}(now)$ changes
    broadcast (upd8, $\mathring{id}$, $\mathring{s}(now)$)   *#Broadcast sensed value and identifier*
    if $d(n) = 0$ for all $n \in \mathring{nbr}$ **then**
      set $rid := u(\mathring{id}, now)$ *#Report region rid appeared*
    else
      if $|\{n \in \mathring{nbr} | d(n) = 0 \text{ and } d(\mathring{cyc}(n)) = 1\}| = 1$ **then**
        send (ident, $\mathring{id}$) to node with identifier $i \in I$
      else
        set $rid := u(\mathring{id}, now)$
        send (regn, rid) to nodes with identifiers $i \in I$ *#Report region merge detected*
    become BNDY
  Receiving (upd8, i, s)
    set $d(i) := s$
    if $\mathring{s} = 1$ **then**   *#Not possible for "inactive" node to change state*
      send (idnt, $\mathring{id}$) to node(s) with identifier(s) $i \in \{n \in \mathring{nbr} | d(n) = 0 \text{ and } d(\mathring{cyc}(n)) = 1\}$
      become BNDY

PNCH

  When $\mathring{s}(now)$ changes
    broadcast (upd8, $\mathring{id}$, $\mathring{s}(now)$)   *#Broadcast sensed value and identifier*
    become IDLE   *#Report region r split*
    for all $i \in \{n \in \mathring{nbr} | d(n) = 0 \text{ and } d(\mathring{cyc}(n)) = 1\}$ **do**
      send (regn, $u(\mathring{cyc}(i), now)$, $\mathring{id}$) to node with identifier $\mathring{cyc}(i)$

BNDY

  When $\mathring{s}(now)$ changes   *Receiving* (regn, r, i)
    broadcast (upd8, $\mathring{id}$, $\mathring{s}(now)$)   **if** $r \neq rid$ **then**
    if $d(i) = 0$ for all $i \in \mathring{nbr}$ **then**   **set** $rid := r$
      Report region rid disappeared   **while** $\neg(d(i) \neq 0 \text{ or } d(\mathring{cyc}(i)) \neq 1)$ **do**
    set $rid = 0$ **set** $i := \mathring{cyc}(i)$
    become IDLE   **send** (regn, r, $\mathring{id}$) to node with id $\mathring{cyc}(i)$
  Receiving (idnt, i)
    send (regn, rid, $\mathring{id}$) to node with id i

PNCH, BNDY

  Receiving (upd8, i, s)
    set $d(i) := s$
    let $I := \{n \in \mathring{nbr} | d(n) = 0 \text{ and } d(\mathring{cyc}(n)) = 1\}$
    if $|I| = 0$ **then become** IDLE   *#Boundary node now in region interior*
    if $|I| = 1$ **then become** BNDY   *#Pinch point now in region interior*
    if $|I| = 2$ **then become** PNCH   *#Boundary node now pinch point*

---

algorithm (such as Protocol 6.8), and then centrally infer what changes must have occurred to lead to the observed changes in state (and it so happens that [58] provides a convenient canonical form for inferring these changes).

We have already addressed many of the assumptions behind this algorithm that an adversary might exploit, including the important assumption of incremental change. One further issue might be exploited by an adversary: delays in messages. When regn messages take a long time to cycle around a boundary, it is possible that a second event occurs to a region before an regn message has had a chance to complete its circuit of the boundary. In such cases it is possible (although not especially likely) that the two regn messages will continue to cycle around the boundary chasing each other, causing the region component to vacillate between two (or more) unique region identifiers. While this would not affect the detection of events, it could lead to increased communication, and depletion of system resources. However, it would not be difficult to introduce a periodic "blocking" mechanism to confound this eventuality.

6.6.3 Summary

Our final algorithm in Part II, Protocol 6.9 is surely amongst the most complex in this book. But this is not accidental. The very same design tools used for our simplest "Crowd!" algorithm all the way back in §3.1.2 have proven adequate to this challenge. In fact, a more sophisticated version of this algorithm was one of the very *first* major decentralized spatial algorithms I ever collaborated on, the focus of (then Mr., now Dr.) Jafar Sadeq's Ph.D. thesis [97]. For me, then, there is a satisfying symmetry to this particular algorithm being the *last* in Part II. The manifest difficulty we encountered in overcoming the substantial complexity of this algorithm *without* the design tools and techniques I have advocated in this book was one of the primary motivations behind seeking a better, more structured decentralized spatial algorithm design technique. This book is itself the culmination of that long search.

6.7 Chapter in a Nutshell

Change is to be expected in geosensor networks. Change in the monitored environment should be assumed for all applications of geosensor networks, since static and unchanging environments are both rare and straightforward to map using conventional spatial data capture techniques such as ground survey. This does not mean that our hard work in Chapters 4 and 5 was pointless. Far from it: decentralized spatiotemporal algorithms can often be built from repeated execution of decentralized spatial (atemporal) algorithms. Further, decentralized spatial algorithms may be integral to initialization of decentralized spatiotemporal algorithms. However, by taking advantage of the inherent correlations in geographic information over time, it is possible to greatly improve the efficiency of truly spatiotemporal algorithms when compared with repeated execution of "snapshot" spatial algorithms.

Spatiotemporal phenomena are also substantially more complex than purely spatial phenomena, and many of the basic ontological classes of spatiotemporal entities are still subjects of active research (e.g., [114]). As a result, achieving clarity on the types of spatiotemporal entities is especially important groundwork that needs to be completed before embarking on any decentralized spatiotemporal algorithm design project. In this book we argue that the top-level distinction for spatiotemporal algorithms concerns whether they must monitor histories (sequences of states) or chronicles (sequences of occurrences) or both. Overlaid on this distinction is the fact that most sensors monitor histories rather than chronicles, and so in applications requiring chronicles, occurrences will typically need to be inferred from states. At the next level, a distinction needs to be made between *what* is changing: the environment or moving objects; and are the nodes themselves static or mobile? Rather like the orthogonality of objects versus fields, and raster versus vector debates in GIS, it is possible for both static and moving sensor nodes to monitor either changing environments or moving objects. To keep things as simple as possible, the algorithms presented in this chapter have generally been chosen to involve *either* dynamic environments or mobile sensor nodes, but not both. But at least one, albeit simple, algorithm (Protocol 6.4) embarks on the challenge of simultaneously managing both sensor mobility and environmental dynamism.

Review Questions

6.1 What adaptations are required to Protocol 6.3 to monitor the movements of mobile objects through planar space, for example, where cell phone towers track the identities of mobile phones while they are in a particular cell?

6.2 Extend Protocol 6.3 to deal with mobile objects that double back on themselves, and so may enter and exit communication graph edges by the same checkpoint.

6.3 The algorithm in Protocol 6.3 stores information about the movements of mobile objects, but never actually uses this information. Let us assume a user wishes to know what objects traverse an edge $\{v_1, v_2\}$ over some time period $[t_1, t_2]$. Extend Protocol 6.3 to enable queries of this type to be satisfied when a user "injects" the query at any node in the network (not just at the nodes v_1 or v_2). .

6.4 Question 6.3 above concerns only simple queries local to a particular node. Thus, question 6.3 is fundamentally a question about routing information between a sink node (which receives the injected query) and a source node (which holds the information about the movement on an edge) and back again. Further extend Protocol 6.3 to respond to queries of the form: "Which mobile objects have completed a path $v_1, v_2, v_3, \ldots$ within time period $[t_1, t_2]$?" Achieving this will require the integration of information from a number of nodes in the network. .

6.5 Extend the epidemic algorithm in Protocol 6.5 with the capability for nodes to use hop-count flooding to reinfect new neighbors with received information.

6.6 The epidemic algorithm in Protocol 6.5 cannot detect new environmental events while nodes are in the SEND state. Using the surprise flooding approach of Protocol 4.4, adapt Protocol 6.5 to allow a node to detect multiple environmental events in quick succession, and infect the neighbors it meets with information about them.

6.7 Adapt Protocol 6.6 to detect only nkr-flocks composed of the *same* nodes, a special case of Protocol 6.6 (cf. Fig. 6.5).

6.8 Complete Protocol 6.7 with events and actions to detect the changes that occur to pits, corresponding directly to the existing events and actions for peaks (see p. 193).

6.9 Protocol 6.7 detects only peaks and pits in a static surface, not passes (saddle points). In fact, it is not so difficult to also detect saddle points using an adaptation of the sweep algorithm introduced in Protocol 4.13, on p. 115. Instead of hops being used as the potential function, the sensed surface itself may be used as a potential function. Then passes can be found using sweeps initiated at the peaks/pits. After an algorithm originally proposed in [100], two sweeps initiated from adjacent peaks or pits will first meet at their connecting pass. Adapt Protocol 6.7 to also detect passes using this technique, outlined in [100].

6.10 Extend Protocol 6.9 to additionally report the sensed topological changes (appearance, disappearance, merging, and splitting) to some known sink node. You may assume the use of any of the routing strategies we have already encountered to achieve this (e.g., routing in a tree, georouting, or even rumor routing).

Part III

Simulating Decentralized Spatial Algorithms

Part III of this book shows how decentralized spatial and spatiotemporal algorithms, designed using the techniques explored in Part II, can be simulated and tested. The chapters in this part do not go as far as implementing our algorithms in *real* geosensor networks: the technology in this area is changing and advancing too quickly to be within the scope of this book. However, all the algorithms in Parts II and III have been implemented in a free and open-source simulation system called NetLogo (and are freely available to download from the Web site that complements this book: http://book.ambientspatial.net).

Chapter 7 begins by introducing the simulation environment and investigating empirically the most important property of any decentralized spatial algorithm: its efficiency. However, efficiency alone cannot guarantee a useful algorithm. Chapter 8 shows how to conduct experiments that go further, testing the robustness and veracity of algorithms, in particular focusing on how well an algorithm deals with uncertainty. Finally, Chapter 9 concludes the book by surveying a few of the many different avenues for further research.

in earlier chapters. We also introduce some of the statistical tests that are important in verifying the algorithms' behavior.

7.1 Simulation with NetLogo

NetLogo is a widely used, free, and open-source simulation system. Every algorithm in this book has been implemented in NetLogo, and can be downloaded from the Web site that complements this book, http://book. ambientspatial.net.

NetLogo was specifically selected for our simulations from amongst a very wide variety of simulation systems that one might choose. Amongst the most commonly used systems for simulating sensor networks are ns-2, TOSSIM, OMNeT++, OPNET, JiST/SWANS, GloMoSim, and MATLAB/Simulink.[1] While these are widely used in computer science and electrical engineering, they tend to focus strongly on the technical aspects of communication in a sensor network, from physical hardware issues to the direct data links between devices to network communication protocols such as TCP/IP. Although important, such considerations are at a much lower level of abstraction than is of interest for decentralized spatial computing. Further, in addition to being notoriously complex, none of these systems provides direct support for efficiently simulating the *geographic* environment being monitored.

In decentralized spatial computing our target is at a higher level of abstraction—away from the details of how a message is physically communicated between two nodes and closer to how an algorithm as a whole behaves and how it interacts with the environment it monitors. Rather few systems provide these capabilities. Ptolemy is an easy-to-use, general purpose simulation system, but with limited support for the spatial aspects of sensor networks. Repast is a popular Java-based simulation system. It can provide many of the required features, but at the time of this writing is relatively complex, realistically only customizable by intermediate or experienced Java programmers.

Thus, NetLogo was the clear choice for a simulation system able to offer the right combination of control over both the computational and the geographic environments; an appropriate level of abstraction for decentralized spatial algorithm design; and remarkable simplicity, with basic simulation results accessible in a few minutes to most users. In addition, NetLogo has an active user community across a range of academic disciplines and, as we shall see, a suite of experimental tools to support rapid empirical evaluation of algorithms.

---

[1] URLs for these systems are not included here, as they tend to change quite frequently. A Web search on any of these names will quickly reveal the latest authoritative Web site for each system.

Simulating Scalable Decentralized Spatial Algorithms

Summary: *This chapter looks in detail at the process of decentralized alg and at how to validate experimentally the efficiency of an algorithm. The cha simulation system used to implement all the protocols in this book, NetLogo. T is used to construct basic scalability experiments. By revisiting many of the prote chapters, the discussion in this chapter demonstrates how to explore experim different aspects of scalability, including by investigating the average-case over complexity, testing hypotheses about the differences between protocols, testing a protocol, decomposing overall scalability into its constituent messages, and, fin the impact of using different measures of communication overheads.*

THE algorithm design technique set out in this book provides a struc for generating, analyzing, and comparing different decentralized sp algorithms. Ultimately, algorithms will need to be deployed in a real de tralized spatial information system, such as a geosensor network. How practical field trials are a major undertaking, requiring time and a team a wider range of skills than those of geographic information scientists al and also including electrical engineers, specialized programmers, and dor experts.

As a result, mistakes at the deployment stage can be costly. Empi testing of algorithms in the lab is therefore a vital intermediate step tow practical deployment. In this chapter we explore empirical testing in the using simulations. This chapter first introduces the simulation system u called NetLogo (§7.1). Then the chapter explores some basic experime designs for testing the efficiency of some algorithms already encount

M. Duckham, *Decentralized Spatial Computing*, DOI 10.1007/978-3-642-30853-6_7,
© Springer-Verlag Berlin Heidelberg 2013

NetLogo is an *agent-based* simulation system. An agent is essentially an autonomous program, situated in a simulated environment and capable of acting upon interactions with that environment and with other agents (cf. [115]). Despite some grander claims of the advantages of agent-based simulation (e.g., [49]), agent-based modeling is, at its root, a mechanism for managing complexity in software development. As a result, agent-based simulation systems are ideally suited for modeling complex decentralized spatial algorithms. There is a direct analogy between sensor nodes and agents, and between geographic and agent environments. Agent-based modeling has much in common with closely related mechanisms for complexity management, such as object orientation [56]. Indeed, the NetLogo agent-based simulation system is built using Java, an object-oriented programming language.

7.1.1 NetLogo Interface

The NetLogo system is thoroughly documented; the documentation is accessible from its home page, http://ccl.northwestern.edu/netlogo. This section provides a brief overview of the key concepts and features relevant to the simulations considered in this book. However, software does evolve rapidly, so the specific interface details are certain to change over time. Interested readers should explore the excellent resources available from the NetLogo team [111] for the latest NetLogo version.

Figure 7.1 summarizes the seven most important elements of the NetLogo system. In explaining the key features of the NetLogo system, we make a distinction between a simulation *model* (the implementation of a specific algorithm within the NetLogo simulation system) and a simulation *run* (a specific execution of a simulation model).

1. *Tabs*: At the top level, NetLogo provides three basic views of a simulation model: the simulation interface itself (shown in Fig. 7.1); information about the simulation model (see below); and a "Procedures" tab for editing the program code behind a NetLogo simulation (§7.1.2).
2. *Simulation environment settings*: The "Settings" button allows editing of a number of settings common to any simulation, in particular the dimensions of the simulation "world" and the wrapping used at its edges (i.e., whether the simulation space is topologically equivalent to a rectangular piece in the plane, a cylinder, or a torus).
3. *Simulation speed and user interface widgets*: The speed at which a simulation executes can be varied before and during a simulation run. In addition, a range of user interface widgets (buttons, text boxes, switches, choosers, plots, etc.) can be added or edited from this section.
4. *Graphical simulation view*: The simulation "world" itself is displayed graphically. In the simulations associated with this book, sensor nodes are depicted as dots of different colors located at certain coordinates in

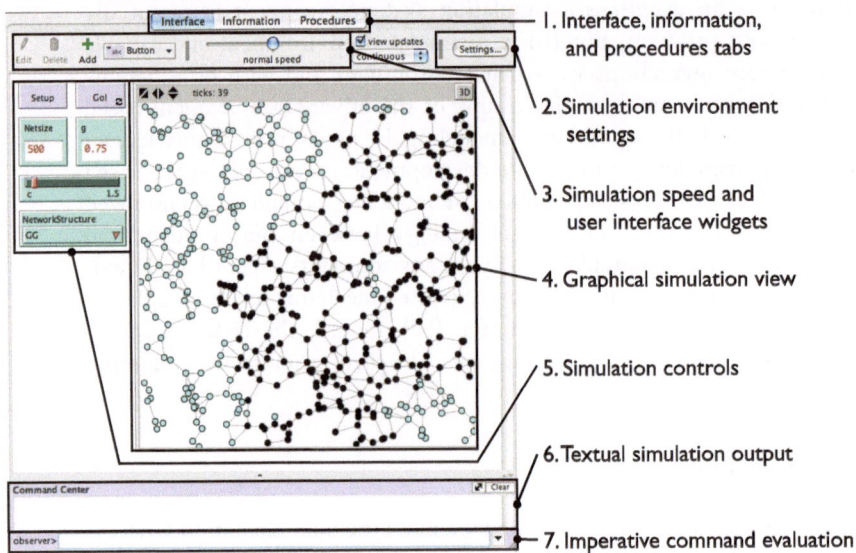

Fig. 7.1. Key elements of the NetLogo interface (version 4.1.3)

space. Direct one-hop communication links between nodes are depicted as lines connecting the dots. In some simulations, nodes or links may have additional labels. Further, the background of the world (white in Fig. 7.1, and termed "patches" in NetLogo-speak) is often used to simulate the monitored geographic environment.

5. *Simulation controls*: The controls for a simulation model are always placed on the left-hand side of the simulation interface view. These controls usually at least contain buttons for setting up and running a simulation; a text box for entering the number of nodes in a simulation; a slider for varying the communication distance c, and a chooser for picking the network structure (e.g., UDG, Gabriel graph, relative neighborhood graph).

6. *Textual simulation output*: On occasion, some output from the simulation may be displayed in the command center window at the bottom of the simulation interface. This feature is especially useful when writing or adapting your own simulations.

7. *Imperative command evaluation*: It is always possible to submit NetLogo commands directly to the interface. These commands are immediately executed by the NetLogo system—an extremely handy feature when debugging or exploring simulations.

The "Information" tab (see 1. above) contains concise documentation for the simulation model. This documentation always includes:

- the protocol name and a brief summary of the protocol;
- simple instructions for operating the protocol simulation;
- any particular effects or features to notice;
- some things to try when running simulations; and
- links to the relevant chapter, credits, and the GNU general public license.[2]

7.1.2 NetLogo Procedures

The NetLogo program code for simulating each protocol can be found in the "Procedures" tab. A full discussion of the NetLogo language is clearly beyond the scope of this book; we present only a few key features. Readers are again encouraged to investigate further the simple and intuitive NetLogo language, in order to be able to adapt and extend the simulation models for this book (see [111]).

The NetLogo program code that complements this book has been written to correspond as directly as possible to the protocols. A specially developed code library provides NetLogo statements that correspond to **become, send,** and **broadcast** operations in protocol specifications. Listing 7.1 provides an example of NetLogo code for Protocol 4.5, gossiping.

The NetLogo program code in Listing 7.1 is separated into procedures. The structure of Listing 7.1 is similar to that of all the NetLogo simulation models for all the protocols in this book:

- The go procedure "asks" each *mote* (simulated sensor node) in turn to execute a single simulation step. While in a real geosensor network motes will operate in parallel, the simulation of this process in NetLogo is serialized, with each mote taking a turn. However, the order in which motes take turns is arbitrary, and will be different each time.
- The step procedure provides the control flow necessary to ensure that each mote responds to events with actions relevant to its current state.
- The events and actions corresponding to each state are contained in the step_INIT, step_IDLE, and step_DONE procedures.
- The step_INIT procedure spontaneously broadcasts an MSGE message before transitioning to state DONE. Comparing this procedure with the corresponding event/action pair in Protocol 4.5 illustrates the close dependence of the NetLogo code on the protocol specification.

---

[2] In short, the GNU general public license (GPL) means you are free to use, adapt, and redistribute the code that complements this book in any way you wish, so long as you continue to acknowledge the source of the code and its authors, and continue to honor the original GNU GPL 3.0 license.

Listing 7.1. Anatomy of protocol simulation in NetLogo (Protocol 4.5)

```
;; Run the algorithm.
to go
  ask motes [ step ]
  tick
end

;; Step through the current state.
to step
  if state = "INIT" [ step_INIT stop ]
  if state = "IDLE" [ step_IDLE stop ]
  if state = "DONE" [ step_DONE stop ]
end

;; Broadcast a message to neighbors.
to step_INIT
  broadcast ["MSGE"]
  become "DONE"
end

;; When a mote receives a message it may rebroadcast the message
;; to its neighbors based on the probability of the g value.
to step_IDLE
  if has-message "MSGE" [ ;; When the mote receives a message
    let msg received "MSGE" ;; Process the MSGE message
    let r random-float 1 ;; Generate random float [0.0,1.0]
    if r < g [
      broadcast ["MSGE"] ;; Rebroadcast MSGE with probability g
    ]
    become "DONE"
  ]
end

;; The DONE state does not respond to messages.
to step_DONE
  set color 0 ;; Nodes in the DONE state are black
end
```

- The step_IDLE procedure defines one action, responding to a received MSGE message. The first two lines of the procedure check for this event, using the procedures has-message and received developed specially for this purpose (and not standard procedures of NetLogo). The remaining lines again correspond very closely to the action for the *Receiving* event in Protocol 4.5.

- As specified in Protocol 4.5, the step_ DONE procedure provides no events or actions, except to change the color of the node in the simulation (to black). Interface commands, such as changing the appearance of motes, patches, or labels, represent the most significant differences between the NetLogo simulation models and the protocols specified in the earlier chapters.

7.1.3 Exploring Simulations with NetLogo

Playing with the NetLogo simulation models can help us acquire intuition into how an algorithm operates, more rapidly than studying the protocol specification alone. All the NetLogo simulation models have at least a communication distance and a network size parameter that can be varied, and usually the network structure can be varied too (e.g., UDG, Gabriel graph, or relative neighborhood graph). Further, the node locations and the message receipt order are randomized for every simulation run. Simulation models often also have protocol-specific parameters and other randomized features that can be modified in different simulation runs.

For example, let us load up the simulation protocol for gossiping, Protocol 4.5. Pressing "Setup" and then toggling "Go" will execute a simulation run. Pressing "Go" again to untoggle it will stop the simulation. The "Information" tab contains a number of suggestions of things to try, including changing the gossip probability g; changing the communication distance c; selecting a different network structure, such as UDG, Gabriel graph, or relative neighborhood graph; changing the simulation run speed to see more clearly the progress of the algorithm. To stop a simulation run that gets stuck or is very slow (say, if you try to execute simulation with a very large number of nodes), you can use the NetLogo "Halt" command (in the "Tools" menu).

Now for a fixed network size (say, 1,000 nodes), communication distance c, say, 1.2 units, and UDG network structure, let us try reducing the gossip probability g from 1.0. What is the lowest gossip probability g for which the message still reaches *most* of the network (shown when nodes turn black, meaning they have transitioned into a DONE state) on *most* runs? Now run the simulation again, this time with a planar network structure (e.g., Gabriel graph). What do you notice? Because of the decreased connectivity in the planar graph compared with the UDG, the messages should now only reach a small proportion of the network (only a few nodes turn black). The simulation in Fig. 7.1 illustrates such a scenario.

Monitoring messages sent and received

One further feature is particularly useful when exploring simulations— monitoring the messages sent and received within a simulation execution. Some procedures have been programmed into the code library to assist with this. To access these procedures, you will need to turn on the trackmsg toggle box in the bottom left of the simulation window. This toggle box is turned off by default to make the simulation run as fast as possible. But when trackmsg is turned on, each node keeps a track of the number and length of all messages it sends and receives.

Execute another simulation run, this time with the trackmsg toggle box set to true. Then try typing the following command into the observer box for imperative command evaluation (item 7. in Fig. 7.1):

show sent-number-msg-totals

In the command center text output window (item 6. in Fig. 7.1) you should now see the total number of messages sent for the simulation run appear. A number of similar procedures have been programmed, including recv-number-msg-totals (total number of messages received) and sent-length-msg-totals and recv-length-msg-totals (total length of messages sent and received). Try running a few simulations and look at how the total number and length of messages sent and received changes.

As with summarizing the total messages sent and received at the end of a simulation, it is not too difficult to plot the increase in these messages over the course of a simulation run. The following steps provide a guide to doing this, although the procedure may differ in the latest NetLogo version:

1. Add a new plot to the right-hand side of the NetLogo interface window (using the "Add Plot" function from the user interface widgets, item 2. in Fig. 7.1). Figure 7.2 step 1 shows a portion of the screen you should see.
2. In the plot settings window that automatically appears, set the plot name to "totals" (Fig. 7.2 step 2). Then create four new plot pens (which will be used to draw the different response curves) called "sent-number," "sent-length," "recv-number," and "recv-length." Finally, set the colors of these four pens to something that will be clear to you (e.g., red, orange, blue, and sky respectively).
3. Next, select (but don't press) the "Go" button with your mouse. Floating the mouse pointer over an empty portion of the interface turns the pointer into crosshairs; then dragging the crosshairs over the "Go" button selects it (Fig. 7.2 step 3).
4. With the "Go" button selected, press the "Edit" function from the user interface widget (item 2. in Fig. 7.1). Add the following code below the go procedure call in the dialog box that should have appeared (Fig. 7.2 step 4):

```
set-current-plot-pen "sent-number"
plot sent-number-msg-totals
set-current-plot-pen "sent-length"
plot sent-length-msg-totals
set-current-plot-pen "recv-number"
plot recv-length-msg-totals
set-current-plot-pen "recv-length"
```

plot recv-number-msg-totals

These lines are very similar, so some cut-and-paste will speed this process up. Essentially, this code tells the NetLogo system to plot each of the different measures of communication resources using a different line on the same plot.

5. Almost there! The penultimate step is to tell the NetLogo system to plot on the widget you just created. The following command when entered into the observer box for imperative command evaluation (item 7. in Fig. 7.1) will achieve this (Fig. 7.2 step 5):

set-current-plot "totals"

6. Finally, set up and run the simulation again. This time you should see the different measures of communication resources increase and plateau in the plot widget you created (Fig. 7.2 step 6).

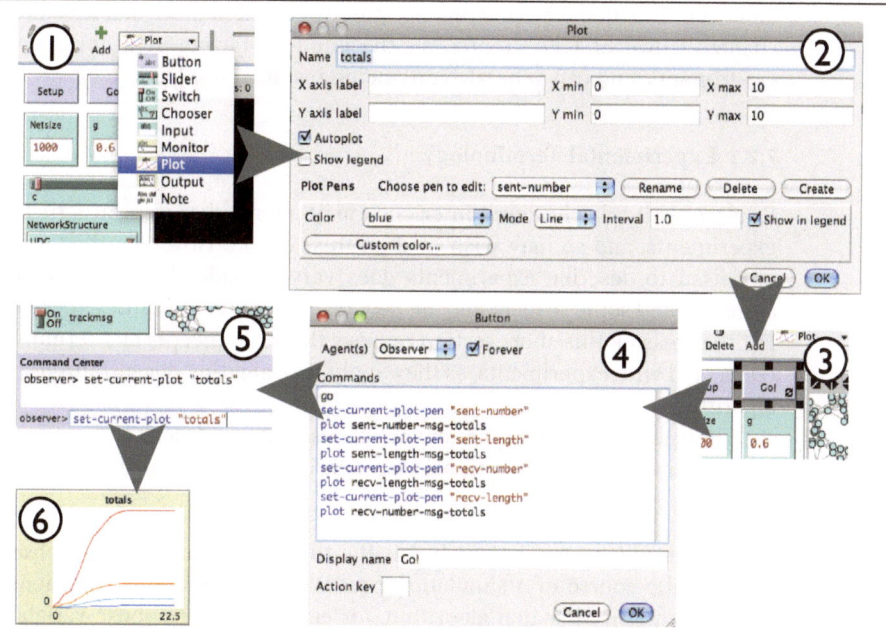

Fig. 7.2. Steps for setting up a plot in the NetLogo system (version 4.1.3)

It is worth investing a little time familiarizing yourself with NetLogo plotting; the plotting functions can provide rapid access to a variety of simulation run characteristics as they emerge.

7.2 Experimental Data from Simulations

The previous section introduced the NetLogo simulation system, and discussed how to gain intuition about the dynamic behavior of protocols by exploring the effects of different factors on individual simulation runs. However, a more structured approach to examining a protocol is required to generate objective, empirical evidence to support an analysis of algorithm efficiency.

The most basic structured approach to gathering objective evidence about an algorithm is one-factor-at-a-time (OFAT) experiments. The OFAT approach can be summarized with the maxim: "only one question should be asked at a time" [96]. The idea is to hold constant all the factors that affect algorithm performance except one, and observe the effects of varying that one factor on the algorithm behavior. When the effects of that one factor are fully understood, it is possible to ask the next question—to vary a different factor.

In fact, OFAT experimental designs have today largely been superseded by more sophisticated experimental designs, discussed in more detail in the next chapter. However, OFAT represents the simplest starting point for designing experiments, and can generate convincing results if used carefully.

7.2.1 Experimental Terminology

Readers with a background in science may already be adept at structuring experiments, and so may wish to skim this section. However, the terminology used to describe experiments does vary considerably across different domains, and some readers may be new to conducting scientific experiments. For that reason, this short section reviews the key concepts and terminology connected with experiments, as they apply to evaluating decentralized spatial algorithms.

There are three main types of variable used in constructing an experiment to investigate the computational characteristics of an algorithm (or indeed any experiment):

- *Response*: A *response* (variable) is the thing being measured or observed over the course of a simulation run. When examining the efficiency of a decentralized spatial algorithm, for example, the response variables of interest are typically the number and length of messages sent or received. Response variables are sometimes also referred to as *dependent* variables.
- *Factor*: A (primary) *factor* is a variable thought to explain the observed effects on the response variable. Factors are purposefully manipulated in an experiment in order to gain information about their effects on the response variable. In an experiment on the scalability of an algorithm, for example, the most important factor will usually be the network size. However, a range of other factors may also be of interest, such as the communication distance or network structure, as well as protocol-specific

parameters. Factors are sometimes also referred to as *explanatory* or *independent* variables.

- *Secondary factor*. In most cases there will be a great many factors that *could* affect the response variable, but only some of these will be of direct interest to the algorithm designer. A *secondary factor* is a factor that is not currently of interest, even though it is expected to lead to a response. For example, node location would normally be a secondary factor when simulating a decentralized spatial algorithm. Although the locations of nodes will often affect the efficiency of an algorithm, this factor is rarely of interest because most algorithms are expected to operate irrespective of the specific locations of nodes. Secondary factors are also referred to as *extraneous factors*.

The *levels* of a factor are the different values it can assume in different simulation runs. For example, in an experiment with network size as a factor, we might choose the levels of 125, 250, 500, 1,000, 2,00, and 4,000 nodes. When experimenting with computational efficiency, the levels are often chosen to increase exponentially (rather than linearly, say 250, 500, 750, 1,000, ...) as this provides better coverage of the wide range of network sizes of interest. A *treatment* is a specific combination of levels of different factors used in an experiment. For example, in a series of experiments varying the factors network size and network structure, the combination of, say, 1,000 nodes and a Gabriel graph network would constitute a treatment.

The essence of OFAT experimental design, then, is to ensure that any observed effects on the response variable can be attributed to a *single* factor. To do this, an OFAT experiment holds constant ("controls") all the (primary) factors except one. Subsequent experiments may vary a different primary factor, but each time all other primary factors are controlled. But what about the secondary factors? If we are unlucky, it could happen that the response observed occurred not because of our manipulation of a primary factor, but because of variations in secondary factors between the treatments.

For example, an experiment on the scalability of an algorithm might first manipulate the network size (factor) at two levels, small and large, while measuring total numbers of messages generated (response). However, if it turns out that the nodes in the small network happened to have a particularly uneven density, while those in the large network happened to have a particularly even density, then it may be that the node location (secondary factor) and *not* the network size (primary factor) are partly or wholly responsible for any observed response.

The answer to this problem involves our final experimental concept: replication.

7.2.2 Replication

Replication involves repeating simulation runs to generate a *population* of results for each treatment. The objective is to ensure that the population of results for each treatment is just as likely to include particular levels or configurations of secondary factors as any other treatment. To achieve this, there are two main techniques that can be used:

- Group secondary factors into a number of levels, and then replicate to ensure that each treatment includes representatives from across each of the different levels (called *blocking*); or
- Randomize the secondary factors, and then perform enough replications to ensure that each treatment is very unlikely to differ systematically from the others with respect to the secondary factors.

In most cases, our experiments will use randomization of secondary factors, rather than blocking, because there are typically a priori no obvious levels to group secondary factors into. For example, in addition to randomizing node locations, our simulations randomize the order in which messages are processed by nodes. For this reason, no two simulation runs will ever be the same, and consequently all the experiments that follow rely on replication of simulation runs.

You may ask how many replications are enough? In general, the more replications, the larger the population of results, and so the greater the confidence we can have in those results. In the natural sciences, where experiments are typically expensive and laborious, determining exactly how many replications is enough is a vexed question: a balancing act between practical constraints and more robust results. However, in the information sciences, simulations are usually cheap and easy to automate, so one would typically perform a great many replications (i.e., hundreds, thousands, or even millions). The size of the data set of results and the time available for running simulations does impose some practical constraints. Consequently, the *actual* number of replications required is often a matter of judgment—ask yourself: "are there enough replications to convince a skeptic that the results could not have occurred by chance?" But when in doubt, do more replications.

A first experiment

Figure 7.3 shows the results of 100 replications of the Protocol 4.5, gossiping, over six different network sizes. The graph is plotted on a log-log scale to better accommodate the wide range of data values on both axes (from 125 to 4,000 on the x-axis and 1 to 3246 on the y-axis).

Summary statistics, such as mode, mean, and standard deviation, can also help in understanding the spread of results from a set of repetitions (e.g., Table 7.1). Of note in Table 7.1 is the difference between mean and mode. The mean is more susceptible to outliers, and so arguably provides a less meaningful measure of central tendency in cases like those in Fig. 7.3, where the bulk of results is skewed towards the maximum value, with a small

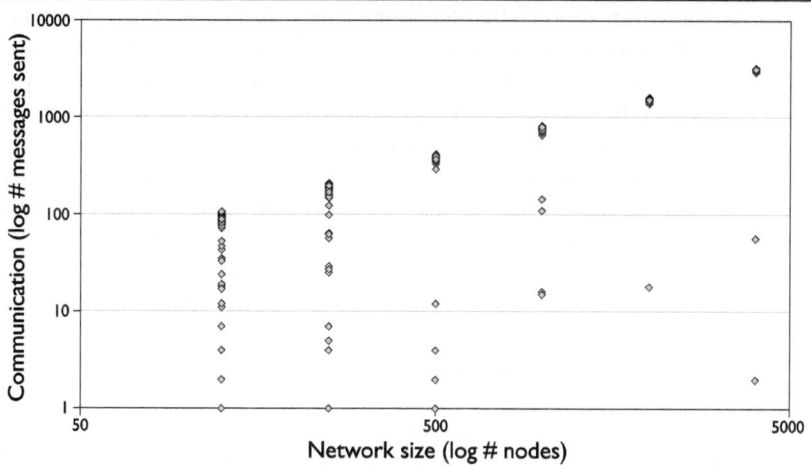

Fig. 7.3. Scalability of gossiping, Protocol 4.5, showing response in overall messages sent for increasing network size. The data comprises 100 repetitions at each network size, each with gossip probability g of 0.8

number of outliers close to the minimum. Further familiar summary statistics, such as quartiles, skew, and kurtosis, can help to describe such variations.

	Repetition set (number of nodes in network)					
	125	250	500	1000	2000	4000
Min	1	1	1	15	18	2
Max	107	210	417	820	1618	3246
Mode	98	194	394	778	1574	3142
Mean	81.87	167.36	367.27	747.98	1551.65	3083.21
Stdev	29.57	56.46	85.47	141.45	158.32	440.74

Table 7.1. Summary statistics for sets of repetitions of the gossiping experiment in Fig. 7.3

The "one question" this experimental data aims to answer is: how does the overall number of messages sent vary as a function of the network size? Overall, the graph appears to indicate a linear increase in messages sent with network size (but more on this later in this chapter). This is to be expected because in the gossiping algorithm (Protocol 4.5) each node rebroadcasts a message at most once. In fact, the gossip probability g was a controlled variable at level 0.8 for the experiments in Fig. 7.3. This also appears to agree with Table 7.1, where the maximum number of messages in each experiment is approximately 80% of the total network size.

However, despite this apparent trend, as noted above the data shows considerable variability, where in several cases the number of messages is very low (instances with less than 20 messages in all sets of repetitions; Table 7.1). In these cases, it appears that the gossip died out relatively soon after the initial message, perhaps because of poor connectivity (e.g., the unique node in INIT state was disconnected from the bulk of the network), or random chance (e.g., it may happen from time to time that by chance neighbors near the INIT node all do not rebroadcast the message). In short, efficiency alone does not normally provide a good characterization of an algorithm; we must also consider issues such as reliability. These issues are picked up again in the next chapter.

One further point is important to note. Even in this simple example, the factor (network size) used to generate Fig. 7.3 was not quite as simple as it at first appeared. When increasing the size of the network (in terms of the number of nodes), it is necessary to either proportionately increase the extent over which the nodes are deployed (i.e., maintain the node density) or proportionately decrease the communication radius (i.e., maintain network connectivity). If the number of nodes *alone* were increased then the network connectivity would change. For example, increasing the number of nodes, but keeping the area over which the nodes were spread and the communication radius constant, would result in individual nodes on average being connected to many more neighbors (i.e., increasing the average degree of nodes in the network). Conversely, decreasing the number of nodes in the network alone will make the connectivity much sparser, eventually leading to a disconnected network. Network connectivity is itself a secondary factor that is expected to result in changes to the response variable. Consequently, when changing the number of nodes in the network, we must also maintain the network connectivity.

The easiest way to increase network size while controlling network connectivity in the NetLogo system is by decreasing communication radius accordingly. Thus every doubling in network size was accompanied by a decrease in communication distance by a factor of $\sqrt{2}$. The same effect would have been achieved by increasing the network extent by a factor of $\sqrt{2}$.

7.2.3 Constructing Experiments with NetLogo

NetLogo provides a convenient tool for replicating experimental simulation runs, called BehaviorSpace. BehaviorSpace is accessed from the "Tools" menu in NetLogo. Opting to create a new (or later edit an) experiment from the BehaviorSpace dialog brings up the window in Fig. 7.4.

Using the BehaviorSpace tool, the process of executing a series of replicated experiments is relatively simple:

1. Give the experiment a name (Fig. 7.4, item 1).

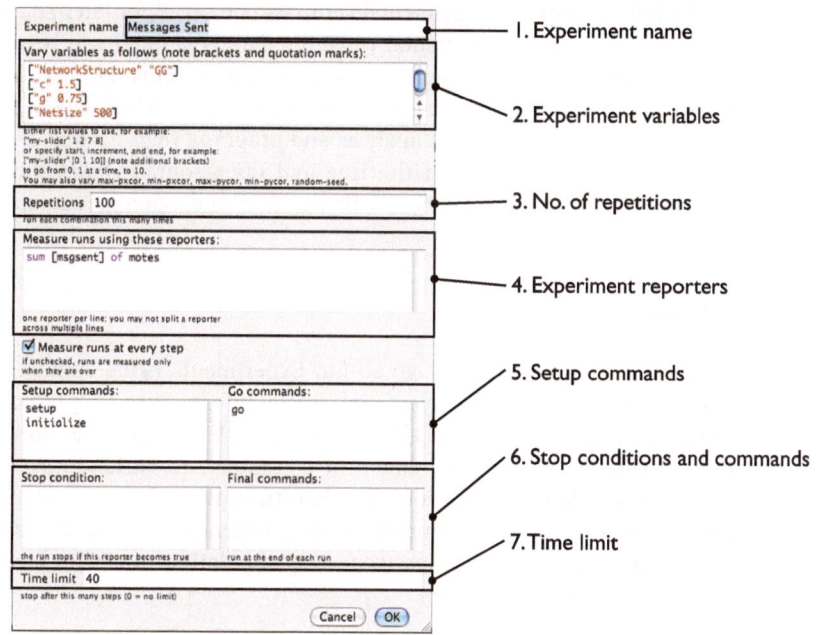

Fig. 7.4. Key elements of the BehaviorSpace experimental setup

2. Specify any factors and controls (Fig. 7.4, item 2). Control variables have only one listed value (e.g., ["Network Structure" "UDG"]). Factors take multiple variables for each level tested (e.g., ["Netsize" 125 250 500 1000 2000 4000]).
3. Specify the number of repetitions (Fig. 7.4, item 3).
4. List the NetLogo reporters (procedures that return a value) that will be used as the response variables (Fig. 7.4, item 4). For example, the total number and length of messages sent and received can be reported using the procedures already encountered (sent-number-msg-totals, sent-length-msg-totals, ...).
5. Specify any set up commands to execute before each simulation run, and the command that runs a simulation (Fig. 7.4, item 5). For most simulation models that accompany this book, the set up commands are simply setup and initialize, with a simulation run using the command go.
6. List terminating conditions (used to detect when a run should end) or final commands (executed after the end of each simulation run; Fig. 7.4, item 6), if any.
7. Finally, specify a time limit, in terms of the number of ticks a run should continue for (Fig. 7.4, item 7; 0 for no time limit).

The BehaviorSpace experimental setup can be saved for later adaptation. When executing the experimental runs, the data generated will be saved in a file you specify (.csv, comma separated variable). This data can then be imported into any statistics or spreadsheet program, such as OpenOffice, R, or Excel, for further statistical analysis and graphing (Fig. 7.3 was generated using NeoOffice, a version of the free and open-source OpenOffice; later experiments were also analyzed with the free and open-source statistical package R).

7.2.4 Summary

Adversarial analysis can only go so far. Experiments using simulation can complement adversarial analysis and create information about the actual performance of an algorithm across a variety of scenarios. At the core of any experiment is the measurement of a response to variation in a factor. To create new knowledge about an algorithm, the effects of the primary factor need to isolated by control (holding other primary factors constant across treatments) or replication (repeating simulation runs with blocked or randomized secondary factors).

7.3 Experimental Investigation of Scalability

The previous sections introduced the NetLogo simulation environment and a basic experimental design. This section steps through a series of experimental investigations of selected algorithms from Part II, with the objective of introducing some useful tools and techniques for drawing solid conclusions from these experiments.

7.3.1 Average Case

To begin with, recall that in §5.1.2 the algorithm for the local minimum spanning tree (LMST) was expected to have in the worst case an overall communication complexity of $O(|V|)$, with the number of messages sent at most $7|V|$ (as a consequence of nodes in the minimum spanning tree having a maximum degree of 6). Let us examine this empirically. The questions in this case are: a. "Is the overall communication complexity of Protocol 5.3 in actuality $O(|V|)$?"; and b. "Is the constant factor less than or equal to 7?"

These questions naturally lead again to an experiment with response variable of total messages sent; factor of network size (again with appropriate adjustments in communication radius to maintain constant connectivity); and all other variables controlled. Six levels were chosen for the factor: 125, 250, 500, 1,000, 2,000, 4,000. The experiments were set up with the BehaviorSpace tool using the NetLogo code for Protocol 5.3 (like all protocols in this book, available from the companion Web site). As with all the experiments in this book, you are encouraged to try this and other experimental investigations yourself.

The results of these experiments are summarized in Fig. 7.5. The results show a markedly tighter distribution than that of the previous experiment, in Fig. 7.3. Over 100 repetitions, the standard deviation of messages sent varies from 5.5 messages for the smallest networks (125 nodes) to 35.5 messages for the largest networks (4,000 nodes).

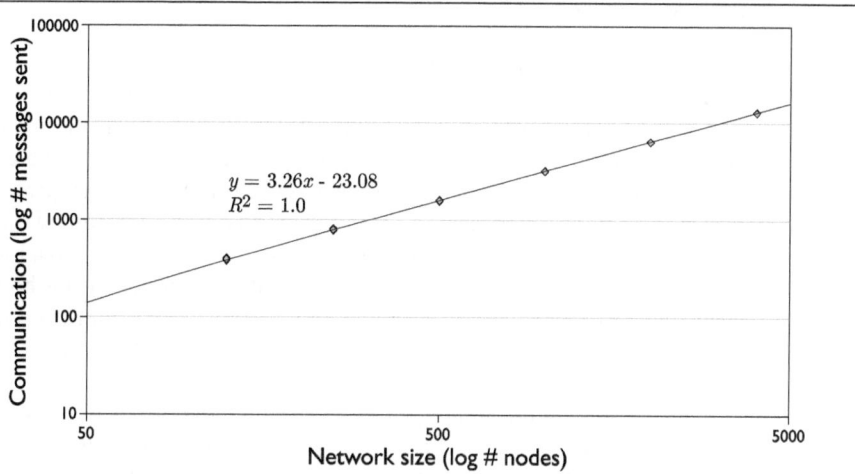

Fig. 7.5. Scalability of LMST protocol, Protocol 5.3, showing response in overall messages sent for increasing network size (100 repetitions per network size)

To verify the order of the communication complexity, one can perform a linear regression of the data. Linear regression is a standard tool for fitting a straight line to a data set (i.e., modeling the response variable as a linear function of the factor). The line fitted minimizes the sum of squares (the sum of squared distances between the regression line and each data point). Any statistics package or spreadsheet (such as R or OpenOffice) will offer this function. Figure 7.5 shows the regression line, along with the line's equation ($y = 3.26x - 23.08$) and R^2 value of 1.0. The determination coefficient R^2 is the sum of the squared difference between the observed data and the regression line as a proportion of the difference between the observed data and the mean value for the observed set. In short, the determination coefficient is an important measure of the goodness of fit of a regression curve. The maximum value for R^2 is 1.0, indicating for Fig. 7.5 a near-perfect fit between the model (i.e., O(n) overall communication complexity) and the observed data.

Finally, the slope of the regression line, 3.26, provides the constant factor for the overall communication complexity. While in the worst case we expect $7|V|$ messages for Protocol 5.3, on average in our experiments we only require about half that number, $3.26|V|$ messages. It is possible that other factors not included in our simulations would in reality alter the average efficiency in an actual deployment of this algorithm. However, the result provides strong support for our analysis, giving us good reason to have confidence in our design.

7.3.2 Multistage OFAT

More complex properties of algorithms can often be tackled experimentally using a sequence of OFAT experiments, each of which yields some information about a single question. For example, back in §5.3.2, Protocol 5.8 provided an algorithm for electing a leader for the boundary of a region, based on a combination of face routing around the boundary and the "as far" protocol for leader election in a ring. The analysis argued that this algorithm was overall $O(|V|)$. Because the "as far" leader election is known to be $O(n \log n)$, a key step in the argument relied on the number of nodes in the boundary cycle being proportionate to $|V|^k$, where $0.5 \leq k < 1$ (or, more specifically, $k = D/2$ where $1 \leq D < 2$ is the fractal dimension of the boundary curve; see p. 155).

The question of how many nodes are in a boundary cycle is important not only to Protocol 5.8. One of the ways to achieve efficient decentralized spatial computing, used in many of the algorithms in this book, is to use spatial structures, such as boundaries, to reduce the number of nodes involved in communication. Consequently, an understanding of the relationship between the size of the network and the expected number of nodes in a boundary cycle is relevant to several different algorithms.

Expected number of nodes in the boundary cycle

Our first question, then, is: "Is there indeed evidence to support the assertion that the number of nodes in the boundary cycle of a region scales with $O(|V|^k)$, where $0.5 \leq k < 1$?" To start to answer this question, we can design an OFAT experiment to measure the number of boundary cycle nodes (response variable) resulting from varying network size (factor). To avoid potentially confounding the results with chance configurations of nodes, node locations will again be randomized. Further, the shape and size of the region itself are expected to affect the response variable. Larger, more complex regions will have longer boundaries; smaller and simpler regions will have shorter boundaries. However, these are *secondary* factors. In practice, we can expect little or no prior knowledge of the size and shape of monitored regions, so our one question really concerns the characteristics of boundary length, ignoring size and shape of the regions.

To make a start on this problem, we shall *control* the region size (fixed at 480 patches) and *randomize* region shape. We shall come back to the implications of this choice a little later. To generate randomized regions, the experimental environment uses a variant of the *dilation* operation often used in image processing. Our region generation procedure starts by randomly selecting an initial patch in the NetLogo world to act as the "seed" region. Then a randomly selected patch (rook's case) neighbor of this seed is added to the region. This process continues, at each step adding a new randomly selected patch neighbor to the region, until the region reaches our predefined size. A final step fills in any "holes" that may have been created while growing the region, to ensure the region is homeomorphic to a disk. Figure 7.6 shows an example of some of the steps in the growth of a small region generated using this randomized procedure.

Fig. 7.6. Example of random region generation procedure, starting with "seed" region (left-hand side), increasing in size (increments of ten patches), and terminating with hole filling (right-hand side)

Using randomly generated regions, the response of the number of nodes in the boundary cycle to varying the network size can be measured. Figure 7.7 summarizes the results of just such an experiment using nine different, randomly generated regions of approximately equal area[3] (about 480 patches). All the experiments used networks with random sensor node locations, connected by the Gabriel graph (GG).

In addition to the data points being represented, a regression curve has been fitted to the data for each region in Fig. 7.7. Instead of the linear regression ($y = ax + b$) used in the previous experiments, this time a power curve of the form $y = ax^b$ has been fitted to each set of data points. The reason for using a power curve as opposed to linear regression is that we expect and wish to test for polynomial orders of scalability. Again, any good statistics or spreadsheet program will provide functions for fitting a range of different growth curves to data.

---

[3] Regions are not guaranteed to be of *equal* size because the final region generation step that "fills in" holes in the region can slightly increase the region area. In our experiments, the generated regions' areas differed by less that 1%.

The results of the regression show that all the regression curves have high determination coefficients ($R^2 > 0.98$ in all cases), indicating a good fit between our regression and the observed data. The power b observed for all curves varied from 0.6 to 0.69 (corresponding to a fractal dimension for the shapes of between dimension 1.2 and 1.38). Indeed, further experiments with more than 30 different, randomly generated regions yielded a range of power curves varying from 0.5 to 0.69 (fractal dimension between 1.0—a non-fractal, Euclidean geometry—and 1.38).

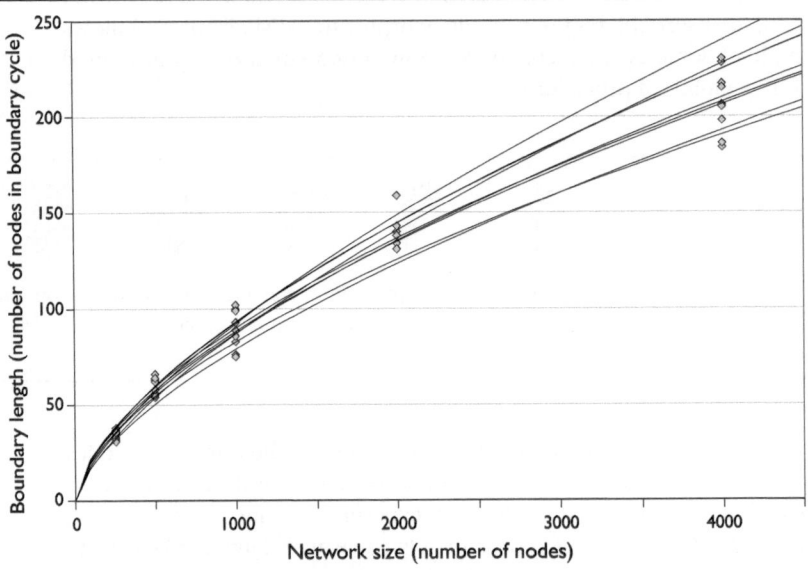

Fig. 7.7. Length of boundary as a function of network size, showing the observed number of nodes in the boundary cycle of nine randomly generated regions with increasing number of nodes in the network

In short, these results do seem to lend support to the argument that the number of boundary cycle nodes scales with $O(|V|^k)$, where $0.5 \leq k < 1$ (in our experiments $k \in [0.5, 0.69]$). Nevertheless, the results need to be treated with a degree of caution. Further experiments should randomize region size, and ideally use different types of regions (e.g., Euclidean circles or squares, of different randomized region growth procedures).

However, in our case, further experiments are not expected to reveal anything particularly new or surprising about the relationship between boundary length and network size. Fundamentally, our experiment only confirms a well-known characteristic of Euclidean and fractal curves in the context of sensor networks. There are enough other studies of the fractal shapes in different application domains for us to have a high degree of confidence that our domain is no different (e.g., [20, 79, 87, 94]). Were we to perform further experiments, the results would show that changing region size has little or no effect on the relationship between network size and boundary cycle length (except for the smallest regions, where the region size is comparable to the spacing between nodes, and so granularity effects take over). Experiments with different types of regions *would* affect the observed value of k, for example, leading to lower k for Euclidean geometries such as circles and squares (because these have a fractal dimension of 1). But observed values for k would always remain strictly less than 1 (question 7.4).

Scalability of leader election in a region boundary

Our second question—"Is the overall scalability of the algorithm for leader election in region boundaries in practice $O(|V|)$?"—could of course be answered directly, without the investigation of boundary cycle length as in the previous section. A straightforward scalability experiment, such as already explored in 7.3.1, could directly provide the evidence required to support the affirmative answer. For example, Fig. 7.8. shows the results of a scalability experiment on Protocol 5.9 (leader election in a boundary cycle). The experiment measured the total number of messages sent (response) with varying network size (factor). Node locations were randomized as usual. Randomized region shapes were again grown and some using regular square regions were also tested. Additionally, blocking was used for region size to ensure each treatment was replicated with each of nine different region sizes. A total of 50 replications at each of three different network sizes (500, 1,000, and 2,000 nodes) led to a total of $2 \times 9 \times 50 \times 3 = 2700$ simulation, runs plotted in Fig. 7.8.

The result does indeed seem to support our earlier analysis that the communication complexity of Protocol 5.9 is $O(|V|)$, with a constant factor of 1.32. However, replotting the results using boundary cycle length as the factor, in Fig. 7.9, reveals further details about the relationship. The figure shows that network size remains an important influence over efficiency (as would be expected from the $O(|V|)$ `ping` messages). But the figure also reveals a lot of structure within the data for each network size: a steady, approximately linear, increase in the number of messages generated with the length of the region boundary. In short, while network size has a determining effect on the length of the boundary of a particular region, *different* boundary lengths (due to different region shapes and sizes) for the *same* network size also partially explain the overall efficiency of the algorithm.

We can quantify the relative contributions of boundary length and network size using *multiple regression*. Multiple (linear) regression is essentially just like the linear regression already encountered, but of the form $y = a_1 x_1 + a_2 x_2 + \cdots + b$. Again, your favorite statistics program should help with the mechanics of performing a multiple regression. In the case of our data in

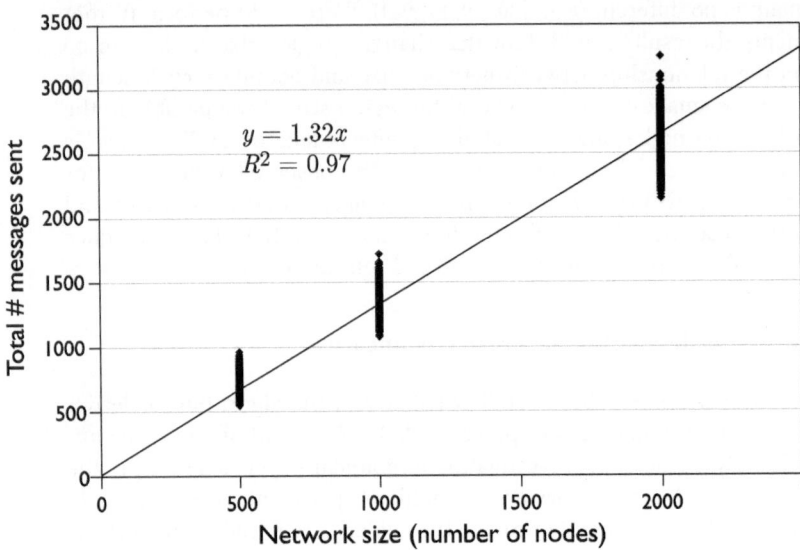

Fig. 7.8. Total number of messages sent as a function of network size for electing a leader in a boundary cycle (Protocol 5.9, 2,700 repetitions)

Fig. 7.9, in the best fit multiple regression has the equation $y = x_1 + 5.45x_2$, where x_1 is the network size and x_2 is the boundary length.

7.3.3 Comparing Overall Efficiency

Quite often it may be important to verify the comparative efficiency of multiple algorithms. For example, in Chapter 3 the very first protocol we encountered was the "Crowd!" protocol (Protocol 3.1). The discussion of the computational efficiency of that algorithm referred to some standard results on the cover time for random graphs. Specifically, the discussion argued that in the worst cases Protocol 3.1 could reach $O(n^3)$ communication complexity, or $O(n^2)$ for planar graphs. The question we might ask is "how do the observed average-case efficiencies for our planar and non-planar graphs compare with each other and with this expectation?"

Figure 7.10 shows the results of an experiment on the overall scalability of Protocol 3.1 repeated for two network structures, the (non-planar) UDG and the (plane) relative neighborhood graph. Because of the relatively low scalability of this algorithm, only ten (rather than the previous 100) repetitions were completed for each factor level.

Again, we have performed a regression of the two data sets, this time with the assumption of a power relationship between the network size and the number of messages sent. Both results achieve relatively high R^2 values (0.94 and 0.92), indicating that we may have reasonable confidence in the

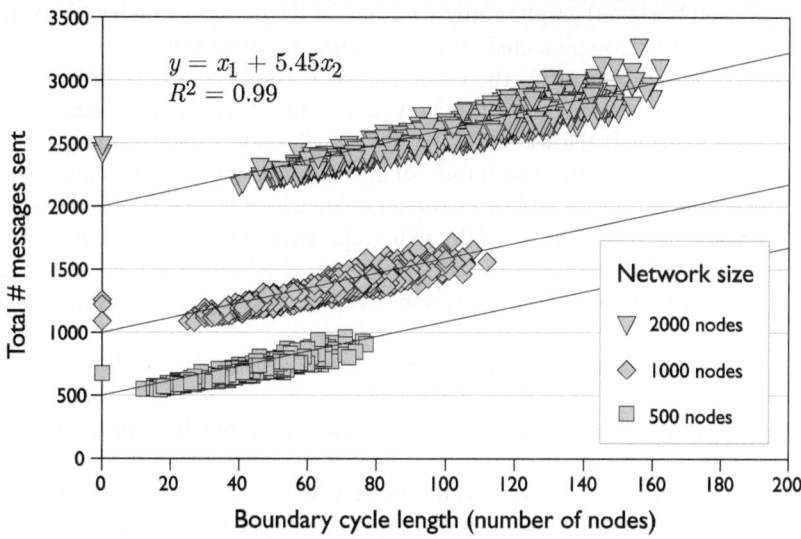

Fig. 7.9. Total number of messages sent as a function of boundary cycle length for electing a leader in a boundary cycle (Protocol 5.9, 2,700 repetitions)

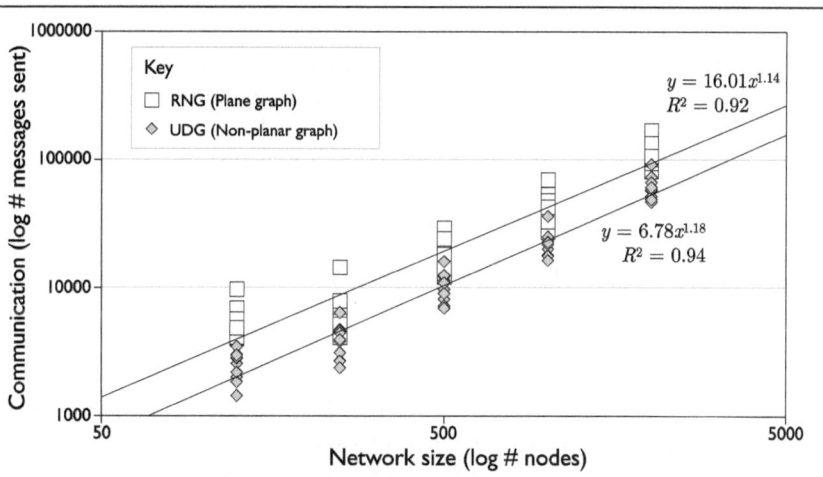

Fig. 7.10. Scalability of the "Crowd!" algorithm, Protocol 3.1, showing response in overall messages sent for increasing network size (ten repetitions per level) for two different network structures, UDG and relative neighborhood graph

two equations as models of the data. Both regression lines also indicate our average case for these algorithms is substantially better than the worst case. The exponent for the regression of the data for the plane (relative neighborhood) graph is only 1.14, and for the non-planar (UDG) 1.18, whereas the worst case indicated $O(n^2)$ and $O(n^3)$ respectively.

The exponent for the UDG data regression curve is slightly higher than that for the RNG data regression curve, indicating lower scalability (as might be expected). However, given the small sample size (ten repetitions per network) and the magnitude of the difference in the exponent (0.04), the results do not provide convincing evidence of any clear difference between the relative scalability of the two graph structures. Indeed, there is considerable overlap between the data points for the UDG and RNG. Does network structure really affect the scalability? Perhaps the apparent differences have occurred by chance?

To answer this question requires *hypothesis testing*. At this point, it would be a good idea to have a trusted text on basic statistics at hand. In this book, we cannot hope to provide a full or thorough account of hypothesis testing. Instead, the following subsection provides the outline of a procedure to address this question as a refresher for those who have studied the topic (but maybe forgotten the details), or as a prompt to those who have not encountered hypothesis testing before (to supplement their reading of this section with a good statistics primer).

Hypothesis testing

The question at hand is whether the two response curves in Fig. 7.10 are really different, or whether the apparent difference might have occurred by chance. One way to approach this question is to look at each level in turn, and ask whether the sample data for each graph structure at that level exhibits any significant differences in sample distribution. More precisely, we can frame the alternatives as two hypotheses:

$$H_0 : \mu_{\text{RNG}} - \mu_{\text{UDG}} = 0 \text{ for some levels tested}$$

and

$$H_1 : \mu_{\text{RNG}} - \mu_{\text{UDG}} \neq 0 \text{ for all levels tested.}$$

The first hypothesis, H_0, asserts that the apparent differences have simply occurred by chance (the null hypothesis), and that the population means μ_{RNG} (population mean of the RNG data) and μ_{UDG} (population mean of the UDG data) are really the same, at least for some levels. The second hypothesis, H_1, asserts that the apparent differences are indicative of a real, underlying difference between these two populations in all cases.[4]

---

[4] It might also be interesting to investigate the hypothesis that $\mu_{\text{RNG}} - \mu_{\text{UDG}} < 0$ for all levels tested, i.e., that the population mean of the algorithm for networks structured as a RNG is *greater* than for networks structured as a UDG. The procedure for testing this hypothesis is almost identical to that described in this section, with the exception that we require a one-tailed rather than two-tailed test (and indeed this test is also significant at the 95% level).

Turning to our statistics textbook, we can now look up the statistic for the testing a hypothesis about the difference between the means of two normal distributions. (We assume that the responses for a particular level are normally distributed). In fact, the usual test statistic in this case is the z-test. However, the z-test is only applicable for *large* sample data sets. Since our data set is relatively small, with each sample containing only ten observations, we require the t-test (which adapts the expectation to the small sample data). If there is any doubt about using a z-test or a t-test, it is always preferable to apply a t-statistic, which degrades gracefully to the z-statistic for large enough sample data sets anyway.

Your favorite statistics package or spreadsheet can now compute the t-test with each pair of sample data sets. The results for the data in Fig. 7.10 indicate that the difference is significant at the 5% significance level for all test network sizes (although not at the 1% significance level for one of the network sizes tested, 250 nodes). Put another way, we can have 95% confidence that network structure, whether UDG or RNG, does indeed affect the observed efficiency of Protocol 3.1, at least across the range of network sizes tested.

7.3.4 Comparing Load Balance

The experiments thus far have investigated the overall communication complexity of an algorithm. But of course the load balance is an equally important (if not more important) feature of any algorithm. It is possible to reuse the experimental structures already encountered, for example, by simply replacing the total number of messages sent with the highest single number of messages sent or received by any node as the response variable. However, in investigating load balance, we are typically interested not only in the node with the *highest* number of messages sent, but also in the distribution of messages sent across all the nodes.

Figure 7.11 compares the load balance of Protocols 3.1 and 3.2 with two frequency histograms. The x-axes in Fig. 7.11 show the numbers of messages transmitted while the y-axes show the total number of nodes that transmitted that number of messages. In interpreting frequency histograms for load balance, tighter distributions with shorter tails towards the larger message counts are more desirable, because they indicate a smaller number of nodes bearing very high loads (e.g., Fig. 7.11b).

In comparing load balance across two algorithms, the load mean and standard deviation are not meaningful because the distributions in the frequency diagrams for load balance are typically strongly skewed. The modal and median values are similarly not especially useful, since when comparing distributions with the same total number of messages, *higher* modal and median values often result from fewer outliers with very large loads (for

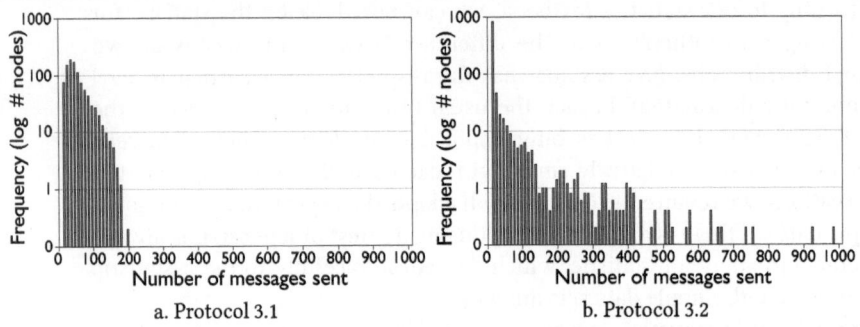

a. Protocol 3.1   b. Protocol 3.2

Fig. 7.11. Load balance for Protocols 3.1 (a) and 3.2 (b), for networks of 1,000 nodes averaged over five repetitions

example, the mode in Fig. 7.11a lies in the class 20–30 messages, whereas in Fig. 7.11b it lies in the class <10 messages).

As a further result of the strongly non-normal distributions in load balance, it is also not so straightforward to perform hypothesis testing. However, while the load balance itself is not expected to be normal, the *proportion* of nodes below or above a certain number of messages sent *is* expected to be normally distributed. The test statistic for the difference between two proportions is

$$z = \frac{P_1 - P_2}{\sqrt{\hat{\pi} * (1 - \hat{\pi}) * (1/n_1 + 1/n_2)}}$$

where P_1 and P_2 are the proportions in the two samples, $\hat{\pi}$ is the proportion in the combined sample (and an estimate of the common population proportion), and n_1 and n_2 are the number of observations in the two samples.

For example, let us test whether there is a significant difference in the proportion of nodes with a load of less than 200 messages for Protocols 3.1 and 3.2 (Figs. 7.11a and b respectively). We start with the following hypotheses:

$$H_0 : \pi_1 - \pi_2 = 0 \text{ as against } H_1 : \pi_1 \neq \pi_2$$

where π_1 is the proportion of nodes with a load of less than 200 messages in the population of executions for Protocol 3.1 and π_2 is the proportion of nodes with a load of less than 200 messages in the population of executions for Protocol 3.2. In our samples of five executions with 1,000 nodes, 5,000 out of 5,000 nodes in Protocol 3.1, as opposed to 4,887 out of 5,000 nodes in Protocol 3.2, had a load of less than 200 messages. The estimate of the common population proportion $\hat{\pi}$ is $9,887/10,000$.

Then the test statistic (for five samples of executions with 1,000 nodes in Fig. 7.11) becomes

$$\frac{5000 - 4887}{\sqrt{0.9887 * 0.0113 * 2/5000}} = 5.34,$$

which is substantially larger than the critical value of 2.33 (for significance at the 99% level). We can therefore reject H_0 and conclude that there is evidence from the data summarized in Fig. 7.11 that the proportion of nodes with a load of less than 200 messages is significantly lower for Protocol 3.1 when compared with Protocol 3.2 (at least for networks of 1,000 nodes).

Viewing the load distribution over a wider range of scenarios is more tricky. However, Fig. 7.12 attempts to depict the load balance across a range of six network sizes, showing how increasing network size alters the load balance.

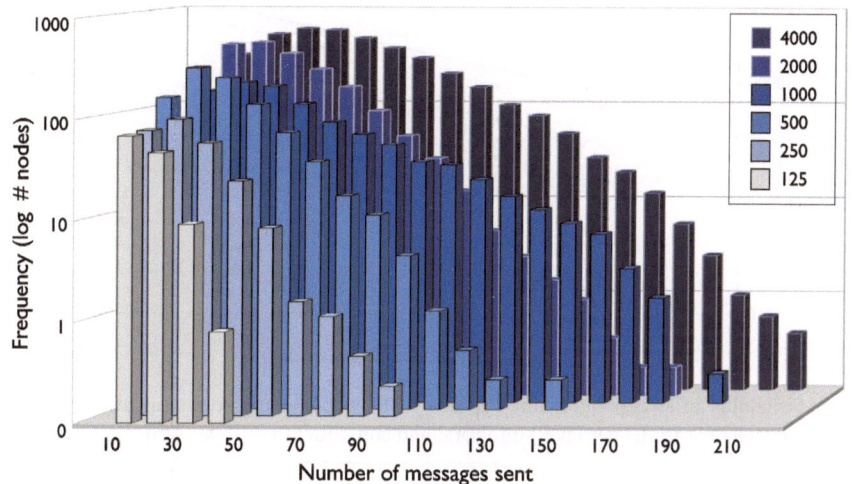

Fig. 7.12. Load balance for Protocols 3.1 across six network sizes (125, 250, 500, 1,000, 2,000, and 4,000 nodes), each averaged over five repetitions

7.3.5 Decomposing by Message Type

One further important empirical technique for understanding the efficiency of an algorithm is to decompose communication into its constituent message components. The NetLogo simulation models constructed for this book are all able to track the breakdown of individual messages for any algorithm.

For example, Fig. 7.13 shows the breakdown of messages sent for the hop-count sweeps, introduced in Protocol 4.13. The protocol has a message structure that uses six different types of messages: ping messages to construct a hop-count potential function; `invt` messages to invite a new node to the sweep front; `unsw` messages for an invited node to check with neighbors if it is in a position to accept an invitation; `acpt` messages from neighbors to inform an invited node that it can join the sweep front (transitioning to FRNT); `swep` messages for a node to check with neighbors if the front has passed; and, finally, `conf` messages for neighbors to inform a node that the front has passed (transitioning it to SWPT). For legibility, the graph in Fig. 7.13 shows only the (linear) regression curves for each message, based on 100 repetitions at each factor level.

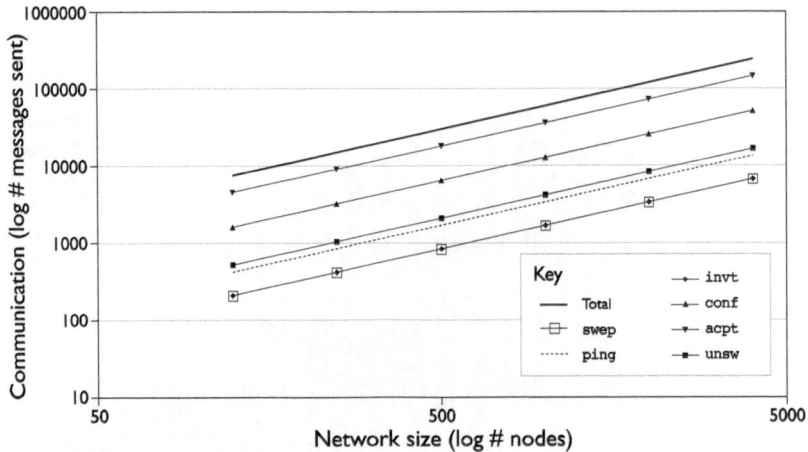

Fig. 7.13. Scalability of sweep algorithm, Protocol 4.13, decomposing communication for changing network size (100 repetitions per level) into total and constituent messages sent

Table 7.2 gives the slopes and determination coefficients for the regression curves shown in Fig. 7.13 (regression curve y-intercept is fixed at 0, the origin). But wait a minute! All the regressions have very high determination coefficients for the linear regression. However, the slope for the `acpt` message regression is 36.26: more than 36 `acpt` messages for every node, while our analysis indicated a load of $3|nbr(v)|$ `acpt` messages sent per node. In all the graphs tested, average node degree was in the vicinity of 7–8 neighbors per node (indicating at the very worst a slope for `acpt` messages of 24).

Further scrutiny reveals another puzzling feature. The regression in Table 7.2 indicates 1.67 `invt` messages per node and 4.17 `unsw` messages, while our analysis in §4.5.1 predicted a load of exactly one `invt` and one `unsw` message per node.

How can this be? In short, it is a mistake. No matter how carefully one performs the pen-and-paper design and analysis process explored in Part II of this book, it is inevitable that from time to time the designer will miss

	Slope	R^2
ping	3.66	0.99
invt	1.67	1
unsw	4.17	1
acpt	36.26	1
swep	1.66	1
conf	12.81	1
Total	59.6	1

Table 7.2. Slopes and determination coefficients for linear regression curves in Fig. 7.13

something, and make an error that will not be uncovered by the analysis. The sweep algorithm in §4.5.1 contains a bona fide example of a couple of such errors that eventuated while writing the book. The mistake was not contrived for this book; I only uncovered it when generating the data for Fig. 7.13. But rather than simply correct it, creating the false impression of a seamless transition from design and analysis to simulation, I decided instead it would be helpful to leave this mistake in place for didactic reasons.

To understand this mistake, it should first be noted that Protocol 4.13 does work (in the sense that it does perform sweeps in overall $O(|V|)$ communication complexity) both in the design and in the NetLogo code developed based on the protocol. This makes the mistake harder for the designer to detect, because it creates the expectation of correctness. Based on the data generated from the protocol and the message breakdown in Fig. 7.13 the consequent investigation uncovered the root causes of the problem:

- The analysis was faulty, failing to note that the swep message broadcast was contained in the action of a *Spontaneously* event (with no state change), which as a result could occur any number of times while the node was in state FRNT. Hence, the analysis should have detected that the number of swep messages sent per node was in actuality *unbounded* (rather than $O(1)$)! Given this error, it was a matter of chance that the protocol, once implemented in NetLogo, ran at all—in other circumstances it is possible that an error like this could have generated enough swep messages to swamp the simulation (rather than have just slightly elevated the load resulting from this message).
- In addition, a coding error in translating the design into the NetLogo implementation further exacerbated the problem, by allowing the occasional additional (rather than exactly one) invt message to be broadcast. This relatively minor error led in turn, via the action for the IDLE/*Receiving* invt state/event pair, to an elevated level of *Receiving* unsw events in the IDLE state, and ultimately multiplied by $|nbr(v)|$, to the additional acpt messages.

In short, a little bit of everything went wrong with this algorithm. The analysis failed to correctly identify the load associated with one message and as a result a design flaw passed into implementation. Further, a mistake in implementation had knock-on effects, increasing the load associated with other messages.

Protocol 7.1 shows a revision of Protocol 4.13, which solves the root design problem. In summary:

- The FRNT/*Spontaneously* state/event pair has been removed. Instead the **broadcast** swep operation has been moved to the action of the FRNT/*Receiving* invt state/event pair (previously empty). This ensures that a FRNT node broadcasts a swep message whenever an important change occurs in its vicinity (i.e., when a new invitation to join the front is overheard, indicating a state change from FRNT to SWPT in the node's neighborhood). As a result, the incorrect expectation of a load of exactly one swep message sent per node needs to be adjusted upwards to at worst $deg(v)$ swep messages sent per node.

Further, revisiting Protocol 4.13 led to one or two further minor changes. In particular, the acpt and conf messages included a Boolean flag b to indicate whether the invitation could be accepted or the front had passed. However, in the case where this flag was set to false, no operations ever eventuated in the corresponding action of the neighbor. As a result, Protocol 4.13 dropped this flag, and instead only broadcast acpt/conf messages when the invitation could be accepted/the front had passed.

Finally, the NetLogo simulation model was reimplemented, with the coding error fixed. Figure 7.14 shows results for the same experiment as in Fig. 7.13, with the revised sweep algorithm and NetLogo simulation model from Protocol 7.1. As shown in Table 7.3, the regression now conforms to the revised analysis expectations, giving us more confidence that the algorithm works as expected.

	Slope	R^2
ping	6.7	0.96
invt	1	1
unsw	1.5	1
acpt	12.47	0.99
swep	1.81	1
conf	14.91	0.99
Total	38.38	0.99

Table 7.3. Revised slopes and determination coefficients for linear regression curves in Fig. 7.14

As a final verification of the analysis, Fig. 7.15 shows a scatter plot of node degree against the number of acpt messages broadcast for a single

---

Protocol 7.1. Hop count sweep (revisited)

Restrictions: $\mathcal{NB}$; identifier function $id : V \to \mathbb{N}$
State Trans. Sys.: $(\{\text{INIT, IDLE, FRNT, SWPT}\}, \{(\text{INIT, SWPT}), (\text{IDLE, FRNT}), (\text{FRNT, SWPT})\})$
Initialization: One or more nodes initiating the sweep in state INIT, all other nodes in state IDLE
Local data: Hop counter h_c, initialized to $h_c := -1$; list of neighbor identifiers of unsw message, N_u, initially empty; list of neighbor identifiers of swep message, N_s, initially empty

INIT
   Spontaneously
      set $h_c := 0$ *#Reset hop count*
      broadcast (ping, h_c)   *#Initiate hop count potential function*
      Wait for a short period
      broadcast (invt, h_c)   *#Initiate sweep algorithm inviting neighbors to join*
      become SWPT

IDLE
   Receiving (ping, h)
      if $(h + 1) < h_c$ **or** $h_c < 0$ **then**   *#Construct shortest path hop count potential function*
         set $h_c := h + 1$ *#Update hop count*
         broadcast (ping, h_c)   *#Continue construction of hop count potential function*
   Receiving (invt, h)
      broadcast (unsw, h_c, $\mathring{id}$)   *#If invited, check for unswept neighbors*
   Receiving (acpt, i)
      set $N_u = N_u \cup \{i\}$ *#Add new neighbor to unswept list*
      if $|N_u| = |nbr|$ **then become** FRNT   *#Accept invitation once all neighbors respond*
   Receiving (swep, h, i)
      if $h = h_c - 1$ **then send** (conf, $\mathring{id}$) to node with identifier i *#Check sweep condition*
   Receiving (unsw, h, i)
      if $h \leq h_c$ **then send** (acpt, $\mathring{id}$) to node with identifier i *#Check sweep front condition*

FRNT
   Receiving (swep, h, i)
      if $h = h_c$ **then send** (conf, $\mathring{id}$) to node with identifier i *#Check sweep condition*
   Receiving (conf, i)
      set $N_s = N_s \cup \{i\}$ *#Store swept neighbors*
      if $|N_s| = |nbr|$ **then**
         broadcast (invt, h_c)   *#Broadcast new invite*
         become SWPT
   Receiving (invt, h)
      broadcast (swep, h_c, i) *#Front nodes can always respond to a swep message from neighbors*

SWPT
   Receiving (unsw, h, i)
      send (acpt, $\mathring{id}$) to node with identifier i *#Swept nodes can always respond to unsw message*
   Receiving (swep, h, i)
      send (conf, $\mathring{id}$) to node with identifier i *#Swept nodes can always respond to conf message*

FRNT, IDLE
   Receiving (unsw, h, i)
      if $h \leq h_c$ **then send** (acpt, $\mathring{id}$) to node with identifier i *#Check if invite can be accepted*

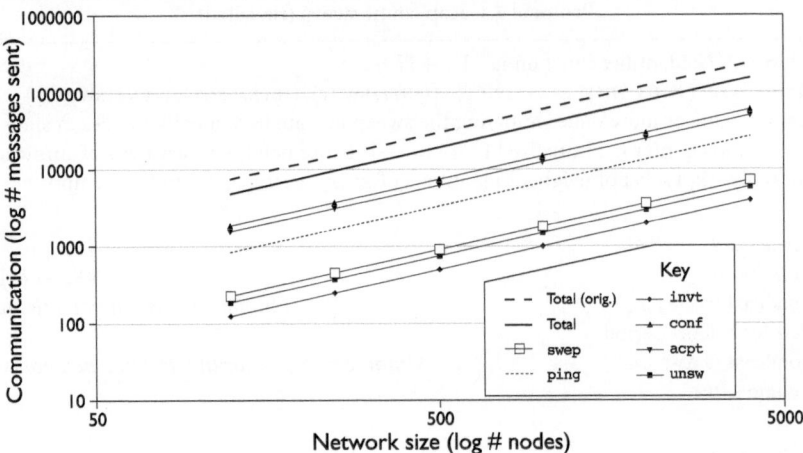

Fig. 7.14. Scalability of revised sweep algorithm, Protocol 7.1, highlighting original total scalability (thick dashed curve; cf. Fig. 7.13)

simulation run of Protocol 7.1 with 500 nodes. The Pearson correlation coefficient, which provides a measure of the linear dependence between these two variables, is 0.87 for this data, indicating as expected relatively high levels of correlation.

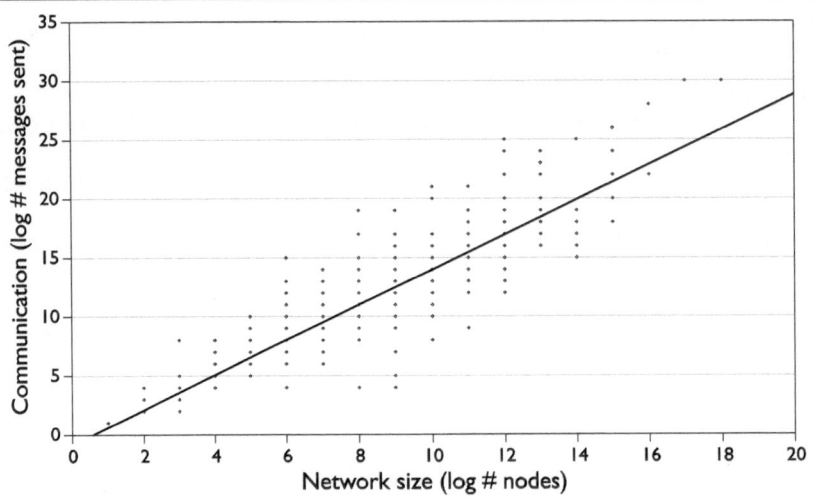

Fig. 7.15. Scatter plot of individual node degree against number of acpt messages broadcast, for single simulation run of Protocol 7.1 with 500 nodes (Pearson correlation coefficient $r = 0.87$)

7.3.6 Measuring Communication Resources

We have used the number of messages sent as the primary measure of the communication resources used by an algorithm. The simplifying rationale behind this default choice is that message transmission is the most energy-intensive operation for a node (§1.2.2), and consequently minimizing the number of transmissions has the greatest impact on the network's energy budget. However, as already indicated in §3.2.2, in reality the situation is more complicated than that. When using package-based communication protocols, longer messages will be broken up into multiple packets, leading to a component of communication resource usage proportional to the *length* rather than simply the *number* of messages sent. Some hardware may also require a substantial amount of power simply to *receive*, and not just transmit, messages. As a result a complete picture of communication resource usage may also require measurement of message count and length for both messages sent and received.[5]

For example, Fig. 7.16 shows the overall breakdown of number (Fig. 7.16a) and length (Fig. 7.16b) of messages sent and received for the greedy georouting algorithm (Protocol 5.4). Each simulation run forwards messages from a (randomly selected) source to a (randomly selected) sink at a known coordinate destination. At each step greedy georouting selects the neighbor that minimizes the Euclidean distance to the destination (see §5.2.1 for a full discussion of greedy georouting). The graph in Fig. 7.16 summarizes 100 repetitions of the simulation at each of six network sizes (125, 250, 500, 1,000, 2,000, 4,000 nodes), with the communication graph structured as the Gabriel graph (and the cost of establishing the Gabriel graph omitted from the experimental investigation). Greedy georouting can fail with unfavorable network configurations (voids; see Fig. 5.3). In this chapter we are interested only in the efficiency of algorithms, and not in whether the algorithm fails (one of the topics covered in the next chapter). Consequently, simulation runs where greedy georouting fails are omitted from the data in Fig. 7.16, leaving at least 70 repetitions in all network sizes. Finally, the overall efficiency of greedy georouting is strongly correlated with the length of the path between source and destination (Pearson correlation coefficient $r \geq 0.99$ in all cases in Fig. 7.16) and much less strongly with network size ($0.68 \geq r \geq 0.7$ in

---

[5] As intimated in §3.2.2, in certain circumstances there are other operations (in particular, taking readings from certain types of sensors) that may even rival the energy costs associated with communication. It would of course be possible to measure the costs associated with these operations in a similar way to communication. However, we do not consider such issues in further in this book for two reasons. First, even in cases where other operations have comparable energy costs to communication, communication typically remains a substantial cost that must be accounted for. Second, it is the constraints to *movement* of information entailed by high communication costs that are specific to decentralized *spatial* computing.

all cases[6]). Hence, the greedy georoute path length is used as the factor in Fig. 7.16 in place of network size.

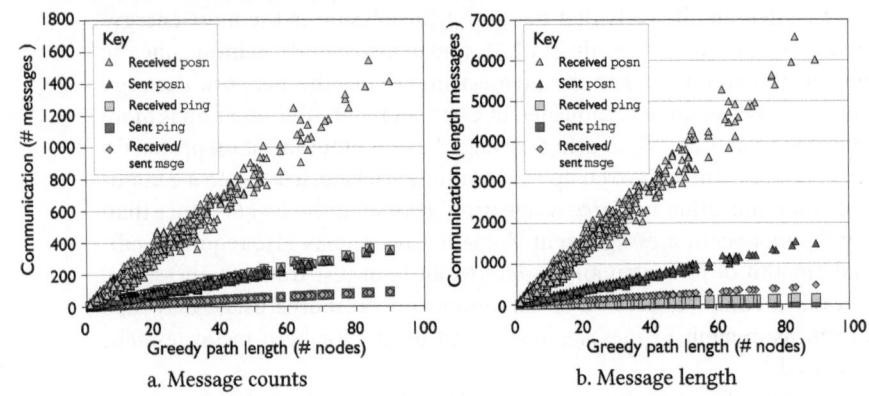

a. Message counts   b. Message length

Fig. 7.16. Comparing response of numbers of messages sent and received a. with length of messages sent and received, and b. with length of route path for Protocol 5.4. Data comprises at least 70 repetitions at each of six network sizes (125, 250, 500, 1,000, 2,000, 4,000 nodes), and excludes cases where the algorithm does not terminate

While all the messages, whether length, number, sent, or received, exhibit strong linear scalability with path length, there are two points worth noting in the patterns in Fig. 7.16:

1. There exist cases which exhibit substantial disparity in the absolute levels of communication resources required for sending and receiving messages, for example, sent and received posn messages (for both message count and length).

2. There exist cases which give a rather different picture of the relative contributions of different messages to the overall scalability, depending on whether message count or length is examined. For example, ping messages sent make the smallest contribution to the overall communication overheads in terms of message length (Fig. 7.16b), but in contrast

---

[6] As the relationship between network size and communication overhead is expected to be non-linear (and is instead logarithmic, see §5.2.1), the Pearson correlation coefficient (which measures linear dependence between two variables) is potentially a misleading guide to the strength of dependence of communication overhead on network size. However, the relationship is *approximately* linear, and so makes a reasonable choice for simplicity. A more correct logarithmic regression of the data yields an R^2 value of 0.47, also providing evidence of the weak dependence between communication overhead and network size.

make the second largest contribution to the overall communication in terms of number of messages sent (Fig. 7.16a).

It should also be noted that message length in Fig. 7.16, and in all the NetLogo simulation models that go along with this book, is measured (rather crudely) as the length of *string* representation of the each message. A more sophisticated and realistic experimental setup would need to measure the number of *bits* required to represent data types. For example, the length of string representation "0.7" is substantially less than that of "0.74855." However, where both data items are represented as floating point numbers, they would be expected to have the same length in transmission (e.g., that of a single-precision floating point number, 32 bits).

7.4 Chapter in a Nutshell

Simulation fulfills three key functions. First, simulation is an important intermediate step between pen-and-paper algorithm design and expensive and complex deployment in the field. Second, no matter how carefully the algorithm design process is conducted, mistakes may occur. Simulation can provide a rapid mechanism for validating the behavior of an algorithm and provide objective evidence to support the reasoned adversarial analysis, ultimately helping to identify and fix design errors if they occur.

Third, simulation enables the performance of an algorithm to be verified through experimentation. Experiments can provide additional information about the scalability of an algorithm that may be difficult or impossible to create purely with pen-and-paper analysis. Average-case information about overall efficiency and load balance, using a variety of different efficiency measures and decomposed by message type, can be created using experiments. In combination with hypothesis testing, experiments can create objective information for comparing different algorithms, network structures, or environments.

However, experiments must also be interpreted with care. Experiments can be used to create *evidence* to support hypotheses. But, as the complexity of experiments increases, so do the chances of misinterpreting results or misusing statistical tests. Well-designed experiments should always tell us *something* about a protocol, but cannot be expected to reveal *everything* we might want to know about how the protocol will perform in every situation.

Review questions

7.1 Using Web search, familiarize yourself with a few of the alternative sensor network simulation systems, such as ns-2 and MATLAB/Simulink. What other alternative simulation systems can you find, in addition to those mentioned at the beginning of §7.1?

7.2 The task set by question 4.2 was to develop an alternative gossiping algorithm that did not suffer from the problem where an INIT node with poor connectivity or few neighbors might result in the premature death of a gossip message [52]. Adapt the NetLogo code for Protocol 4.5 to implement your solution to this problem.

7.3 Question 5.8 asked you to adapt the greedy georouting (or GPSR) protocols to geocast a message from a single source to a specified axis-parallel rectangle. Taking your answer and the existing NetLogo code for Protocol 5.4 as a basis, implement your solution. What other answers to questions from Chapters 4–6 can you implement? With practice and building on the existing protocols and NetLogo simulation models that accompany this book you should be able to implement all of them.

7.4 Adapt the existing NetLogo procedures for region evolution to construct randomized Euclidean shapes (circles, squares, etc.). Using this procedure, repeat the boundary length experiments in §7.3.2 to confirm that they result in the same order of communication complexity, but with a lower constant factor than the regions grown using the original procedure.

7.5 Design and run an experiment to investigate the average-case overall communication complexity and average-case load balance for Protocol 4.11, the algorithm for establishing a rooted tree with multiple initiators.

7.6 Design and run an experiment to investigate the load balance of Protocol 5.3, the algorithm for constructing the LMST (local minimum spanning tree). As discussed in §5.1.2, in the worst case the algorithm should generate at most seven messages per node. What is the average-case load per node, and how evenly is this load distributed throughout the network?

Simulating Robust Decentralized Spatial Algorithms

8

Summary: Evaluating efficiency helps us understand how much of a system's resources will be consumed by an algorithm. But efficient algorithms must also be robust: able to reliably generate useful information under a range of different circumstances. Most algorithms must strike a balance between efficiency and robustness; it is frequently possible to increase an algorithm's robustness at the cost of decreased efficiency. Robustness becomes especially important in the practical contexts of algorithms for geosensor network deployments. Geosensor networks are also expected to operate in environments of uncertainty, for example, where sensors are inaccurate or network coverage is sparse. Further, it is in many cases desirable that decentralized algorithms continue to operate at some level even if the the assumptions upon which they are founded are violated, for example, when communication becomes unreliable.

E FFICIENCY was the measure used for evaluating and comparing algorithms in the previous chapter. In most cases, this involved an investigation of the scalability of an algorithm in terms of its communication complexity. However, measures of efficiency are rarely the *only* feature of an algorithm we are interested in; rather, we are usually also interested in how *well* an algorithm works, to what extent it can complete a useful task. Imagine a decentralized spatial algorithm that *never* communicates any information, where each node simply processes and hoards its own sensed data. Such an algorithm would surely be highly scalable in terms of communication complexity—$O(1)$! However, it would also not be of any *use*, at least not in the context of spatial computing tasks.

In practice we almost always need to balance computational efficiency with other measures of the efficacy or usefulness of an algorithm. The question we would like to ask is to what extent does an algorithm do what it is supposed to? Specifically, the main focus of this chapter is on the algorithm's *robustness*: its capability to generate useful information, and to continue to do so under a range of different circumstances and within different contexts.

This chapter approaches robustness from three perspectives. Evaluating the intrinsic balance between efficiency and robustness, where increased robustness can sometimes be attained at the cost of decreased efficiency, is tackled first. Then, two further extrinsic factors are investigated, outside the control of the algorithm itself: uncertainty and faults. Uncertainty is an unavoidable feature of computing with geographic information—our information about the geographic world is never perfect. Faults occur when system components or operations fail. Common faults, such as unreliable communication, may also violate the specified restrictions on our algorithms. It is highly desirable that algorithms degrade gracefully even in the presence of extrinsic factors such as uncertainty and faults.

8.1 Balancing Efficiency and Robustness

In some of the algorithms presented in Part II of this book, there is just one answer we are searching for. For example, in generating the Gabriel graph or relative neighborhood graph in §5.1.1, the graphs themselves are uniquely defined for a given UDG. Assuming all our restrictions are met (and our design procedures have eliminated all errors from our protocols), these algorithms will generate the correct answer.

However, in many other cases there may be degrees to which an algorithm may perform better or worse. For example, we might be interested not simply in the scalability of an algorithm, but its in operational latency (how long it takes it to generate the answer). In cases where an algorithm may approximate the true answer, it will be important to evaluate how close the answer generated is to the *correct* answer. In other cases we have seen, where the algorithm is not always guaranteed to work, it can be important to gather information about how often and under what circumstances it fails to terminate or to generate an answer in practice.

In all these cases, it is often important to understand the balance between efficiency and robustness—what are the costs in terms of efficiency for achieving or improving robustness.

8.1.1 Example: Path Length in Rooted Trees

Protocol 4.7 demonstrated how to construct a rooted tree using an algorithm with overall $\Theta(n)$ communication complexity. Protocol 4.8 then showed how a similar algorithm could generate the *shortest* path rooted tree, at the cost of reduced scalability, $O(n^2)$, but with the expectation that the average case would be much closer to Protocol 4.7. Indeed, a quick check of the average-case scalability of Protocol 4.8 (Fig. 8.1) shows that in our simulations the

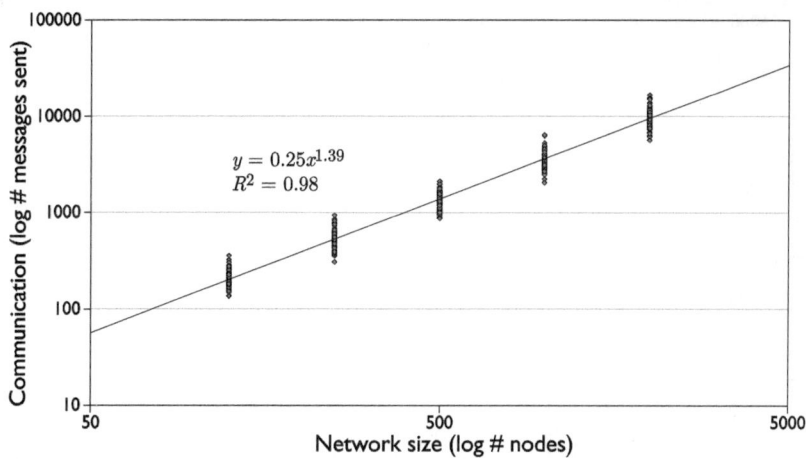

Fig. **8.1.** Scalability of shortest path tree, Protocol 4.8, for 100 randomized repetitions across six network sizes

number of messages scale approximately in proportion to $|N|^{1.4}$ (regression $R^2 = 0.98$ for 100 experimental runs repeated across six network sizes).

Reducing the length of the paths in a rooted tree is important as it may increase the scalability of other algorithms that must route information using that tree. However, path length is also important for robustness. Reducing the length of routes through the overlay network may in turn reduce the chance of a random communication breakdown, causing the algorithm to fail, or simply reduce the latency of an algorithm using the tree. In short, it may be useful to understand the balance between how efficient Protocols 4.7 and 4.8 are and how long the paths in the resulting overlay network are.

Our question then becomes "What extra communication costs are incurred in achieving the shortest path lengths?" Figure 8.2 shows a set of data that offers one perspective on the answer to this question. In Fig. 8.2, the dependent variable of communication cost is measured as the additional number of messages required by Protocol 4.8 when compared with Protocol 4.7. Other measures of communication cost are possible, such as the proportion of messages required by Protocol 4.7 to those required by Protocol 4.8, but tend to yield a similar picture.

Path length is shown as the independent variable in Fig. 8.2, measured as the reduction in hops of the length of the longest path generated by Protocol 4.8 when compared with Protocol 4.7. Here again, other measures are possible, such as the median or mean path length. However, in terms of robustness, we may be interested in single points of failure, so the *longest* path (i.e., the one most likely to lead to failure or to high latency) seems a sensible choice.

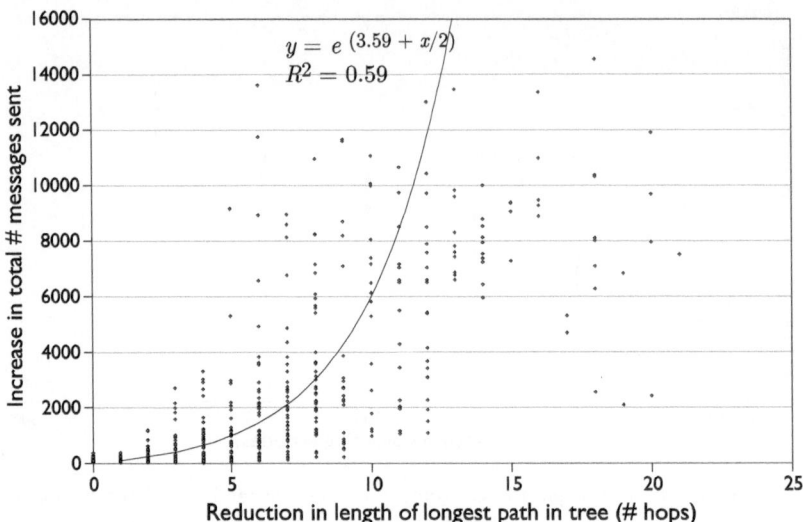

Fig. 8.2. Increase in number of messages sent for reduction in length of longest path in overlay network tree for Protocol 4.8 when compared with Protocol 4.7 (500 repetitions)

For this comparison to be meaningful, each simulation run was paired— Protocols 4.8 and 4.7 were simulated on the *same* 100 randomized networks at each of five network sizes. For each pair of simulation runs, the total number of messages and the length of the longest path generated by Protocols 4.8 and 4.7 were compared. If the simulations had not been paired, it would have become much harder to attribute any change in path length or communication overheads to the protocols, as opposed to attributing them to random differences in the node locations or in the root location. For even the smallest networks (125 nodes), the standard deviation of messages sent by Protocol 4.8 was 41.46, indicating high levels of variability due solely to the randomized factors.

From Fig. 8.2, there does broadly seem to be an increase in messages sent with reductions in path length, as would be expected. As we look for larger reductions in the length of the longest path in the tree, the magnitude of the increase in the total number of messages increases accordingly. However, there is no obvious pattern or relationship between these changes. The length of the longest path in *balanced* trees (where branches are all of approximately the same length) is a function of the log of the number of nodes in the tree (an important computational feature of trees, used for efficiently indexing databases). As a result, a conjecture about the relationship underlying Fig. 8.2 would be that any reduction in path length is related to the log of the increased number of messages sent. A regression fitted to the data where $x = 2\ln y + 7.18$—or equivalently $y = e^{(3.59 + x/2)}$—indicates some weak support for this idea ($R^2 = 0.59$). However, the main message of Fig. 8.2 is in the high levels of variability (e.g., a reduction of maximum path length of six hops required anywhere between 100 and 13,000 additional messages). In short, Protocol 4.8 can be relied upon to reduce path length, but cannot be relied upon to do so at low or even predictable computational costs— other randomized factors (such as node and root location) must be strongly influencing the communication overhead.

8.1.2 Evaluating Multiple Factors

Although the "one question at a time" approach is simple and powerful, it has one important drawback: it does not allow for the exploration of the interaction *between* factors. In examining one factor at a time, and controlling other factors, the implicit assumption is that each factor is independent and does not alter the effect of other factors. For instance, in our example of the multistage OFAT in §7.3.2, the size of the network directly affected the number of nodes located at the boundary of a fixed region; and the number of boundary nodes directly affected the efficiency of the algorithm for leader election at the boundary. However, these two factors (network size and number of boundary nodes) were independent in that changing the network size did not change the effect of the number of boundary nodes. The number of boundary nodes was directly related to the efficiency of the algorithm whether or not the network size was big or small.

However, factors are not always independent. Instead, they may interact, with changes in one factor altering the magnitude of its effects. In such cases the OFAT approach will fail to adequately characterize the behavior of an algorithm, since altering one factor at a time will not yield information about the effects of one factor on another. The answer to this problem is a more sophisticated approach to experimental design, called simply "design of experiments" (DOE). In DOE every possible combination of factors is tested, and the results analyzed to determine which factors have effects on the response variables. The disadvantage in examining all possible combinations of factors (termed a *full factorial* experiment) is that problems can arise from a combinatorial increase in experimental size. For example, for an experiment with three factors each of which has four levels, it would be necessary to run $4^3 = 64$ simulation models, each of which would require sufficient repetitions to ensure a representative sample. As a result, the number of levels in a full factorial experiment is kept to a minimum (often only two or three levels). Alternatively, it is also possible to test only a proportion (typically half) of the possible combinations of factors (termed a *fractional factorial* experiment).

8.1.3 Example: Greedy Georouting

Section 5.2.1 introduced the greedy georouting protocol (Protocol 5.4). To recap, in greedy georouting information is routed between a source node and a sink node at a known coordinate destination using purely local decision making, at each step the information being forward to the neighbor closest to the destination. Greedy georouting can sometimes fail because of voids (local minima) in the network (i.e., at a non-destination node that is closer to the destination than any of its neighbors; see Fig. 5.3).

It is expected that the distance between the source and sink nodes will affect both the communication efficiency (because the further the message must travel, the more the number of hops required) and the failure rate of the algorithm (because the longer the route, the more the chance of encountering a void). However, it is similarly expected that the type of network (whether it be UDG, RNG, or GG) will also affect the communication efficiency (because longer paths will be necessary in sparser graphs) and the failure rate (because the changing network structure is expected to affect the likelihood of voids occurring).

To disentangle the effects of these different factors, we can run a full factorial experiment with two factors and three levels for each factor (UDG, RNG, and GG for network structure, and close, middling, and far for the distance between source and sink nodes). NetLogo is fully compatible with DOE, and so setting up BehaviorSpace to run such an experiment is no different that it was with earlier experiments (except that multiple factors must be varied using the experimental variables setup; see Fig. 7.4).

Experiment	1	2	3	4	5	6	7	8	9
Network structure	UDG	UDG	UDG	RNG	RNG	RNG	GG	GG	GG
Source/sink distance	Close	Mid	Far	Close	Mid	Far	Close	Mid	Far

Table 8.1. A full factorial experimental design for testing the effects of network structure and distance between source and sink on the performance of the greedy georouting algorithm (Protocol 5.4)

The first output from these experiments is the failure rates of the algorithm. Table 8.2 shows the percentage of successful repetitions based on 100 runs for each of the treatments (recall: a "treatment" is a specific combination of levels tested in an experiment). The values appear to clearly indicate decreasing reliability with both increasing delivery distance and sparser graph structure (from UDG to GG to RNG). This impression can be statistically verified again using the test for difference between two proportions (§7.3.4, p. 234). Doing so for all treatments reveals that all the differences are indeed significant at the 1% level, except when comparing the UDG and the GG for close sources and destinations.[1]

---

[1] More precisely, the null hypothesis H_0 that there is no difference between the proportion of successfully delivered messages in the UDG and the GG *cannot* be rejected in the case of close messages. The corresponding hypotheses can be rejected at the 1% significance level in all other cases.

	UDG	GG	RNG
close	93	86	64
mid	67	40	18
far	47	29	6

Table 8.2. Percentage of experimental runs of greedy georouting where the message successfully reaches the destination

Turning to the communication efficiency, we are primarily interested in the efficiency when the message is successfully delivered. When delivery fails, the algorithm will not terminate, with messages being passed backwards and forwards between nodes at a local minima indefinitely. Further, the two factors are expected to interact, since a change in the network structure may change the *magnitude* of the effects of the distance between source and sink. Examination of these effects can be achieved by using an important statistical technique called ANOVA (analysis of variance). Using ANOVA is similar to performing individual t-tests between multiple pairs of treatments, but considerably less laborious. It can also provide useful information about the interaction between factors, which would be missed with a simple pairwise t-test.

However, an important assumption of ANOVA is that the variances of the samples are equal. Variance is a measure of spread of the results, and numerically simply the square of the standard deviation. Unequal variances, especially in combination with small sample sizes, can increase the chance of incorrectly rejecting the null hypothesis (termed a type I error) or of failing to reject a null hypothesis that is false (a type II error). The sample sizes for successful message delivery are indeed unequal. There also exist a number of hypothesis tests for unequal variances, such as Levene's test, which are often included with the results of an ANOVA test. An initial analysis of the total number of messages generated by the successful algorithm repetitions revealed that there was evidence that the associated variances are indeed unequal (the null hypothesis H_0 that the variances are unequal cannot be rejected at the 1% level using Levene's test).

Consequently, before the ANOVA was run, the two least populous treatments (RNG/mid and RNG/far) were rerun to generate enough successful results to allow an ANOVA test for equal sample sizes of 25 repetitions (by randomly selecting subsamples of 25 runs from treatments with larger samples). The formats of the results generated by ANOVA differ depending on which statistics software you use, but the results in Table 8.3 are typical of the outputs often found. The first two rows concern the effects of the two factors (distance between source and sink and network structure) on the total number of messages generated (termed the *main* effects); the third row concerns the *interaction* effects between the two factors.

	Df	Sum Sq	Mean Sq	F value	$P(> F)$
A: Distance	3	9549651.725	3183217.242	7134.167	0.0
B: Network	3	34383.836	17191.918	38.530	0.0
A/B: Interaction	3	11253.179	2813.295	6.305	0.0

Table 8.3. Results of ANOVA on distance between source and sink (close, mid, far) and network structure (UDG, GG, RNG) for greedy georouting, Protocol 5.4

The ANOVA output often includes information about associated statistics, such as the degrees of freedom (df) and the sum of squares and mean squared difference. However, the vital information is the F-statistic associated with the main and interaction effects. This provides information about three null hypotheses: H_{0A}, that factor A (distance between source and sink) has no effect on the response variable (total messages sent); H_{0B}, that factor B (network structure) has no effect on the response variable; and H_{0AB}, that the interaction between factors A and B have no effect on the response variable. Looking at the test statistic (F value) and the associated probability that this result could occur by chance ($P(> F)$) in Table 8.3 indicates that the null hypothesis can be rejected in all cases. In other words, the experiments provide evidence that both factors individually and in combination have a significant effect upon the efficiency of Protocol 5.4.

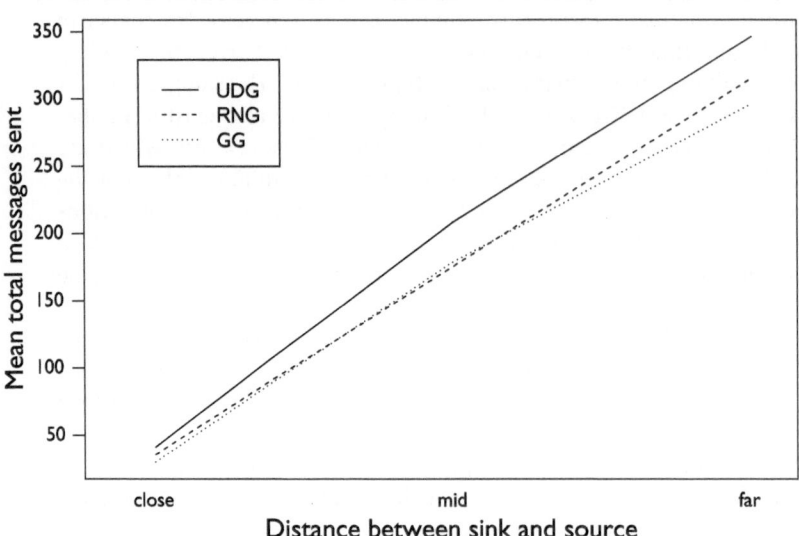

Fig. 8.3. Interaction diagram for total messages generated for greedy georouting, Protocol 5.4, with two factors: distance between source and sink (close, mid, far) and network structure (UDG, GG, RNG)

A convenient way of depicting the results of ANOVA is with an interaction diagram, shown in Fig. 8.3. The figure summarizes the response (mean total messages sent for each treatment), showing one factor (the distance between sink and source nodes) on the x-axis and the second factor (network structure) as different response curves. Each curve shows the responses for the different treatments. For example, the curves show that as expected the total number of messages generated by the algorithm increases with distance between sink and source nodes, as would be expected. The curves also show, perhaps surprisingly, that the UDG network structure generates *more* messages than the sparser networks structures (GG and RNG), even though the paths between sink and source are expected to be longer in the sparser network structures. This result can be explained because the analysis investigated *total* messages sent. Because nodes in the UDG will have on average higher degree than nodes in the GG or RNG, there are on average more neighbors of nodes on the route from source to sink. In turn, this means a greater number of nodes have to respond to ping messages with their own posn messages (see Protocol 5.4). The plot of the interaction diagram for msge messages (as in Fig. 8.4) shows a different pattern, as expected with increasing numbers of msge messages with increasing network sparseness.

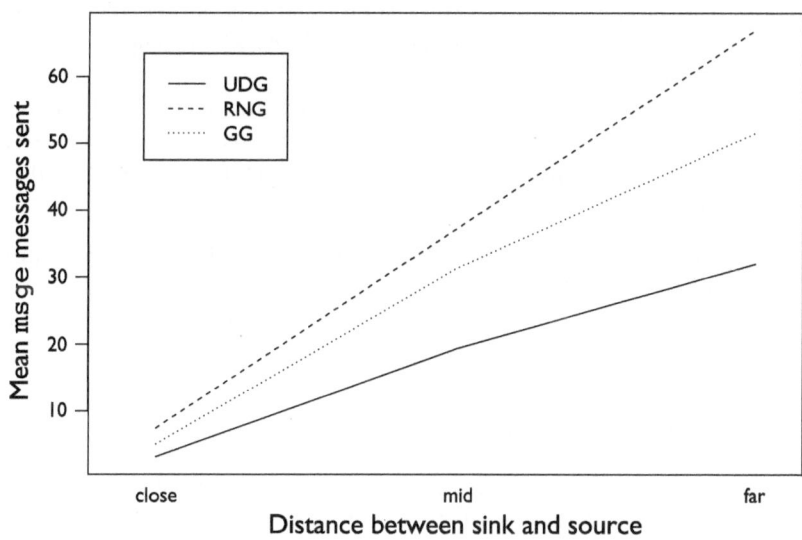

Fig. 8.4. Interaction diagram for msge messages generated for greedy georouting, Protocol 5.4, with two factors: distance between source and sink (close, mid, far) and network structure (UDG, GG, RNG)

As one might guess, interaction diagrams such as Figs. 8.3 and 8.4 also show the interaction between the two factors. Interaction can be seen in the different slopes of the response curves. For example, the different slopes of the curves in Fig. 8.4 indicate that while increased distance between sink and source leads to more msge messages, the magnitude of this effect is *increased* by sparser network structures (i.e., the slope of the response curve is steeper for RNG than for GG than for UDG). The interaction in Fig. 8.3 is clearly less marked, with only slight (although still statistically significant) slope differences between the three curves. Based on this observation, a follow-up ANOVA considering only two levels of network structure (the two closest curves in Fig. 8.3, GG and RNG treatments) reveals that they exhibit no significant effects or interactions. Thus, even though we rejected the null hypotheses H_{0B} and H_{0AB} when considering all *three* levels of network structure (UDG, GG, and RNG), closer inspection reveals that there is no significant evidence to reject the null hypotheses H_{0B} and H_{0AB} when looking only at GG and RNG together (ignoring the effects of the UDG). This type of exploration, where an analysis prompts further significance testing for specific relationships within the data set, is typical of the process of applying and understanding the results of an ANOVA test.

To recap, our empirical investigation of Protocol 5.4 has revealed:

- increasing both distance between source and sink and network sparseness significantly decreases the reliability of routing;
- increasing distance between source and sink decreases the efficiency of successful routing, as does using the UDG when compared with GG and RNG; and
- increasing network sparseness significantly increases the path length (total number of msge messages) between particular source and sink, as well as magnifying the effect of increasing distance between the source and sink.

While none of these results on their own may be especially surprising, in addition to introducing some important experimental techniques, the example does illustrate how careful experimental work can support understanding of the detailed operation and characteristics of an algorithm.

8.1.4 A Note on Deadlock and Termination

Two further intrinsic characteristics of decentralized algorithms are frequently of interest when discussing algorithm robustness: deadlock and termination. Both deadlock and termination are major topics in the field of distributed systems more broadly, and are not specific to the design of decentralized spatial algorithms. Consequently, in this section we only sketch a brief overview of the key issues, rather than provide more detailed empirical examples.

Deadlock occurs when all nodes of a set are blocked, unable to generate any events, waiting for events triggered by other nodes in the set.

Because nodes are all waiting for each other to trigger an event, deadlocks are sometimes also called *circular waits*. Ideally, the algorithm design and analysis process should identify any potential for deadlocks in an algorithm. However, even when the potential for deadlocks is acknowledged, it may not be possible to guarantee deadlock avoidance, at least not without decreasing the scalability of an algorithm. In these cases, a range of distributed algorithms do already exist for detecting deadlock. As deadlocks are already an established and well-studied research problem, without any intrinsically spatial characteristics, we leave further exploration of this topic to the reader (e.g., [99]). In the context of this chapter, it is enough to acknowledge that deadlocks are another potential behavior of decentralized spatial algorithms that may be important to consider, balanced against efficiency and other concerns.

Termination is another a well-studied problem in distributed systems. Termination concerns whether a node has completed its activity in a protocol. Not all our algorithms are required to terminate. Many of the spatiotemporal protocols in Chapter 6 remain resident in the network, continually monitoring and reporting on changes (termed long-running or "push" queries, discussed in more detail in the next chapter, §9.4). In these cases, the algorithm is never expected to terminate. However, in other algorithms termination can be an important property, especially where an algorithm is composed of multiple protocols being operated in sequence on nodes (i.e., when one protocol terminates, a different protocol begins).

For those algorithms that are expected to terminate, detecting *local* termination, where an individual node must determine whether it has completed its actions, is usually trivial. In most cases, the state transition system for each protocol will define a terminal state (often called DONE, frequently with no or empty actions for all system events). Nodes entering that state can be said to have locally terminated. By contrast, detecting *global* termination, where nodes can determine when a protocol has completed across the entire network, is often challenging and usually more computationally intensive. The issue of detecting termination already arose, for instance, in the context of Gabriel and relative neighborhood graphs, in §5.1.1. However, as for deadlocks, we leave further exploration of this issue to the reader, with a note that the detection of termination also usually imposes efficiency costs on an algorithm.

8.1.5 Summary

We have now seen that some algorithms may fail under certain circumstances. Empirical investigation can yield useful information about how likely or serious these failures can be (e.g., Table 8.2). Surely, it would be preferable if our algorithms *always* worked; but, in practice there will often be an important role for *approximate* algorithms that can generate the correct answer *most* of the time (especially when we have a good understanding of when these algorithms will fail). Approximate algorithms can be useful in a number of circumstances, and indeed we have already encountered several approximate algorithms. For example, when monitoring dynamic environments for salient spatiotemporal events, it may not be essential to capture *every* event; this premise was behind rumor routing, back in §4.2.2.

8.2 Uncertainty

Information captured about our geographic world is never perfect; imperfection is unavoidable. In the presence of imperfection, a decentralized spatial algorithm may not be *certain* to generate a correct answer. We use the term *uncertainty* to refer to this situation, where we can't have complete confidence in the information generated by an algorithm. The term *imperfection* is reserved for those features of information that give rise to uncertainty.

8.2.1 Types of Imperfection

Much has been written about imperfection and uncertainty. Different fields including computer science and geographic information science have a plethora of different conventions for classifying imperfection and quantifying uncertainty. A few of the many terms used to describe aspects of imperfection and uncertainty include: error, granularity, currency, reliability, quality, ambiguity, completeness, fuzziness, and consistency (cf. [119]).

Despite this fog of words, the essence of the concepts underlying these different terms can largely be captured using just two basic types of imperfection [118]:

- *inaccuracy*, which concerns a lack of *correctness* in information, where a node's knowledge about its geographic environment does not accord with reality; and
- *imprecision*, which concerns a lack of *detail* in information, where a node's knowledge about its geographic environment lacks specificity in some way.

Figure 8.5 illustrates the concepts of imprecision and inaccuracy in (non-spatial) temperature information sensed by a node. Note that imprecision and inaccuracy are *orthogonal*: imprecision and inaccuracy can occur together, or entirely independently of each other. Being independent does not mean that imprecision and inaccuracy are unrelated. One can usually make information more accurate by making it less precise. For example, at the time of this writing the statement "I am in Sydney" is inaccurate (I am in truth in Melbourne). However, the statement "I am in Australia" is at the same time less precise but more accurate. Similarly, *false precision* occurs when information is rendered inaccurate by an unwarranted level of precision. For example, when I was at school I learned that dinosaurs died out 65 million years ago. While this information can be considered accurate, it would be *inaccurate* (false precision) to say now, three decades later, that dinosaurs died out 65,000,030 years ago.

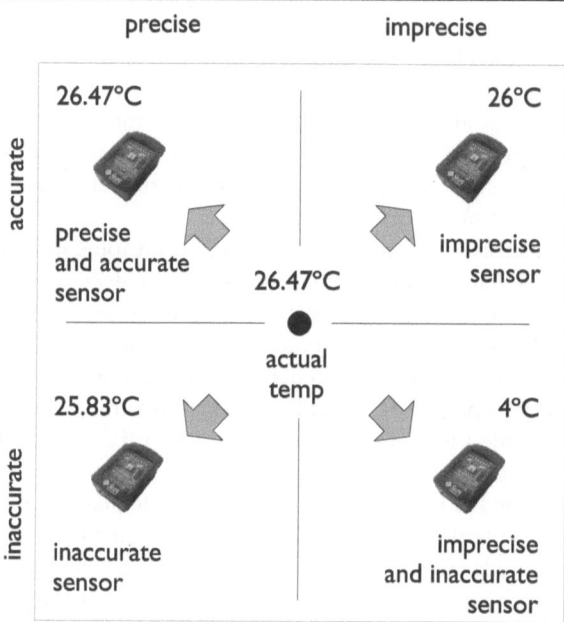

Fig. 8.5. Imprecision and inaccuracy are unavoidable features of information about the world

Returning to the common terms for describing uncertainty and imperfection mentioned above, these can almost always be mapped to some aspect, feature, or combination of inaccuracy and imprecision. Error, for example, is a synonym for inaccuracy—it concerns a lack of correspondence between reality and our information about reality. Granularity is the existence of *grains* in information, where individual elements cannot be discerned apart. Granularity arises therefore as a result of imprecision. Currency is in essence a statement about temporal accuracy: how close is an observation in the past to what would be observed now. Ambiguity concerns information that does not adequately distinguish between multiple possibilities, a form of imprecision.

An important example where a *direct* mapping to imprecision and inaccuracy may not be obvious is the issue of *vagueness*. Paradoxically, the term vagueness has a very precise meaning—it concerns the existence of *borderline* cases, where it is not always possible to say whether a statement is or is not true. Classic examples of vague spatial objects and relations include "mountain" and "near." For example, if I say "I am near the Eiffel Tower," there will exist some locations which we can regard as definitely near the Eiffel Tower (say, the Pont d'Iéna and the Champ de Mars). Other locations we can be sure are not near the Eiffel Tower, such as the CN tower in Toronto or Tower Bridge in London. But crucially there will exist locations for which it is indeterminate whether they are or are not near the Eiffel Tower (such as the Gare Saint Lazare). However, vagueness too is arguably another form of imprecision: this time not an imprecision in measured information, but an imprecision in categories used to classify information. The problem of vagueness arises because concepts like "mountain" and "near" are not precisely defined. In turn, this gives rise to uncertainty as to where the precise boundaries of those categories lie, and whether specific boundary instances are members of the category or not.

Vagueness in information is an important topic in its own right (cf. [62] for an excellent introduction to vagueness) and does from time to time arise in the context of decentralized spatial computing. In this book, we have skirted around the topic at several junctions. For example, geographic regions like "hot spots" are often vague. In many cases, it is not possible or not meaningful to identify a precise threshold that separates things inside the region from those outside the region. Similarly, the crisp (i.e., not vague) definition of a flock in §6.4 (as a group of n mobile objects that remain within a disk of radius r for a period of time k) arguably does not adequately capture the concept of flocking. In both cases it might be more reasonable to admit indeterminacy at the boundaries (i.e., locations for which it is not possible to say whether they are in or out of the hot spot, or movement patterns for which it is indeterminate whether or not they constitute a flock).

Despite the tantalizing possibilities, from the perspective of this book (designing decentralized spatial algorithms) it makes sense to leave further investigation of the issues surrounding vagueness to one side, and concentrate instead on the more basic forms of imperfection in sensed data: imprecision and inaccuracy.

8.2.2 Computing under Uncertainty

Imprecision and inaccuracy are endemic features of information captured, used, and generated by geosensor networks. No matter how closely nodes in a network are spaced, there will still be places at which no node is located. Thus, a geosensor network will always provide a lack of spatial detail about the monitored environment. Similarly, no matter how frequently a sensor is set to sample, changes in the monitored environmental parameter will occur in between consecutive sensor readings. In short, we also always lack temporal detail about the monitored world.

Turning to inaccuracy, the information generated by environmental sensors at different locations will normally be subject to inaccuracy (e.g., Fig. 8.5). Similarly, sensed spatial information, such as coordinate location from a

GPS, will itself be to a greater or lesser extent inaccurate. Distinguishing between these two inaccuracies in location sensors and environmental sensors may not always be possible (e.g., an accurate environmental sensor at an inaccurate location will typically be indistinguishable from an inaccurate environmental sensor accurately located).

Here again, we can expect an interplay between imprecision and inaccuracy. For example, localization techniques that rely on anchors are typically highly accurate. Your cell phone network provider can determine what your location is in terms of the nearest cell phone tower with high levels of accuracy; but typically this information is not especially precise and can only resolve your location to within a few hundred meters, in some cases maybe to within tens of meters. By contrast, your GPS-enabled smartphone can tell you where you are located to within a few meters; but typically this information may be subject to substantial inaccuracy, at times placing you several hundred meters from your actual location.

A further important feature of spatial information, when compared with some other types information, is that its inaccuracies are normally autocorrelated. Errors tend not to be randomly distributed through space. Instead, just like geography itself, inaccuracy in geographic information tends to be more closely related when the locations that the information refers to are closer together. Inaccuracies are similarly usually temporally autocorrelated.

It is worth highlighting that these characteristics are true for *any* geographic information, not just that used in a geosensor network. All geographic data capture technologies have inherent, inbuilt spatial imprecision and inaccuracy. Indeed, at least in the case of spatial and temporal granularity (i.e., the level of detail available in space and time), one might expect data captured by a geosensor network to be *more* precise than that captured using more conventional methods, such as remote sensing or ground survey (see §1.2.1).

Happily, we can again use the experimental tools we have already developed to investigate for a particular algorithm the likely effects of imprecision and inaccuracy.

8.2.3 Example: Topological Relations Between Regions

As an example of how to incorporate imperfection into an experiment, we will use Protocol 4.15: topological relations between regions. In this algorithm, the sensor nodes deployed across a region sense environmental variables in their vicinity (e.g., low soil moisture, high temperature), and use this to infer the 4-intersection topological relation between the regions exhibiting those characteristics (e.g., whether the region of low soil moisture was contained in the region of high temperature).

There are two types of imperfection these experiments will investigate: sensor inaccuracy, where a sensor can misdetect in which region it is located; and spatial imprecision, where the limited number of nodes covering the regions results in limited detail about the topological relation that holds be-tween the two regions. As is often the case, both sensor inaccuracy and spatial imprecision can lead to the same outcome: inaccuracy in the topological relation detected. For example, in Fig. 8.6 the square region is in actuality inside the circular regions. In Fig. 8.6a, however, the coarse spatial granularity of the network means that there happen to be direct one-hop connections between nodes inside both the square and the circular regions, and nodes outside both regions. Consequently, the network will detect the square region as covered by the circular region. Similarly in Fig. 8.6b, a node actually located in the circular region that inaccurately detects both the circular and the square regions gives rise to the same inaccurate detected topological relation, even though the granularity problem in Fig. 8.6a does not occur in Fig. 8.6b.

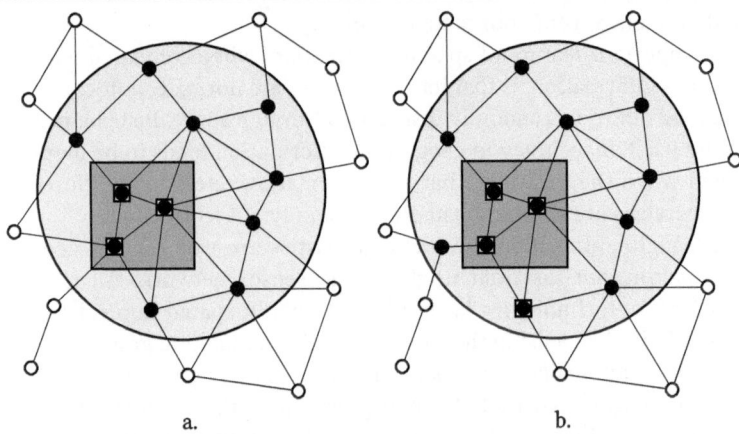

a.   b.

Fig. 8.6. Inaccuracy in detected topological relations (square covered by circle instead of square inside circle) resulting from a. node imprecision, and b. sensor inaccuracy

To investigate more systematically these kinds of effects resulting from sensor inaccuracy and spatial imprecision, a full factorial experiment using the following factors and levels was set up in NetLogo:

- *Sensor inaccuracy*: The level of sensor inaccuracy was varied across three levels, from none (probability of an inaccurate sensor reading $P(inaccurate) = 0.0$); low (probability of an inaccurate sensor reading $P(inaccurate) = 0.05$); and high (probability of an inaccurate sensor reading $P(inaccurate) = 0.2$). Further, the level of autocorrelation in the inaccurate readings was varied from absent (inaccurate readings were randomly selected from all possible sensor readings) to present (inaccurate readings were randomly selected from the actual readings of nearby locations).

- *Spatial imprecision*: The level of spatial imprecision was varied by changing the number of nodes monitoring the regions, from low (250 nodes),
  through medium (500 nodes), to high (1,000 nodes); by varying the network structure, over UDG, relative neighborhood graph (RNG), or Gabriel
  graph (GG); and by randomly generating regions themselves at a range of
  sizes, small, medium, and large.
- *Input topological relation*: Although the region generation process was
  randomized, the process was also blocked to ensure the generated regions
  were representative of the eight possible topological relations: overlap,
  meet, disjoint, equals, covers/covered by, contains/inside.

Each simulation setting was repeated 100 times. The experiments were
also paired such that the same sets of randomized node locations and region
configurations were used for the different levels of inaccuracy and network
structures. Altogether, this required a large experiment with six factors and
a total of 3 (probability of sensor inaccuracy) $\times$ 2 (autocorrelation) $\times$ 3
(network size) $\times$ 3 (network structure) $\times$ 3 (region size) $\times$ 8 (input topological
relation) = 1,296 treatments, with 100 repetitions for each treatment, leading
to 129,600 simulation runs.

When faced with a large, multifactor experiment like this one, there are
many possible effects that one might look for and find. In addition to the
effects of the six factors, we might also have effects for the $^6C_2 = 15$ possible
interactions between two factors, for the $^6C_3 = 20$ possible interactions
between three factors, and so on. In total there are $2^6 - 1 = 63$ possible
interactions we might look at! Consequently, the key to making sense of such
a large data set is to identify the subset of effects that are most important to
the behavior of the algorithm.

A good starting point is to look at the six factors individually, shown
using box plots in Fig. 8.7. In a box plot, five summary statistics are shown:
the sample minimum and maximum (the lines at the top and bottom of the
"whiskers" for each box); the upper and lower quartiles (delimiting the upper
and lower 25% of the data, shown by the extremes of the box itself); and the
median (shown by the band near the middle of the box).

In Fig. 8.7, the actual topological relation (Fig. 8.7e), the probability of
sensor errors (Fig. 8.7f), and the autocorrelation in sensor errors (Fig. 8.7a)
show clear effects on the response variable (the accuracy of the identified
topological relation). The region and network sizes (Fig. 8.7b and d, respectively) appear to show lesser effects, while no clear effect is evident
for network structure (Fig. 8.7c). In some cases, there may be data points
that are more than 1.5 times the interquartile range (i.e., the height of the
box), considered to be outliers. If so, outliers are indicated with circles (as in
Fig. 8.7e).

The next step is to perform an ANOVA (two-way with replication) for all
pairs of factors, and focus on understanding those where statistically significant patterns arise, and in particular two-way interactions. In general, more

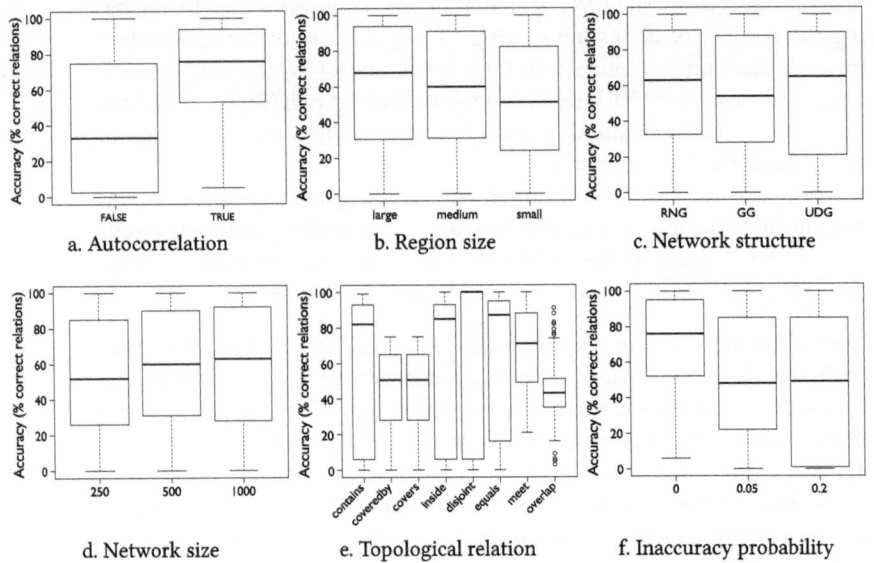

Fig. 8.7. Summary box plots for six factors used to explore accuracy of topological relations between regions (Protocol 4.15)

than two-factor interactions are rare, statistically problematic, and difficult to interpret. Consequently, restricting the investigation to at most two-factor effects is reasonable in most circumstances. The main effects are dominated by four factors: the probability of sensor inaccuracy, autocorrelation in sensor inaccuracy, region size, and the input topological relation. More specifically, the main effects of the six factors can be summarized as:

- *Probability of sensor inaccuracy*: Main effects in all cases, showing increased rates of error with increasing sensor inaccuracy.
- *Autocorrelation*: Main effects in all cases, showing reduced error rates with autocorrelation.
- *Region size*: Main effects in all cases, showing increased error rates with smaller region sizes.
- *Network size*: No significant main effects in any cases.
- *Network structure*: No significant main effects in any cases.
- *Input topological relation*: Main effects in all cases, showing overlap, covers, and covered by amongst the least reliable input relations; inside and contains as more reliable; equals the next most reliable; and disjoint and meet as the most reliable (cf. Fig. 8.7e).

However, turning to the interaction between factors, the picture does get more complex. Figure 8.8 shows interaction diagrams for those pairs of factors where the main effects and the interaction effects are statistically significant at the 1% level. In summary, these interaction effects are:

- between probability of sensor inaccuracy and autocorrelation in sensor inaccuracy, with autocorrelated inaccuracy masking the effect of increasing probability of inaccuracy (Fig. 8.8g and h).
- between input topological relation and autocorrelation in sensor inaccuracy (Fig. 8.8a and d). Although in most cases increasing autocorrelation in sensor inaccuracy improves robustness, in the case of the meet and overlap relations, increasing it increases algorithm error rates.
- between input topological relation and probability of sensor inaccuracy (Fig. 8.8b and e). As in the interaction between input topological relation and autocorrelation in sensor inaccuracy, increasing probability of sensor inaccuracy bucks the trend in the cases of meet and overlap relations, surprisingly *improving* overall robustness.
- between input topological relation and network size (Fig. 8.8f). Although network size shows no main effects, it does interact with the input topological relation, leading to improvements in robustness for covers/covered by, inside/contains, and overlap; little or no effect upon disjoint or equals relations; and an increase in error rates associated with the meet relation.
- between input topological relation and network structure (Fig. 8.8c). Network structure seems to produce a strong effect in the meet relation, leading to the highest accuracy for the UDG and lowest accuracy for the RNG; in the cases of most other topological relations, the sparser graphs appear to perform better.

At first sight the interaction effect between autocorrelation and probability of sensor inaccuracy might appear surprising, but perhaps also should be expected. In effect, spatial autocorrelation in sensor inaccuracy tends to make the algorithm substantially *more* robust to sensor inaccuracy. When the sensor inaccuracy is spatially autocorrelated (i.e., nearby sensors nodes are more likely to make the same errors), then the topological relation observed is more likely to be unchanged. Like everything else in geographic space, errors too tend to be autocorrelated, with higher errors clustering in similar locations. This inherent structure in information in space actually helps our algorithm, ensuring that greater autocorrelation in errors in most cases leads to greater algorithm robustness.

Looking into the data further, we see that three of the key interaction effects hinge around the meet relation. Unlike most other relations, meet exhibits decreasing accuracy with increasing network size, with sparser network structure, with increasing autocorrelation, and even with decreasing probability of sensor error. As already discussed, meet is a problematic topological relation when moving from continuous models of space to limited spatial granularity observations of space (cf. the discussion of the meet relation and discrete sensors on p. 124).

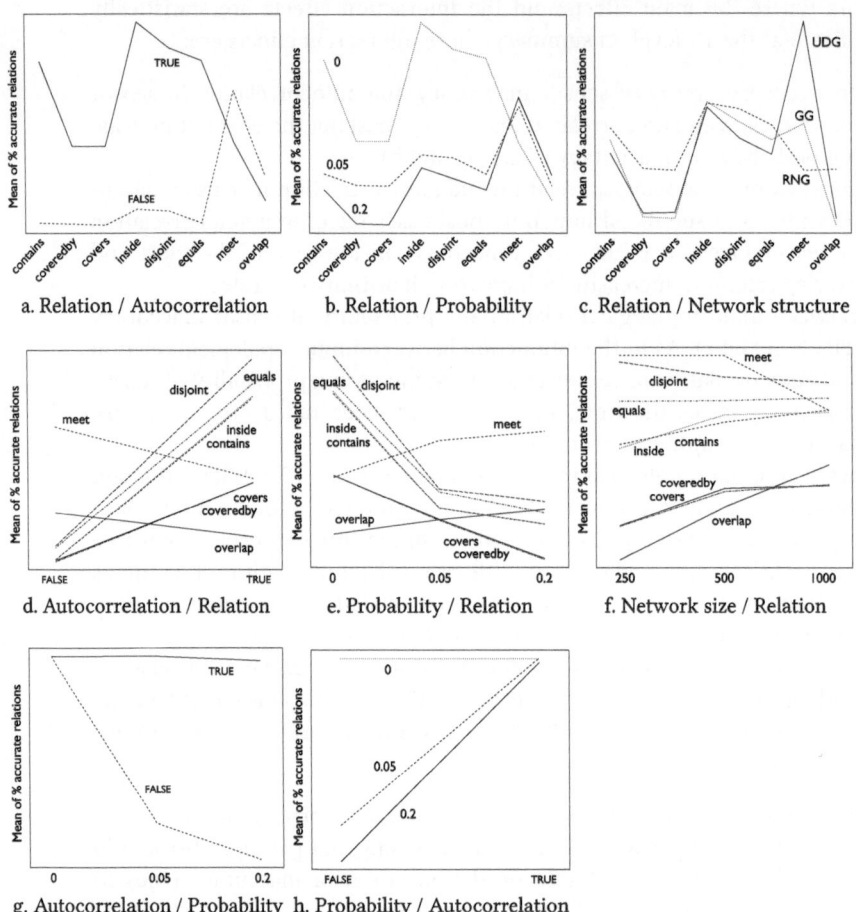

Fig. 8.8. Significant interaction diagrams for accuracy of topological relations between regions (Protocol 4.15)

More detailed investigations into the structure of errors between the different topological relations can be assisted by the use of a misclassification matrix, shown in Table 8.4. In the misclassification matrix, each row represents the topological relation *input* to the algorithm, whereas each column represents the topological relation *identified* by the algorithm. The number in each cell shows the percentage of the input topological relations (row) that the algorithm identified for each of the different possible topological relations between regions (column heading).

Over all the experiments, Table 8.4 shows that almost 25% of overlap, covered by, and covers relations were misclassified as meet. One potential cause of these errors is granularity effects. For example, at smaller (i.e., coarser spatial granularity) networks, it is more likely that some smaller sensed regions have no interior nodes. Correspondingly, it seems reasonable to infer it is also less likely that meet will be misclassified with smaller network sizes and smaller regions.

	overlap	disjoint	meet	equal	inside	contains	covered by	covers	Total
overlap	41.60	4.59	24.78	4.52	1.62	1.92	9.93	11.04	100
disjoint	0.20	68.72	30.56	0	0.28	0.19	0.03	0.01	100
meet	11.04	16.90	68.73	0.02	1.45	1.60	0.19	0.06	100
equal	16.93	1.27	10.83	63.6	0.00	0.00	3.73	3.64	100
inside	11.78	5.93	16.19	0.00	58.54	0.01	7.55	0.00	100
contains	12.05	6.51	15.91	0.01	0.00	58.45	0.02	7.06	100
covered by	12.86	4.50	24.60	9.60	4.16	0.00	43.68	0.60	100
covers	13.00	4.50	24.52	9.60	0.00	4.17	0.73	43.48	100
Total	119.46	112.92	216.13	87.35	66.06	66.34	65.86	65.89	

Table 8.4. Overall misclassification matrix for percentage of input topological relation (rows) identified as each of the possible topological relations (columns) between regions in experiments with Protocol 4.15

Similar effects can also account for the interaction between autocorrelation of sensor inaccuracy and meet and overlap topological relations (Fig. 8.8d). If the sensor inaccuracy is not autocorrelated, it is relatively more likely that any inaccurate sensors that falsely detect $A \cap B$ will be in locations that are remote from its true location. In turn, this will lead to regions that appear to have disconnected parts, something that Protocol 4.15 is not designed to deal with. Careful examination of the algorithm reveals that such disconnected regions are much more likely to be identified as meet or overlap. Thus, the algorithm has an in-built bias to detect, by chance, meet and overlap as opposed other relations as sensor inaccuracy increases and becomes less correlated.

8.2.4 Designing for Robustness

Given that imperfection is an endemic feature of all geographic information, in particular information captured by a geosensor network, how should we design decentralized spatial algorithms to allow for uncertainty? At root, there are three distinct strategies the designer can adopt when attempting to design robust algorithms:

1. Accept and report uncertainty

The first strategy is to accept the imperfection inherent in information, and simply report on the expected levels of uncertainty in our algorithms that result from it. For example, the experiment in the previous section yielded some empirical information about how reliable Protocol 4.15 is expected to be across a range of scenarios. From Table 8.4 we can infer that if the *detected* relation between two regions is "contains," then we can have high confidence that the actual relation between the two regions is also "contains" (in Table 8.4 "contains" is misclassified as some other relation in only $1 - 58.45/66.34 = 11.89\%$ of cases). By contrast, we may have relatively low confidence that something that is *not* detected as "contains" is in actuality not "contains" (in Table 8.4, regions that actually contained another were misclassified as something else in $1 - 58.45/100 = 41.55\%$ of cases). These kinds of figures could be used in applications of Protocol 4.15 as a guide to how reliable the algorithm is expected to be in different cases.

2. Resolve uncertainty

Different sensors in a geosensor network may often generate closely related observations of the same phenomenon. Although each of these observations will normally be subject to imperfection, combining multiple observations can potentially increase our certainty in the information about that phenomenon, effectively providing corroboration for observations. This approach is an example of *information fusion*, where the goal is to integrate multiple information sources to generate new information with increased usefulness or reliability [24]. For example, Protocol 4.15 uses data aggregation to discard all observations that duplicate known information. This approach eliminates any redundancy in communication, and so improves algorithm efficiency (see p. 123). However, at the cost of some efficiency, the algorithm could be adapted to forward some or all of the *repeated* observations to the sink. In turn, the sink node that collates the partially processed information from across the network could additionally use the *frequency* of a particular observation to improve the accuracy of the algorithm in some cases (see question 2).

3. Allow approximate and imprecise answers

A third strategy that may be used by the designer in the pursuit of robustness is to allow approximate and imprecise answers. As we have already seen, reducing precision will in general increase accuracy: the more is the detail in information, the harder it is to ensure correctness. Further, in many cases *less* precise answers are actually *more* useful, particularly when the less precise categories are chosen to coincide with meaningful concepts. A great many of the algorithms in this book already rely on this principle—that less is more. Characterizing continuous fields using thresholded regions (e.g., Protocols

4.14, 4.15, 5.7, and so on), surfaces using peaks and pits (e.g., Protocol 6.8), and groups of moving objects using flocks (e.g., Protocol 6.6), for example, was partly motivated by efficiency concerns, restricting communication to spatial structures such as boundaries and ridges. However, one can argue that information about the qualitative spatial structures is more useful than precise information about the entire environment, especially in the context of the risk of information overload inherent in geosensor networks. Even in cases where exact and detailed information is required, it is unlikely that this will be needed all the time. Instead, imprecise and approximate answers can be used to efficiently and robustly determine *when* the system may need to switch from low-detail filtering to high-detail monitoring. Thus, there is an attractive serendipity to allowing for imprecision and approximation in the answers generated by a decentralized spatial algorithm. Doing so can lead not only to increased efficiency, but also increased robustness and usefulness.

8.2.5 Example: Structure of Complex Areal Objects

As an example of the robustness that can arise from imprecision, we return to Protocol 5.12 for determining the structure of complex areal objects. The algorithm does require information about the coordinate locations of nodes, both to ensure a planar overlay network and to compute the area of region components. However, the algorithm is imprecise in the sense that it only generates the containment relationships between different region components in a complex areal object (see Fig. 5.13). The quantitative coordinate locations required by the area computation are only used to generate qualitative (imprecise) information about the orientation of the boundary cycle (a negative area for a clockwise orientation, a positive area for a counterclockwise orientation; cf. §5.4.2). The actual magnitude of the area is discarded by the algorithm.

Discarding area information may seem a waste after our having computed it. However, the magnitude of the area of each region component is expected to be highly sensitive to inevitable errors in the coordinate positions of nodes. Positioning systems are inherently inaccurate, and in the case of low power, low-cost geosensor networks are notoriously so. By contrast, the sign of the area is expected to be relatively robust to coordinate inaccuracy. As a result, Protocol 5.12 should continue to operate to high levels of reliability even as nodes' coordinate positions become more and more inaccurate.

To investigate this expectation further, Fig. 8.9 shows the results of an experiment on the robustness of Protocol 5.12 to inaccuracy in nodes' coordinate positions. The response variable in these experiments is the percentage of simulation runs for which the answer generated by the algorithm is in some respect incorrect (i.e., for which at least one containment relationships is incorrect or absent). A response of 100% of runs containing errors means that every simulation run in a treatment was incorrect in some way; a response variable of 0% of runs containing errors means that every simulation run in a treatment generated a perfectly correct query response.

The primary factor tested by the experiment summarized in Fig. 8.9 was the level of simulated inaccuracy in the nodes' positioning systems. Unlike in other experiments in this book, in this case the node locations were fixed in a grid pattern to ensure comparability between multiple simulation runs. The level of inaccuracy was then measured as the ratio of the maximum possible random perturbation in each simulation run to the distance between neighboring nodes in the grid (termed the "internode distance"). For example, a simulation run where the internode distance was 40 units and the maximum possible (randomized) perturbation in a node's knowledge of its coordinate position was 20 units led to a level of inaccuracy of $20/40 = 0.5$.

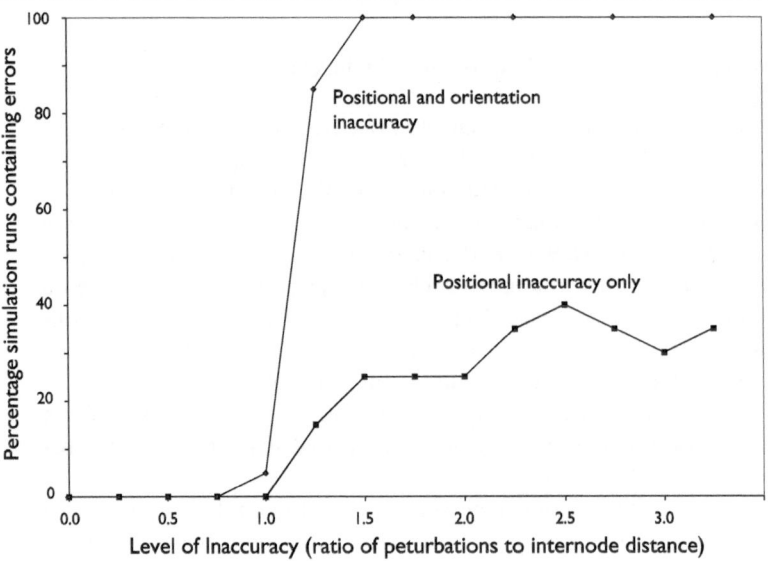

Fig. 8.9. Robustness to imprecision in Protocol 5.12, structure of complex areal object [35]

This primary factor was varied across 14 levels of inaccuracy, from a ratio of perturbation to internode distance of 0.0 (no inaccuracy) to 3.5 (inaccuracy of coordinate location up to three times greater than the distance between a node and its neighbors). At each level, 20 repetitions of a simulation run for a fixed complex areal object were performed. In addition, the entire experiment was repeated for two distinct cases. In the first case, only the coordinate position for a node could be in error, but the cyclic ordering of neighbors around each node was assumed to be correct (simulating the case where cyclic ordering of neighbors around a node is sensed independently of the positioning system, for example, using ultrasound direction finding technology). In the second case, both coordinate position and cyclic ordering could be incorrect (simulating the case where the cyclic ordering of neighbors around a node is computed from each neighboring node's coordinate position).

The results in Fig. 8.9 show remarkable robustness to inaccuracy. In the case where inaccuracy could affect both coordinate position and cyclic ordering, the algorithm continues to perform perfectly until the level of inaccuracy rises above 0.75 (i.e., the maximum random perturbation is 75% of the spacing between nodes, the internode distance). It then degrades rapidly until at an inaccuracy level of 1.5 the algorithm completely fails, providing no correct results. In the cases where inaccuracy only affects coordinate position (and not cyclic ordering), the algorithm is even more robust, degrading more slowly still.

In summary, these results illustrate the potential for increased robustness that can follow from decreased precision. In our specific example, monitoring only imprecise information about the containment relations between regions, as opposed to, say, the precise areas or centroids of regions results in substantial tolerance to inaccuracy in a node's knowledge of its coordinate position. As we have already seen, the algorithm is also efficient as a result of computing primarily at boundary structures. Further, one can argue that qualitative information about the spatial structure of a region, such as the containment relations between region components, may in many applications be more *meaningful* than quantitative information about each region's location or area.

8.3 Fault Tolerance

To this point, we have assumed a high degree of *reliability* underpinning our algorithms. Communication was explicitly assumed to be reliable (in the sense that messages sent would eventually be received) back in Chapter 4. Consequently, communication reliability is fundamental to all our algorithm designs. Implicitly, we have also assumed that *nodes* are reliable: our analyses did not consider that they might die, or, say, unexpectedly start executing a new protocol. In reality, of course, faults do occur. A *fault* here means some behavior other than that specified in our algorithm and its associated restrictions.[2]

It is normally worth distinguishing between the two distinct locations where faults occur:

- *Link faults* occur when communication between nodes fails in some way, for example, with messages becoming lost or corrupted.

---

[2] Note that by this definition we exclude the issue of *design* faults. Design faults may indeed occur, but identifying and correcting them is fundamentally the focus of our algorithm specification and design procedure, in particular the algorithm adversarial analysis and subsequent implementation and testing.

- *Node faults* occur when nodes themselves fail, for example, when some function of a node, or even the entire node, becomes inoperable.

Orthogonally, there are also two main *types* of faults that can be distinguished:

- *Stop faults* occur when a component (link or node) or function temporarily or permanently stops operating.
- *Byzantine faults* occur when a component is allowed to experience arbitrary failures, such as message payloads or headers becoming corrupted, or nodes arbitrarily changing state or triggering incorrect actions.

A fault in a single component, or even in a group of components, will not necessarily result in an algorithm failing. However, it is to be expected that at *some* level of faults *any* algorithm will fail. *Fault-tolerance*, then, is concerned with the extent to which an algorithm can continue to operate even in the presence of faults. While, as we shall see, it is possible to design highly fault-tolerant algorithms, it is worth beginning with a cautionary tale which illustrates how the introduction of faults, and specifically unreliable communication, can fundamentally change the nature of decentralized computing: the Two Generals' Problem.

8.3.1 Two Generals' Problem

The ramifications of unreliable communication are often illustrated using the "Two Generals' Problem." As the story goes, long ago two allied armies had trapped an enemy army in a valley. However, the enemy's forces were strong enough to defeat the two allies individually; only if the allies attacked together could they hope to win the battle. Both the allied generals knew this, and would only to attack if they were sure the other would too.

The two allied armies were camped on opposite sides of the valley. Being in a time before telegraph or radio communication, and not wanting to reveal their position with fires or lights, the two allied armies' generals had to rely on messengers for communication and coordination of their attack. But of course there is always the chance that messengers, who must run the gauntlet of enemy scouts and sentries, could be intercepted and killed. The situation is depicted in Fig. 8.10. The question is, then, how can the two generals coordinate their attack?

The first general sends out a runner to the second general with the message "Let us attack at sunrise." Happily, this message gets through, and the second general sends another runner back with the message "Message confirmed! We will attack at sunrise." Again, the message gets through to the first general without mishap. We might hope at this point everything is set for the battle. However, the second general does not *know* that her message got through, and so can't be certain that the first general will attack as planned. Not wanting to risk a massacre, she will want to hold back her

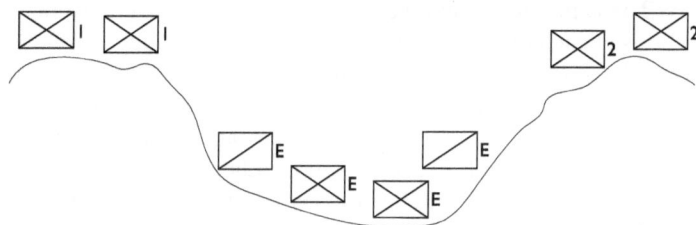

Fig. 8.10. The Two Generals' Problem. The smaller allied armies (1 and 2) of the two generals, on either side of the valley, can only hope to defeat the enemy (E) if they coordinate their attack using messengers that must get past the enemy sentries and reconnaissance (diagonal boxes)

forces until she is sure the first general has received the confirmation of the attack. Anticipating this, the first general sends out a third runner out with the message "Confirmation received! We will attack at sunrise." Again, the enemy scouts are sleeping at their posts, and this message gets through to the second general. But now the *first* general cannot know whether *his* new confirmation got through or not. So the second general must now dispatch *another* runner with *another* confirmation message.

The surprising conclusion of this story is that as long as the two generals need to be *sure* that the other will attack, no matter how many messages are sent, the attack will never occur. The two generals can never be certain they are in possession of the same knowledge. The analogy shows how common knowledge in a distributed system can never be guaranteed in the presence of unreliable communication. The striking feature of this analogy is that it doesn't even matter if all the messages are received—merely the *possibility* of unreliable communication is enough to prevent the establishment of common knowledge.

8.3.2 Example: Faults in Boundary Tracking

To illustrate the impact of faults, and the empirical examination of that impact, this section uses the example of the algorithm given in Protocol 6.1. This protocol is a naïve boundary tracking algorithm, with nodes periodically polling neighbors to determine whether they are at the boundary of a region. The experiment in this section investigates the response of the algorithm, in terms of boundary detection accuracy, to increasing link faults (both stop faults and Byzantine faults). Given a fault probability $P(fault)$ (primary factor), link stop faults can be simulated by randomly "dropping" messages with probability $P(fault)$. To simulate a Byzantine fault, our experiment randomly switches the payload of ping messages, again with probability $P(fault)$. The response variable is measured as the proportion of *actual* boundary nodes correctly *detected* by the algorithm as such.

The experiments summarized in Fig. 8.11 show the average of 100 repetitions at each of 11 different levels of fault probability (from 0.0 to 1.0). The error bars added to the data points in Fig. 8.11 show the magnitude of the standard deviation for each treatment. Our stop faults lead only to errors of omission, because in the absence of a message caused by a link failure, nodes in Protocol 6.1 default to detecting no boundary, and remain in the IDLE state. The Byzantine faults only lead to errors of commission. Errors of omission would require that all the neighbors of a true boundary outside the region suffer payload switches; but none of the neighbors inside the region suffer any payload switches. Because payload switches are random, such chance correspondences between node state and payload switches are very unlikely to occur (and indeed did not occur in any experiment).

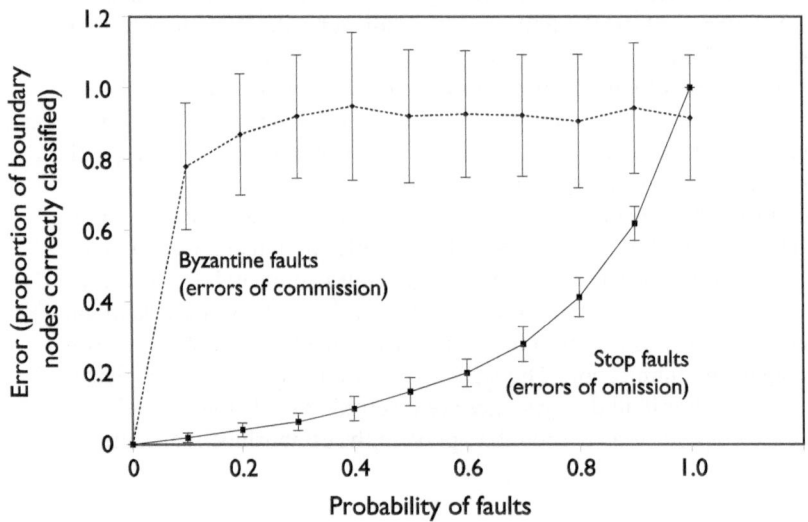

Fig. 8.11. Impact of stop and Byzantine link faults on Protocol 6.1. Response in proportion of correctly classified boundary nodes to failure probability, based on 100 repetitions per data point, for a 1,000-node UDG

Figure 8.11 shows that Protocol 6.1 is robust to stop faults, with relatively high link stop fault probability of up to $P(fault) = 0.4$ leading to on average less than 10% of boundary nodes undetected. By contrast, the Byzantine faults cause a total breakdown of the algorithm: with even relatively low payload switch probabilities, $P(fault) = 0.1$, as many non-boundary as boundary nodes are detected as boundary. Further increasing the fault probability has only limited impact on the algorithm errors. The size of the region and network used in the experiments resulted in the total number of nodes in the region being about double the number of nodes located at the boundary.

Hence, the proportion of errors caused by Byzantine faults is capped at around 1.0 (i.e., where all the nodes in the region are identified as boundary nodes, only about half of which are actually boundary nodes).

In summary, the example illustrates how different *types* of faults can lead to dramatically different algorithm behaviors and robustness.

8.4 Chapter in a Nutshell

Useful algorithms will normally need to strike a balance between efficiency on the one hand and reliability and robustness to uncertainty and faults on the other. Increasing robustness and reliability frequently requires additional computation and communication that reduces efficiency. An important exception to this rule occurs when approximate and imprecise answers are acceptable. Many of the algorithms in the book make full use of this fact. Focusing on *qualitative* spatial structures, such as boundaries, topological relations, and events, can often reduce communication overheads for an algorithm and increase the accuracy of results even in the presence of sensor inaccuracy and imprecision. The price of using qualitative spatial structures is that, by definition, qualitative spatial information is less precise, and so provides less information than quantitative counterparts. However, carefully chosen qualitative spatial structures that correspond to objects and relations that are meaningful in the application domain can nullify even this drawback—less is sometimes more.

Finally, simulation and experimentation becomes essential when exploring not just with the efficiency of an algorithm, but an algorithm's robustness to random errors or faults. Moving beyond OFAT to DOE enables this exploration, because the latter is well adapted to gathering information about the balances between multiple factors, and even the interaction between those factors.

Review questions

8.1 Design and execute an experiment to investigate the efficiency of the algorithm for flocking, Protocol 6.6. Specifically, your experimental design should investigate the influence of the factors of flock size, flock radius, flock duration, movement pattern, and overall network size on the response in terms of total messages sent.

8.2 In Table 8.4, regions that are in actuality disjoint are frequently misclassified as meet by Protocol 4.15 as a result of random sensor inaccuracy (i.e., when an isolated sensor happens to erroneously sense a part of $A \cap B$). Adapt Protocol 4.15 to filter out these isolated sensor misreadings, and rerun the experiment to determine empirically the effect of your adaptation. Perform an appropriate hypothesis test to ensure that any observed changes associated with the new protocol could not have happened by chance.

8.3 It is to be expected that the fault-tolerance of Protocol 6.1 is strongly related to the network connectivity (e.g., average node degree). Experimentally explore the fault-tolerance of Protocol 6.1 further, as in §8.3.2, to investigate this expectation (e.g., examine the effect on fault rates of varying the network structure or changing the network density by manipulating the communication radius c).

8.4 Design and execute an experiment to investigate the tolerance of Protocol 4.15 to different stop and Byzantine node and link failures. What are some of the most and least robust characteristics of that algorithm? Indeed, one could select any algorithm in this book and similarly investigate fault-tolerance—one might consider designing a "test suite" to check the fault-tolerance of any new algorithm.

8.5 Design and implement an extension to Protocol 4.11 to deal with the issue raised by the adversary in §4.4.1: that the root node cannot detect when the algorithm has terminated (i.e., when all nodes in the network are correctly updated with knowledge of the identity of the tree root). Determine either analytically or experimentally the additional communication overhead of your extension.

8.6 Design and run an experiment to investigate the scalability and robustness of rumor routing, Protocol 4.6, when compared with gossiping, Protocol 4.5. Your comparison should focus on the balance between scalability (i.e., the communication overheads associated with the two protocols) and robustness (i.e., how likely it is that the source node actually receives information about some environmental event). It may help to refer to [16] for further ideas on how to design the experiment.

Further Topics and Technologies

9

Summary: *The coming years will no doubt yield numerous technological advances that will help to make decentralized spatial computing more commonplace. This chapter addresses a selection of important, ongoing challenges that arise from the increasing ubiquity of geosensor networks and decentralized spatial computing environments. Some of these are challenges to the decentralized spatial algorithm designer, such as algorithm design with greater modularity, synchronous network design, and biologically inspired algorithm design. However, we end with a look at the challenges in supporting users of geosensor networks, both human users as well as automated devices and information systems.*

THIS book represents for me a beginning; there are many different endings still to be explored. This chapter aims to tie up a few loose ends, such as decentralized computing in heterogeneous or synchronous networks. But more importantly, it aims to show where some of the interesting "endings" might lie, including in the design of algorithms inspired by natural systems, in computing with data streams, and in human interaction with spatiotemporal sensor data.

9.1 Modularity

Chapter 5 introduced modularity into the protocol specification, so that a sophisticated algorithm could be constructed from simpler components. After its introduction, many of the subsequent protocols rely on modularity to decompose complex protocols into more manageable fragments. Indeed, it is difficult to imagine a useful algorithm design technique without modularity; breaking hard problems into smaller solvable parts is a prerequisite to effective algorithm design. Consequently, it is worth having a section

M. Duckham, *Decentralized Spatial Computing*, DOI 10.1007/978-3-642-30853-6_9,
© Springer-Verlag Berlin Heidelberg 2013

examining the characteristics of our pragmatic approach to modularity, and to considering where other approaches could lead.

The approach to modularity used in this book is simple: to allow specialization of protocols but never generalization. More specifically, in writing a new protocol that extends an existing protocol, we can always:

- Add more restrictions: If a protocol works in one computational environment, it will also work in another more restrictive environment that is a special case of the original. For example, a protocol that assumes only neighborhood-based ($\mathcal{NB}$) restrictions can always be extended by requiring location-based ($\mathcal{LB}$) restrictions, since $\mathcal{LB}$ restrictions are a special case of $\mathcal{NB}$ restrictions with the *addition* of the restriction of a planar communication graph (as opposed to a possibly non-planar communication graph). Conversely, a derived protocol can never safely remove or relax the restrictions of the protocols it extends (e.g., a protocol that assumes $\mathcal{LB}$ restrictions can never be extended with one that assumes $\mathcal{NB}$ restrictions).

- Add new local data: Adding new local data to a derived protocol will have no impact on the communication complexity of the protocol being extended, and so is always allowable.

Also, in the approach to modularity adopted in this book, we additionally allow:

- Add new states and transitions: Derived protocols are allowed to add states and/or transitions to the original state transition system. Conversely, removing states or transitions is not permissible. Further, it is not, in general, allowable to change the initialization conditions of a protocol, except where the derived initialization can be regarded as a special case of the original initialization conditions. For example, Protocol 5.12 (for determining the topological structure of a complex areal object) extends the initialization conditions of Protocol 5.7 (for determining the inner boundary nodes and cycle for a region). The derived protocol changes the original initialization conditions from "all nodes in state INIT" to "all nodes in state INIT except one node in state ROOT." The ROOT node performs one spontaneous action before transitioning directly into the INIT state (i.e., satisfying the original initialization conditions of the protocol being extended).

- Add new events and new actions: In our simple model of modularity, new events and actions can be added. By default, the actions for any events not specified in a protocol are assumed to be the empty action (i.e., "do nothing"). Therefore, adding a new action can also be regarded as replacing an empty action with a nonempty action for a particular event.

It is at this point worth noticing that these latter modifications, adding new states, transitions, events, and actions, can fundamentally change the operation of an algorithm. As soon as new events and actions are permitted, the analysis and properties of the protocol being extended may no longer hold true. For example, we might add to each of the existing states a spontaneous event with an associated action that simply transitions the node into a terminal state. In this case, the derived protocol will perform no useful task and have $O(1)$ communication complexity, regardless of the complexity of the original protocol. At the other extreme, we might add a spontaneous event with an associated action that transmits a msge message; and a *Receiving* msge event with an associated action that retransmits the received msge message. In this case, the derived protocol will most likely again perform no useful task, but may now have unbounded communication complexity, never terminating.

Thus, the approach to modularity taken in this book is purely syntactic: providing a sensible framework that makes *writing* extended protocols more manageable. The approach, however, does not assist in *analyzing* extended protocols. It is not possible to infer the properties of a derived protocol with new events and actions from the protocol it is based on. Instead, such a derived protocol must be analyzed and tested *as if it were a new protocol.* It is *likely* that the prior analysis of an existing protocol will help in the analysis of a derived protocol, but this cannot be *relied* upon.

There do already exist some alternative, formal approaches to modularity that could assist in analyzing modular decentralized algorithms. For example, the formal notion of object-oriented data types and object-oriented inheritance in [3] has direct parallels with modularity in decentralized algorithm design. Potentially, such formal models could help us define rules for modularity that go beyond purely syntactic mechanisms for modularity, additionally safeguarding the properties of the derived protocol. However, as already discussed in §3.1, there is a trade-off in decentralized algorithm design between formal rigor and practical ease of use. The design techniques set out in this book, including the approach to modularity, have consistently preferred practicality to formal power where the two cannot easily be reconciled.

Adopting a syntactic, as opposed to a formal, analytical approach to modularity, we also allow derived protocols to override existing nonempty actions with new actions (using the *Updates* keyword). For example, in routing around a boundary cycle, Protocol 5.8 extends the action for the IDLE/*Receiving* ping state/event pair in Protocol 5.7 with the addition of a new msge message sent to neighboring boundary cycle nodes. As in the case of adding new events and actions (which as mentioned can be treated as a special case of updating an empty action), overriding actions can fundamentally change the behavior and properties of an algorithm.

This syntactic approach to modularity can also be used for merging existing protocols. In these cases, it may be necessary to rename states, messages, or local data where namespace clashes exist (e.g., where two protocols have two distinct IDLE states). However, it is usually necessary to override some actions when merging protocols at those points where the original protocols must interact.

9.1.1 A Note on Node Heterogeneity

All of the algorithms explored in this book have been homogeneous, in the sense that the behavior of the system is specified with a single protocol. Even in cases where the nodes have different capabilities and take on different roles in the computation, a single protocol has sufficed. Yet, heterogeneity is a strong assumption. It seems certain that the future will see highly heterogeneous networks, comprising nodes with different capabilities, taking on different roles, and, in practice, based on different platforms and technologies. For example, §6.3.1 explored the movement of objects through a transportation network monitored by checkpoints. The associated algorithm, Protocol 6.3, only involved the checkpoints in that computation. But more sophisticated algorithms could also enlist the assistance of the mobile objects, for example, taking advantage of the mobility diffusion principles (see §6.3.2). In such cases, don't we need a more sophisticated algorithm design process, able to combine different protocols operated by different types of nodes?

Happily, the answer to this question is "no!" An important result from distributed systems tells us that no matter how many different types of nodes we have in a network, how many different types of behaviors are required, it is always possible to define a single, homogeneous protocol that can satisfy those requirements [99]. The intuition behind this result can be glimpsed by noting that the state transition system need not be *connected*, in the sense that some states may only be accessible from certain initialization states. Thus, it may be that two nodes starting in different states may exhibit completely different, heterogeneous behaviors based on the same homogeneous protocol. In short, as we have already seen in several cases, having a *single* protocol does not mean that all nodes executing that protocol must behave in the *same* way.

9.2 Synchronous Networks

All of the algorithms we have explored in the previous chapters make only minimal assumptions about the *time* taken by a protocol. The design process does require that actions be atomic and "terminate within a finite amount of time" (p. 61). And all the algorithms we have encountered assume *reliable* communication "where messages sent will always arrive within a finite amount of time, although necessarily in the order they were sent" (see p. 86, although of course the empirical investigation of fault-tolerance in §8.3 did illustrate how the effects of unreliable communication can dramatically affect the behavior of a protocol).

Networks that make no assumptions about time, other than that messages and actions are delivered within a finite amount of time, are termed (fully) *asynchronous*. Thus, in a fully asynchronous system, the clock time at which communication occurred, how much time communication or processing required, or whether two nodes performed actions at the same time may be of *interest*, but do not fundamentally affect the process of computation. In contrast, some problems do require algorithms that rely on stronger assumptions about time, termed *synchronous* networks. Some of the increasingly restrictive assumptions that can also be made about time (summarized in Fig. 9.1) include:

- *Message ordering*: The assumption of reliable communication implies that messages may overtake one another en route: it is possible that a message that arrived first may have been sent *after* one that arrives later. The protocols in this book allow this possibility. Indeed, to ensure robustness, the NetLogo simulations of the protocols are all programmed so that message ordering is not preserved, and overtaking will be a frequent occurrence. In some cases, it may be useful to make a stronger assumption about message ordering, that messages arrive in the order in which they were sent—in other words that the communication channels between pairs of nodes are FIFO (first in first out).
- *Bounded communication delays*: A closely related assumption about timing in networks is that the time taken to deliver a message is at most some known value Δ. By ensuring that no node ever sends two messages closer that Δ, a distributed system with bounded communication delays can also guarantee message ordering.
- *Synchronized clocks*: In addition to making assumptions about the communication order and delays, it is sometimes useful to assume that the clocks across the network are all synchronized. Clock synchronization requires either that nodes in the system all increment by one unit at the same time (often termed synchronized clock *rates*—note that in this case the actual *times* shown on the clocks may all be different) or that all system clocks tell the same time (full clock synchronization) [75].

Together, a distributed system that assumes bounded communication delays and synchronized clocks is termed a (fully) *synchronous* system. The need for precise timing and synchronization is sometimes a little overstated in the research literature on the topic (for instance, "unless the clocks in each machine have a common notion of time, time-based queries cannot be answered" [108]). As we saw in Chapter 6, queries about the changes over time are clearly answerable, even though the algorithms specified were fully asynchronous, with no common notion of time. In much the same way that neighborhood-based algorithms can detect spatial properties without any geographic coordinates, spatiotemporal algorithms can monitor change without requiring quantitative timestamps.

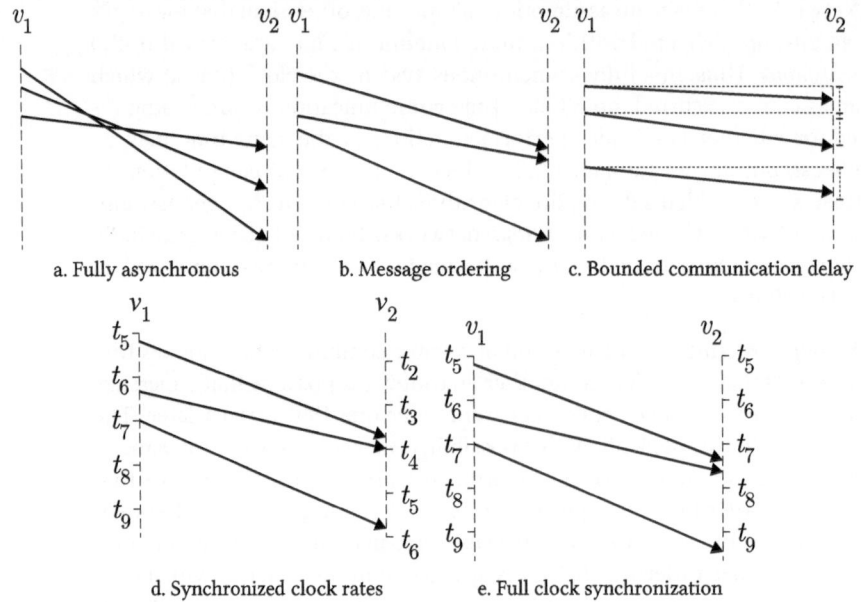

a. Fully asynchronous　　　b. Message ordering　　　c. Bounded communication delay

d. Synchronized clock rates　　　e. Full clock synchronization

Fig. 9.1. Increasingly restrictive assumptions about synchronization in a network

However, there are reasons why precise times are desirable or even essential to some applications, including those in which queries explicitly require time (e.g., "At what clock time did the region split" or "What was the speed in mph of the vehicle") or in order to assist with efficient duty cycling or scheduling [95, 103]. Perhaps more significantly, synchronization can afford important opportunities for coordination amongst nodes, helping with issues such as data consistency and efficiency (e.g., with at-most-once message delivery schemes [75]). Synchronous algorithms are in fact not subject to the same fundamental complexity constraints as asynchronous algorithms. For example, a synchronous version of the "as far" algorithm for leader election in a ring (§4.4.2) could preferentially speed messages with smaller identifiers around the ring, delaying messages with larger identifiers. Depending on the precise cost function used to delay messages, it is possible to use this approach to design (worst case) $O(n)$ communication complexity synchronous variants of the "as-far" protocol, in contrast to the $O(n^2)$ asynchronous version [99].

Achieving efficient clock synchronization in sensor networks is an ongoing research challenge. The overwhelming majority of computing devices, such as the nodes in a sensor network, rely on a crystal oscillating at known frequency to measure time. However, these devices are all subject to *clock drift* (more correctly, if less transparently, also termed *clock skew*), where over time the rate of oscillation varies slightly. Even for clocks that were initially operating at exactly the same frequency, changes in supply voltage and the environment (e.g., temperature) will lead to clock drift [95]. As a result of clock drift, geosensor nodes cannot usually be assumed to have synchronized clock rates. In turn, even if all clocks were synchronized before deployment, it would usually not take long for the clocks to fall out of synchronization.

GPS can of course be used to generate accurate times. GPS satellites have on board atomic clocks, which rely on a completely different physical mechanism than crystal oscillators in electrical circuits, and are highly accurate. Consequently, in applications where sensor nodes are fitted with GPS receivers, clock synchronization comes "for free." However, as we have already seen, cost, energy limitations, and adverse environmental conditions (such as signal attenuation in indoor environments, under water, or in dense vegetation) mean that GPS cannot be assumed to be available in many cases.

Where clock synchronization is required and GPS is unavailable, many alternative approaches and algorithms have been proposed. A very simple example of a synchronization protocol for two neighboring nodes is *round-trip synchronization*. In round-trip synchronization, a node n_1 sends a message to a neighbor n_2 asking for its current timestamp, t_2^c. When node n_1 receives the response to this message, it can at least bound $t_2^c \in (t_1^s, t_1^r)$, where t_1^s and t_1^r are the timestamps at n_1 of sending and receiving the message, respectively. Repetition of this process yields a set of samples of time differences, which in turn can be used to estimate the relative clock drift of the two nodes (for example, using regression). More sophisticated protocols still can be used to synchronize parts of or entire networks, rather than just neighbors.

The example of round-trip synchronization illustrates two important principles of clock synchronization in sensor networks. First, clocks in a sensor network are never *precisely* synchronized. The variable delays in communication between nodes mean that the goal of synchronization is to reduce the time differences to a point where they can no longer affect the execution of an algorithm, but not to completely eliminate time differences. Second, clock synchronization is typically expensive in terms of communication efficiency. Round-trip synchronization requires $O(|E|)$ communication complexity (recall: in the worst case $|E|$ is proportional to $|V|^2$, although $|E|$ is proportional to $|V|$ often in sparse networks and always in planar networks). Repeated rounds of clock synchronization are of course necessary for nodes to remain in synchronization, and prevent clock drift from again taking over.

Rather than adopting traditional approaches to clock synchronization, sensor network synchronization often adopts the principle of *post-facto* synchronization. In post-facto synchronization, routines for synchronizing nodes are only run when needed, in response to some event or stimulus. In normal operation, nodes run at their own pace, only synchronizing timestamps of received or stored retrospectively, when it becomes clear this extra effort is necessary [39].

Clearly, all the (asynchronous) algorithms in this book will operate just as well in synchronous systems—our asynchronous algorithms make fewer assumptions about the computational environment in which they operate than their synchronous counterparts. But, as we have seen, it may be that more efficient synchronous variants of many algorithms exist. Given that many spatial algorithms do rely on coordinate location, synchronization can also be assumed for nodes that derive these coordinates from GPS.

9.3 Biologically Inspired Computing

One of the fundamental questions posed by decentralized spatial algorithm design was how to design local protocols that exhibit desirable global behaviors (see p. 80). There is a major class of systems that provides a wealth of examples where the local interactions between distributed agents can lead to robust and highly sophisticated global behaviors: biological systems [13]. *Biologically inspired computing* is made up of a collection of computational problem-solving heuristics that are inspired by biological systems such as social insect behavior and biological processes such as evolution.

Biologically inspired computing is closely related to artificial intelligence (AI). Like artificial intelligence, biologically inspired computing is well suited to addressing computationally intractable problems, where the solution space is too large to explore exhaustively. In genetic computing, for example, large solution spaces are efficiently explored using a process of evolution. The most successful solutions from a population are repeatedly combined to form new populations of solutions, which become increasingly well adapted over multiple generations.

Within biologically inspired computing, the technique with the greatest relevance to decentralized spatial computing, and in particular to geosensor networks, is *swarm intelligence* (SI). Swarm intelligence has much in common with agent-based computing and cellular automata. It is concerned with the emergent behavior of large numbers of individual agents that can react to their immediate environment and communicate with nearby neighbors. The biological inspiration for swarm intelligence can be seen most dramatically in social insects, such as bees, ants, and termites.

For example, ant colony optimization (ACO) takes ant behavior as the inspiration for a range of problem-solving algorithms. In an unexplored space, ants move in a relatively random fashion (and are therefore called *blitz* ants). However, as ants move they leave a trail of pheromones that evaporate over time. Ants that subsequently stumble across pheromone trails are more likely to follow the trail than to strike out on a new path. This process where agents leave a trace in the environment that can then influence the behavior of other agents is termed *stigmergy*. For large numbers of ants over time, stigmergy leads to more direct paths being reinforced, with less direct paths being progressively weakened.

Using these principles, ACO has been applied to a range of combinatorial optimization problems, including to the traveling salesperson problem (TSP) and other vehicle routing and scheduling problems [30]. To illustrate the potential use of ACO in geosensor networks, Protocol 9.1 presents a biologically inspired algorithm for identifying *convoys* in groups of moving objects. A convoy is defined as a set of n or more objects, where there exists some sequence of nodes each separated by a distance of less than r that connects any pair of nodes in the set [57]. Equivalently, consider the UDG with unit distance r induced on some set of objects in the plane. A convoy is then any connected component of the resulting UDG containing at least n nodes. Like their movement pattern cousins, flocks, convoys are usually spatiotemporal objects that can be defined to persist for some period of time k. Figure 9.2 shows an example of an nkr-convoy with direct comparisons to the flocks discussed in §6.4.

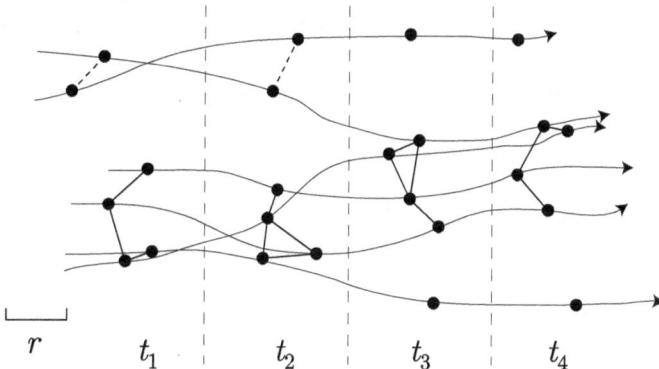

Fig. 9.2. An example nkr-convoy, with convoy size $n = 4$, convoy duration $k = 4$ time steps, and convoy distance r (cf. flocks in Fig. 6.5)

As with flocks, the combinatorial nature of convoys makes them expensive to compute. Instead, Protocol 9.1 uses a simplified version of ACO to identify convoys. The general idea behind the algorithm is:

- Blitz "ants" are messages that explore the network, keeping track of which nodes they have visited in a table. At each step, the next move of a blitz ant is randomly selected, but biased towards one-hop neighbors with higher pheromone levels.
- When a blitz ant reaches n hops, where n is the required size of the convoy, it returns to its origin along the path on which it arrived. At each

step on its return trip it increments the pheromone level of nodes along its return path by 1.

Over time, dead-end routes and cycles should become less favored; more direct routes along the spine of a convoy will tend to be more favored. By identifying more direct routes, the approach is expected to be substantially more efficient in terms of messages sent than a brute-force approach (for example, at each timestep build a tree using multiple initiators as described in §4.4.1, and then count the number of nodes in each tree using, say, the TAG algorithm in Protocol 4.10).

For simplicity, Protocol 9.1 assumes that the communication distance c equals the convoy distance r, and the required convoy duration k is just one timestep. However, it is relatively straightforward to relax these assumptions, at least for the former case where $c > r$ and nodes have positioning or range-finding systems that can provide information about the distance to neighbors.

In understanding Protocol 9.1, it is worth highlighting:

- At each time step, every node broadcasts to its neighbors its pheromone level in a pher message. Neighbors receiving this pher message update their stored data about neighboring nodes' pheromone levels in their np table. Nodes also delete from their np table any information about past neighbors that have moved out of their neighborhood since the last time step.
- The blitz ant chooses a neighbor to move to based on the pheromone information in the np table using the *choose* function. The *choose* function takes as input the neighbor node identifiers and the pheromone levels of neighbors contained in np not already visited by the blitz ant (if any). The *choose* function returns the identifier of one of those unvisited nodes in np, randomly selected based on a distribution biased by the pheromone levels.

A range of *choose* functions might be designed, but a simple example is given below:

$$choose(b) \mapsto m \in \text{SELECT } nid \text{ FROM } b \text{ where } P(m) =$$
$$\text{SELECT } pher/(\text{SELECT SUM } pher \text{ FROM } b) \text{ FROM } b \text{ WHERE } nid = m$$

where b is a nonempty table of the form

$$\langle nid : \mathbb{N}, pher : \mathbb{R} \rangle$$

and $P(m)$ is the probability of selecting node m from table b.

In short, given a set of N unvisited neighbors, which together have the total pheromone level of p_t, the probability of an ant visiting a single neighbor n with pheromone level p_n will be p_n/p_t. Many other *choose* functions are possible, and indeed may be more desirable.

Protocol 9.1. Ant colony optimization for monitoring convoys

Restrictions: reliable communication; $G(t) = (V, E(t))$; identifier function $id : V \to \mathbb{N}$; distance $dist : E \times T \to \mathbb{R}$; neighborhood function $nbr : V \times T \to \mathbb{N}$, where $nbr(v, t) \mapsto \{id(v') | \{v, v'\} \in E(t)\}$; convoy parameter n (minimum number of objects in a convoy); $choose : \langle nid : \mathbb{N}, pher : \mathbb{R} \rangle \to id_*$

State Trans. Sys.: $(\{\text{IDLE}\}, \varnothing)$

Initialization: All nodes in state IDLE

Local data: Pheromone neighbor table $np = \langle nid : \mathbb{N}, pher : \mathbb{R} \rangle$, initialized with zero records; level of pheromone on node lp, initialized $lp := 0$

IDLE

  When time trigger elapsed
    set *pher* = *pher* * 0.8   *#Pheromones degrade over time*
    broadcast (pher, lp,$\overset{\circ}{id}$)   *#Broadcast current pheromone level to neighbors*
    DELETE FROM np WHERE nid NOT IN $nbr(\overset{\circ}{now})$ *#Discard info about old neighbors*
    let $q := choose(np)$ *#Pick next neighbor*
    let b be CREATE TABLE b (*step* $\mathbb{N}$, *nid* $\mathbb{N}$)   *#Create new blitz table*
    INSERT INTO b VALUES (0,id)   *#Populate blitz table with first record*
    send (bltz, b) to neighbor with identifier q *#Start blitz ant travels*
  Receiving (pher, l, i)
    DELETE FROM np WHERE $nid = i$ *#Remove old record for neighbor i if any*
    INSERT INTO np VALUES (i, l)   *#Insert new record for neighbor i with pheromone level l*
  Receiving (bltz, b)
    let h be SELECT MAX (*step*) FROM b *#Retrieve current length of blitz ant path*
    if $h \geq n - 1$ **then**   *#Check if blitz ant traveled > n hops*
      let q be SELECT nid FROM b WHERE $step = h$ *#Find final node on the path of this ant*
      set $lp := lp + 1$ *#Increment pheromone levels*
      send (antx, b, h) back to neighbor with identifier q *#Return ant along path*
    else
      if EXISTS SELECT * FROM np WHERE nid NOT IN (SELECT nid FROM b) **then**
        let $q := choose$(SELECT * FROM np WHERE nid NOT IN (SELECT nid FROM b)) *#Pick neighbor*
        INSERT INTO b VALUES ($h + 1$, id)   *#Add this node to blitz ant path*
        send (bltz, b) to neighbor with identifier q *#Continue blitz ant travels*
  Receiving (antx, b, h)
    DELETE FROM b WHERE $step = h$ *#Remove last record from b*
    if $h = 0$ **then**
    Convoy found by ant $\overset{\circ}{id}$ at time *now*!
    else
      set $h := h - 1$ *#Decrement hops to start*
      let q be SELECT nid FROM b WHERE $step = h$ *#Find next node on return path of this ant*
      set $lp := lp + 1$ *#Increment pheromone levels*
      send (antx, b, h) back to neighbor with identifier q *#Return ant along path*

In summary, there are natural synergies between biologically inspired and decentralized spatial computing. This section has only attempted to scratch the surface; but there are a great many other opportunities still to be explored (and see questions 9.3–9.5).

9.4 Users

This book has focused on the design of efficient and reliable decentralized spatial algorithms. In keeping tightly to this focus, the question of who or what *uses* this information, and how those uses are supported, has deliberately not been directly tackled. However, this question is clearly of fundamental importance in the practical application of decentralized spatial computing to real-world problems, and more broadly the vision of ambient spatial intelligence (§1.4). In this section, we start at the bottom by looking specifically at the role of *queries* in a decentralized spatial information system, such as a geosensor network. But we also take a longer view, pointing the way to several other strands of ongoing research that have relevance to how data from sensor networks can be shared and used by human users or other automated processes.

9.4.1 Query Models

Several of the algorithms in this book take the form of a query—an unambiguous statement of the information to be retrieved from the information system. For example, Protocol 4.15 can answer queries about the 4-intersection topological relation between two regions. Similarly, Protocol 5.12 can satisfy queries about the containment relations between connected components of complex areal objects. In these cases, one can imagine a person submitting the query to a sink node in the network, in much the same way that the same basic queries might be submitted to a traditional, centralized spatial information systems, such as a GIS or a spatial database.

However, many of the algorithms we have encountered do not fit this traditional idea of a query—a request for information—so closely. Protocol 4.14 enables nodes at the boundary of a region to identify themselves. But the protocol does not specify how this information might be communicated to a sink node or a user. Building on these earlier techniques for identifying boundaries, Protocol 6.9 enables specific nodes to identify the events that occur in monitored regions. But again, the protocol does not specify how the information generated is to be *used.*

The lack of a traditional query is a frequent feature in decentralized spatial computing. As first indicated in §6.3.1, it often makes sense to separate the algorithms for *maintaining* information about the changing state of the environment from the explicit retrieval and querying of that information. One reason for this is that in a decentralized spatial information system, such as a geosensor network, there is often no single, simple model of how the information will be used. Consider the following scenarios for the use of information captured by a geosensor network monitoring environmental parameters such as temperature, humidity, and wind speed as part of a bushfire warning system in a high-risk part of rural Australia:

- An area commander sitting remotely at his computer in a Melbourne Country Fire Authority (CFA) office submits a query for information about wind speed in the monitored area;
- A fire fighter located in the area monitored by the network requests information about humidity in her immediate vicinity;
- A remote fire modeling system requests information about temperatures across the monitored region, to combine with other weather and environmental data sets as part of a simulation of bushfire hazard; and
- A digital camera attached to the network and located in the monitored region queries the network for information about environmental conditions in order to initiate video surveillance in response to atypical readings.

These examples were chosen because they illustrate that queries can originate remotely, external to the network, or within the network itself. Queries may also come from human users, or from other information systems or devices. In fact, the situation can be more complex still. In the examples above an implicit assumption is that the entity submitting the query is also the entity that requires the response. However, it is also possible that queries initiated by one entity may require responses delivered to another entity, such as when the area commander requests information from the network to be sent directly to the fire fighter in the field. Further, the types of queries submitted in these examples could fall into one of two different categories [14]:

- Pull queries: In a pull query, a one-off request for information about the current or past state of the monitored environment requires an immediate response. For example, the area commander might submit the query: "Show me on a map the current wind speeds." Alternatively, the fire fighter might submit the query: "Show me on a map the average humidity over the past week."
- Push queries: In a push query (also called a *long-running* query) the query remains resident in the network, proactively generating responses based on predefined triggers. For example, the area commander's query might be "On the hour, show me the latest wind speed" or "Alert me when any node senses winds stronger that a 'high wind' on the Beaufort scale."

These issues are not unique to decentralized spatial information systems. The balance between push and pull strategies, for example, is an important issue in location-based services [41]. Nevertheless, there is no reason to assume a priori that queries to a decentralized spatial information system will be either push or pull; originate from humans or automated devices; or have sources and sinks that are in-network or remote. Instead, each of these different possibilities is likely to be relevant in some situations. Consequently, it is in general sensible to decouple the mechanisms for responding to queries from the mechanisms for maintaining processed information about the monitored environment (cf. the discussion of tracking mobile objects in a cordon-structured network, Protocol 6.3, on p. 182).

Despite these inherent challenges, it should be noted that active research is making progress in defining consistent query languages for sensor networks, using the analogy of conventional, centralized databases. The concept of the "network as a database" [50] is intended to help manage the complexity of large and heterogeneous sensor networks by treating networks as "virtual databases" capable of being queries using extended versions of standard query languages such as SQL. Of course, if the queries specified are to be satisfied efficiently, in-network decentralized algorithms for computing atomic query operators are still necessary. Indeed, the challenges in developing the atomic decentralized (spatiotemporal) query operators required to underpin a simple and consistent (spatiotemporal) query language contributed to the genesis of this book.

9.4.2 Data Stream Processing

At this point, it is worth introducing an alternative approach to querying sensor data that starts from a radically different conceptual basis to decentralized spatial computing: data stream processing. While the "network as a database" idea aims to apply tried-and-tested database concepts to sensor networks, data stream processing starts from the premise that traditional database concepts are inadequate for managing data from sensor networks. Specifically, in a sensor network, new data is continually being generated, whereas traditional databases are typically updated relatively infrequently; queries in a sensor network are typically long-running (push) queries, while databases are well adapted to one-off (pull) queries; and different sensors and sensor networks are highly heterogeneous data sources, whereas databases are most useful when a fixed and precise database schema can be defined.

Given these differences, data stream processing uses the analogy not of a database, but of continuous and unbounded streams of data channeled through a purpose-built information system for managing and querying these streams, termed a data stream management system (DSMS). The data flowing into a DSMS is assumed to be timestamped tuples. The timestamp associated with each tuple will ideally indicate the time at which the data was captured (e.g., a tuple of sensor readings from a particular sensor node with an associated timestamp of the when those readings were taken). However, as we have seen, timing and synchronization in sensor networks are problematic, so in many cases DSMSs may need to juggle multiple timestamps and time constraints, such as the timing and the ordering at which tuples are received at the DSMS. The DSMS then provides access to the ability to construct long-running queries over the data streams from one or more sensor networks. A typical long-running query over a sensor network monitoring, say, a marine environment might be "Continuously report on the average dissolved oxygen in the monitored area over the previous 1 hour."

For further information about the principles behind DSMS, as well as specific DSMS implementations, the reader is referred to existing research on the topic (e.g., [2, 22]). However, from the perspective of this book, two characteristics of data stream processing are worth emphasizing. First, while distributed DSMSs have been developed, data stream processing is inherently more centralized than the approaches and algorithms pursued in this book. A data stream is, at root, a centralized concept, combining the data generated by a large number of real-time data sources. As a consequence, data stream processing is not well adapted to the constraints on the *movement* to information that are fundamental to decentralized spatial computing. Second, the inherent disaggregation of the data in data streams presents special challenges to answering spatial queries when data from closely related locations cannot be relied upon to arrive at the same time and spatial relationships between tuples may not be explicitly represented (e.g., see [85]).

9.4.3 Sensor Web

In introducing data stream processing, we took a step away from the central topic of this book: that computing happens somewhere. If we continue to imagine a world where sensors in the environment are all interconnected, with little or no spatial constraints on the movement of information, we might ask "what happens then?" How should we organize and coordinate heterogeneous sensors and sensor nodes across larger-scale spaces, integrating data from these different sources? These sorts of questions are encapsulated by the idea of a *sensor web*, a term first introduced in the late 1990s by Kevin Delin [26].

By approaching sensors as an amorphous mass of interconnected sensing and computing devices, akin to computers connected by the Internet, the aim of the sensor web is to treat them as a single distributed instrument. The idea is immediately attractive: a sort of "reverse microscope" for monitoring fine-grained environmental changes over large spatial extents. (Indeed, the term "macroscope" has already been suggested in this context [23].)

As might be anticipated for such a grand vision, a great many disparate topics have become associated with the sensor web, from the fine-grained features of the sensors and sensor nodes (termed "sensor web pods" in Delin's terminology) to the establishing and maintaining of network connectivity between devices to the large-scale architectures and processing required to support resource discovery, event detection, notification systems, and integration with predictive models.

However, one specific topic under the umbrella of the sensor web where progress is surely being made is that of information sharing. The large number of heterogeneous sensors, nodes, and communication technologies that must be integrated in a sensor web can present a major impediment to collating and combining the diverse data. In tackling this problem, the Open Geospatial Consortium (OGC) has developed the Sensor Web Enablement (SWE) standard which aims to specify how data and metadata from sensors should be described, as well as standardized mechanisms for retrieving data, retasking sensors, and publishing alerts generated from the data [17]. Figure 9.3 shows an example of how two sensor measurements can be encoded in SensorML, the XML-based specification language used in SWE. Being XML-based, SensorML specifications are designed to be both computer- and human-readable (albeit only barely readable to a human).

```
<swe:DataRecord definition="urn:ogc:def:property:OGC:atmosphericConditions">
  <swe:field name="AirTemperature">
    <swe:Quantity definition="urn:ogc:def:property:OGC:AirTemperature">
      <swe:uom code="Cel"/>
      <swe:value> 35.1 </swe:value>
    </swe:Quantity>
  </swe:field>
  <swe:field name="AtmosphericPressure">
    <swe:Quantity definition="urn:ogc:def:property:OGC:AtmosphericPressure">
      <swe:uom code="hPa"/>
      <swe:value> 950.0 </swe:value>
    </swe:Quantity>
  </swe:field>
</swe:DataRecord>
```

Fig. 9.3. An example data record in SensorML, showing the encoding of a sensor measurement of temperature and atmospheric pressure, after [88]

Unfortunately, standardizing the specification of geographical information is rarely the end of the story. Frequently, the many different organizations and end users that generate and consume data from geosensor networks have rather different interpretations and *meanings* associated with data items. A major ongoing research challenge is to capture these meanings in a way that can enable automated reasoning about high-level, human queries (e.g., [18]).

9.4.4 Geovisualization

The last stop on our journey through this book is arguably the one that matters most of all, as it concerns humans and human understanding of sensor data. The approach set out in this book is predicated on the assumption that any human user already has a clear idea of what information he or she requires from a decentralized spatial information system. Given a clear specification of the information required, the techniques in this book should help us design a robust decentralized algorithm that can efficiently generate that information. For example, if you need to know the area of a hot spot or the topological structure of a complex areal object or to be alerted to the formation of a flock, then the tools in this book can help you generate a decentralized spatial algorithm to meet those requirements.

But what if you don't have a clear idea of what you are looking for? An ecologist, for example, will posses considerable expertise in the principles governing the relationships between organisms and their environment. But when presented weeks' or months' worth of fine-grained, spatially and temporally detailed sensor data about a range of environmental parameters, he or she is unlikely to know in advance which if any principles are expected to explain the variations observed. In such cases, what is required is a mechanism for the ecologist to interact with the data and try to identify and even understand the patterns and events that may be occurring.

Geovisualization concerns the design of user interfaces that support humans in interacting with, discovering, and ultimately understanding the patterns in geographic information. Although the term "geovisualization" conjures up the idea of graphical user interfaces (and indeed geovisualization most frequently uses graphics), in fact geovisualization need not, or not only, involve graphical user interfaces. The term "visualization" derives from the objective of engaging humans' "visual thinking" processes—thinking in pictures—as opposed to their verbal and logical thinking process. To illustrate, Fig. 9.4 shows census data concerning change in population across the state of Texas. Discerning any pattern from the tabular text would be extremely difficult, even if you had intimate local knowledge of where the counties were located. In contrast, the map engages our visual thinking capabilities, immediately helping us identify patterns in the data, such as the densely populated (urban) and sparsely populated (rural) areas.

In fact, maps are surely a first step to engaging visual thinking. However, such simple map-based representations are poorly suited to visualization of data from a sensor network for at least two reasons:

- maps are static, whereas sensor data is highly dynamic; and
- maps present limited opportunities for going beyond mere data display, and instead interacting with data.

Consequently, research over the past decade or more has focused on exploring key principles for constructing representations of geographic data that are dynamic and interactive. Corresponding principles for static maps are already well established. The effectiveness of certain *visual variables* (for example, position, size, shape, orientation, color, texture of symbols) for graphically communicating to humans different characteristics has been thoroughly explored [12]. For instance, a change to the *value* of a symbol (i.e., the lightness or darkness of colors of the same hue, e.g., lighter versus darker greens) is associated with certain well-understood effects. Value is

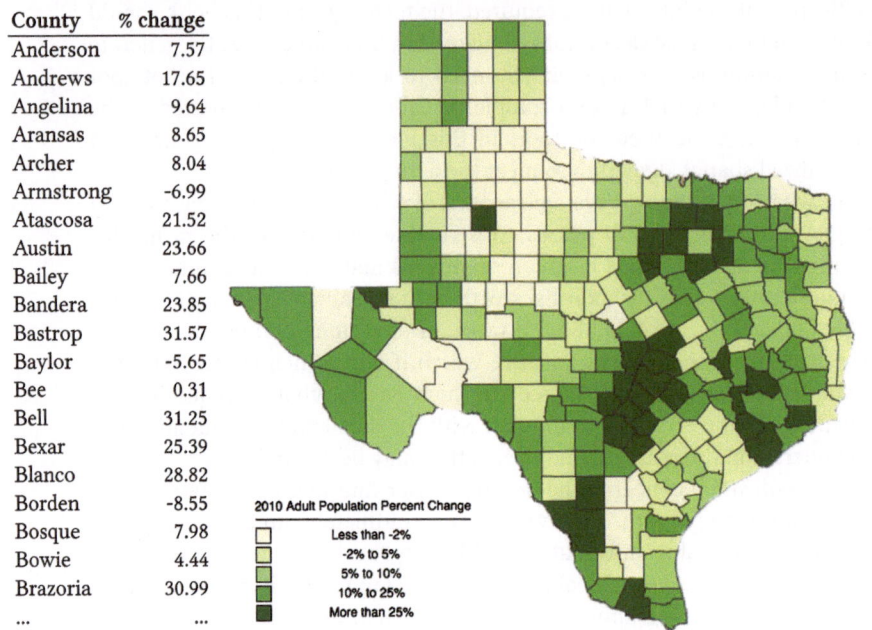

County	% change
Anderson	7.57
Andrews	17.65
Angelina	9.64
Aransas	8.65
Archer	8.04
Armstrong	-6.99
Atascosa	21.52
Austin	23.66
Bailey	7.66
Bandera	23.85
Bastrop	31.57
Baylor	-5.65
Bee	0.31
Bell	31.25
Bexar	25.39
Blanco	28.82
Borden	-8.55
Bosque	7.98
Bowie	4.44
Brazoria	30.99
...	...

2010 Adult Population Percent Change

- Less than -2%
- -2% to 5%
- 5% to 10%
- 10% to 25%
- More than 25%

Fig. 9.4. Thinking in pictures: engaging visual thinking can dramatically change human understanding of patterns in data (Data source: US Census Bureau; map source: *The Texas Tribune*)

particularly effective when representing ordinal data, such as the bushfire risk across an area (e.g., increasing darkness from low, moderate, through high, very high, to severe, extreme). By contrast, value is confusing when representing purely categorical data, such as land cover classes (e.g., urban, agricultural, forest, water).

Moving from static to animated maps, similar effects can be observed in dynamic visual variables, such as the speed of the animation and the order in which scenes appear and for how long they are visible [28]. Similar efforts have yielded corresponding principles for interactive maps (e.g., [37]), for depicting uncertainty in maps (e.g., [31]), for using three-dimensional views (e.g., [113]), and even for non-visual representations such as sound (e.g., [67]). To illustrate, Fig. 9.5 shows a typical interactive and three-dimensional interface to geographic information from the NASA World Wind project [84].

Designing effective interfaces to support visualization of data from a geosensor network is especially challenging. As we have already seen, data from a geosensor network is typically highly dynamic, even varying in real time; it is often inaccurate and imprecise, with low-cost and poorly calibrated sensors; and it typically concerns a large number of point observations in space and time, and requires additional effort to relate those observations

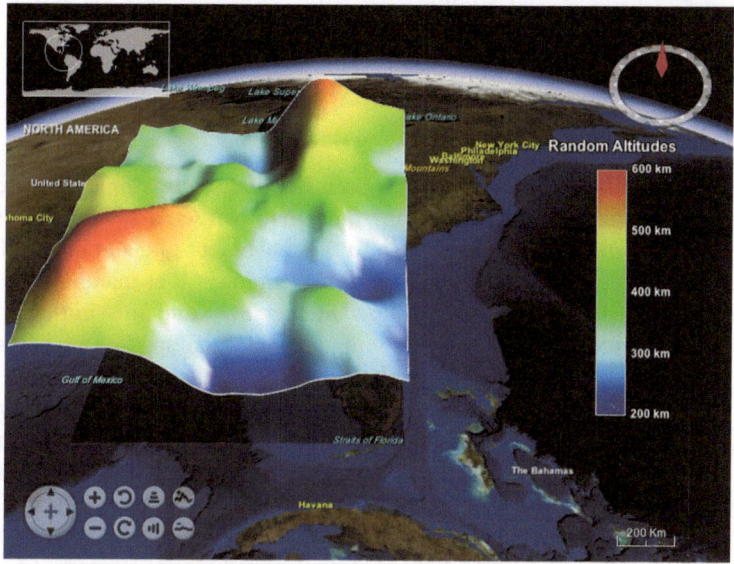

Fig. 9.5. Example interactive, animated, and 3D map in NASA World Wind [84]

to construct meaningful objects and events (such as hot spots and traffic jams). Thus, the objective of geovisualization for sensor data is frequently to explore data, to formulate and test hypotheses about the processes that might be driving the observed changes, and to ask "what if?" questions.

These objectives are the particular focus of an area of geovisualization called *(geo)visual analytics*. Visual analytics aims to combine our existing highly developed human abilities to identify patterns and gain insights, with the ability of computers to generate rapidly diverse, interactive, and engaging views of large and complex sensor data sets [4]. Visual analytics hopes to provide the tools that can make it easier for human users to spot trends, outliers, associations, clusters, and distributions that are meaningful in the relevant application domain: in short, to "answer questions you didn't know you had" [89].

Many challenges remain. Amongst the most pressing challenges is that of *evaluating* new styles of interface. Classical usability evaluation techniques focus on the measurable characteristics of user interfaces, such as how quickly users complete tasks, how often they make errors, and even subjective satisfaction measures [102]. However, these measures are not especially helpful in assessing whether an interface helped in making a decision, solving a problem, or even gaining insight into a phenomenon [89]. Nevertheless, the topics of geovisualization and geovisual analytics are amongst the most active areas of geographic information science, and the outcome of research into them is expected to be every bit as important to realizing the vision of ambient spatial intelligence as the design of efficient algorithms is for geosensor networks.

9.5 Coda

This book began with a simple premise: that, increasingly, "computing happens somewhere." Computing somewhere presents spatial constraints not only to the generation but also to the movement of geographic information. The models, algorithms, and design techniques that make up decentralized spatial computing, to me follow ineluctably from this initial premise. Of course, we should not throw out our GISs and spatial databases just yet! As both the first chapter and this final chapter have attempted to argue, the idea of ambient spatial intelligence—embedding in the environment the intelligence to monitor geographical spatial change—relies on a combination of both decentralized and centralized spatial data sources and models of computing. However, since first embarking on this topic in 2004, I have been repeatedly surprised by the wide range of spatiotemporal queries and operations that *can* be supported in a decentralized spatial information system. If I have managed to convey enough of my enthusiasm for this topic to encourage others to work in this area, then I shall regard this book as an unqualified success.

Review questions

9.1 Look back through the algorithms in Chapters 4–6. Which algorithms can be considered "traditional" queries, in the sense of §9.4.1 ("an unambiguous statement of the information to be retrieved from the information system")? Which algorithms adopt some nontraditional model, for example, lacking a single node to initiate the query or collate the response?

9.2 Looking back again through the algorithms we have seen, briefly write down example applications of selected algorithms that can illustrate all 12 possible combinations of the following distinctions: an in-network versus a remote query; a human-generated versus an automated query; and a push versus a pull query (see §9.4.1).

9.3 Design and execute an experiment to test empirically the computational complexity of the ant colony optimization version of the convoy algorithm, Protocol 9.1.

9.4 Design a protocol to detect convoys using the brute-force approach suggested in §9.3 (i.e., build a tree using multiple initiators as in §4.4.1, and then count the number of nodes in the tree using the TAG algorithm in Protocol 4.10). Can you implement this algorithm? If so, design and run an experiment to compare the computational complexity of the brute-force algorithm with that of the ant colony optimization algorithm in question 9.3.

9.5 How might either the brute-force or the ant colony convoy detection algorithm be extended to work in cases where the communication distance is *larger* than the convoy distance, $c > r$? Design a protocol to deal with this scenario. Can you sketch an algorithm design to approximate convoys in the other case, when $c < r$?

References

1. AARTS, E. Ambient intelligence: A multimedia perspective. *IEEE Multimedia 11*, 1 (2004), 12–19.
2. ABADI, D., CARNEY, D., ÇETINTEMEL, U., CHERNIACK, M., CONVEY, C., ERWIN, C., GALVEZ, E., HATOUN, M., MASKEY, A., RASIN, A., SINGER, A., STONEBRAKER, M., TATBUL, N., XING, Y., YAN, R., AND ZDONIK, S. Aurora: A data stream management system. In *Proc. ACM SIGMOD International Conference on Management of Data* (2003), p. 666.
3. ABADI, M., AND CARDELLI, L. *A Theory of Objects.* Springer, Berlin, 1996.
4. ANDRIENKO, G., ANDRIENKO, N., DEMŠAR, U., DRANSCH, D., DYKES, J., FABRIKANT, S., JERN, M., KRAAK, M.-J., SCHUMANN, H., AND TOMINSKI, C. Space, time, and visual analytics. *International Journal Geographical Information Science 24*, 10 (2010), 1577–1600.
5. ARAUJO, F., AND RODRIGUES, L. Single-step creation of localized Delaunay triangulations. *Wireless Networks 15*, 7 (2009), 845–858.
6. AU YEUNG, C., LICCARDI, I., LU, K., SENEVIRATNE, O., AND BERNERS-LEE, T. Decentralization: The future of online social networking. In *W3C Workshop on the Future of Social Networking* (2009).
7. AUGUSTINE, D., AND FRANK, D. Effects of migratory grazers on spatial heterogeneity of soil nitrogen properties in a grassland ecosystem. *Ecology 83* (2001), 3149–3162.
8. AVIN, C., AND ERCAL, G. On the cover time and mixing time of random geometric graphs. *Theoretical Computer Science 380*, 1–2 (2007), 2–22.
9. BACHARACH, J., AND BEAL, J. Building spatial computers. Tech. Rep. 2007-017, MIT CSAIL, March 2007.
10. BEAL, J., AND SCHANTZ, R. A spatial computing approach to distributed algorithms. In *45th Asilomar Conference on Signals, Systems, and Computers* (2010).
11. BENKERT, M., GUDMUNDSSON, J., HÜBNER, F., AND WOLLE, T. Reporting flock patterns. *Computational Geometry 41*, 3 (2008), 111–125.
12. BERTIN, J. *Semiology of Graphics.* University of Wisconsin Press, Madison, 1983.
13. BONGARD, J. Biologically inspired computing. *IEEE Computer 42*, 4 (2009), 95–98.
14. BONNET, P., GEHRKE, J., AND SESHADRI, P. Querying the physical world. *IEEE Personal Communications 7*, 5 (2000), 10–15.
15. BOSE, P., MORIN, P., STOJMENOVIC, I., AND URRUTIA, J. Routing with guaranteed delivery in ad hoc wireless networks. In *Proc. 3rd International Workshop on Discrete Algorithms and Methods for Mobile Computing and Communications* (1999), pp. 48–55.

M. Duckham, *Decentralized Spatial Computing*, DOI 10.1007/978-3-642-30853-6,
© Springer-Verlag Berlin Heidelberg 2013

16. Braginsky, D., and Estrin, D. Rumor routing algorthim for sensor networks. In *Proc. 1st ACM International Workshop on Wireless Sensor Networks and Applications (WSNA'02)* (New York, NY, USA, 2002), ACM Press, pp. 22–31.

17. Bröring, A., Echterhoff, J., Jirka, S., Simonis, I., Everding, T., Stasch, C., Liang, S., and Lemmens, R. New generation sensor web enablement. *Sensors 11*, 3 (2011), 2652–2699.

18. Bröring, A., Maué, P., Janowicz, K., Nüst, D., and Malewski, C. Semantically-enabled sensor plug and play for the sensor web. *Sensors 11*, 8 (2011), 7568–7605.

19. Burrell, J., Brooke, T., and Beckwith, R. Vineyard computing: Sensor networks in agricultural production. *IEEE Pervasive Computing 3*, 1 (2004), 38–45.

20. Burrough, P. Fractal dimension of landscapes and other environmental phenomena. *Nature 294* (1981), 240–242.

21. Caelli, T., Lam, P., and Bunke, H., Eds. *Spatial Computing: Issues in Vision, Multimedia and Visualization Technologies*, vol. 24 of *Series in Machine Perception and Artificial Intelligence*. World Scientific, Singapore, 1997.

22. Chen, J., DeWitt, D., Tian, F., and Wang, Y. NiagaraCQ: A scalable continuous query system for Internet databases. *SIGMOD Record 29*, 2 (2000), 379–390.

23. Culler, D. Toward the sensor network macroscope. In *Proc. on Mobile Ad Hoc Networking and Computing (MobiHoc)* (2005), p. 1.

24. Dasarathy, B. Information fusion—what, where, why, when, and how? *Information Fusion 2*, 2 (2001), 75–76.

25. de Berg, M., Cheong, O., van Kreveld, M., and Overmars, M. *Computational Geometry: Algorithms and Applications*, 3rd ed. Springer, Berlin, 2008.

26. Delin, K., and Jackson, S. The sensor web: A new instrument concept. In *Proc. International Society for Optical Engineering (SPIE)* (2001), vol. 4284, pp. 1–9.

27. Demers, A., Greene, D., Hauser, C., Irish, W., Larson, J., Shenker, S., Sturgis, H., Swinehart, D., and Terry, D. Epidemic algorithms for replicated database maintenance. In *Proc. 6th Annual ACM Symposium on Principles of Distributed Computing* (1987), PODC '87, ACM, pp. 1–12.

28. DiBiase, D., MacEachren, A., Krygier, J., and Reeves, C. Animation and the role of map design in scientific visualization. *Cartography and Geographic Information Systems 19*, 4 (1992), 201–204, 265–266.

29. Dodge, S., Weibel, R., and Lautenschütz, A.-K. Towards a taxonomy of movement patterns. *Journal of Information Visualization 7*, 240–252 (2008).

30. Dorigo, M., and Gambardella, L. Ant colony system: A cooperative learning approach to the traveling salesman problem. *IEEE Transactions on Evolutionary Computation 1*, 1 (1997), 53–66.

31. Drecki, I. Visualization of uncertainty in geographical data. In *Spatial data quality*, W. Shi, M. F. Goodchild, and P. F. Fisher, Eds. CRC Press, 2002, ch. 10, pp. 144–163.

32. Ducatel, K., Bogdanowicz, M., Scapolo, F., Leijten, J., and Burgelman, J.-C. Scenarios for ambient intelligence in 2010. Tech. Rep. IST Advisory Group Final Report, European Commission, 2001.

33. Duckham, M., Jeong, M., Li, S., and Renz, J. Decentralized querying of topological relations between regions without using localization. In *Proc. SIGSPATIAL ACMGIS* (2010), ACM, pp. 414–417.

34. Duckham, M., and Kulik, L. Location privacy and location-aware computing. In *Dynamic & Mobile GIS: Investigating Change in Space and Time*, J. Drummond, R. Billen, E. João, and D. Forrest, Eds. CRC Press, Boca Raton, FL, 2006, pp. 35–51.

35. DUCKHAM, M., NUSSBAUM, D., SACK, J.-R., AND SANTORO, N. Efficient, decentralized computation of the topology of spatial regions. *IEEE Transactions on Computers 60*, 8 (2011), 1100–1113.

36. DUCKHAM, M., AND REITSMA, F. Decentralized environmental simulation and feedback in robust geosensor networks. *Computers, Environment, and Urban Systems 33* (2009), 256–268.

37. DYKES, J. Exploring spatial data representation with dynamic graphics. *Computers and Geosciences 23*, 4 (1997), 345–370.

38. EGENHOFER, M., AND FRANSOZA, R. Point-set topological spatial relations. *International Journal of Geographical Information Science 5*, 2 (1991), 161–174.

39. ELSON, J., AND ESTRIN, D. Time synchronization for wireless sensor networks. In *International Parallel and Distributed Processing Symposium (IPDPS '01)* (Los Alamitos, CA, USA, 2001), vol. 3, IEEE Computer Society.

40. EMILIANI, P., AND STEPHANIDIS, C. Universal access to ambient intelligence environments: Opportunities and challenges for people with disabilities. *IBM Systems Journal 44*, 3 (2005), 605–619.

41. ESPINOZA, F., PERSSON, P., SANDIN, A., NYSTRÖM, H., CACCIATORE, E., AND BYLUND, M. GeoNotes: Social and navigational aspects of location-based information systems. In *Ubicomp 2001: Ubiquitous Computing* (Berlin, 2001), G. Abowd, B. Brumitt, and S. Shafer, Eds., vol. 2201 of *Lecture Notes in Computer Science*, Springer, pp. 2–17.

42. FREY, H., AND STOJMENOVIĆ, I. Geographic and energy-aware routing in sensor networks. In *Handbook of sensor networks: Algorithms and architectures*, I. Stojmenović, Ed. Wiley, Hoboken, NJ, 2005, ch. 12, pp. 381–415.

43. GALLAGHER, R., HUMBLET, P., AND SPIRA, P. A distributed algorithm for minimum-weight spanning trees. *ACM Transactions on Programming Languages and Systems 5*, 1 (1983), 66–77.

44. GALTON, A. *Qualitative Spatial Change*. Oxford University Press, Oxford, UK, 2000.

45. GALTON, A. Fields and objects in space, time, and space-time. *Spatial Cognition and Computation 4*, 1 (2004), 39–68.

46. GALTON, A. Dynamic collectives and their collective dynamics. In *Proc. Conference on Spatial Information Theory (COSIT)*, vol. 3693 of *Lecture Notes in Computer Science*. Springer, 2005, pp. 300–315.

47. GEORGOUDAS, I., SIRAKOULIS, G., AND ANDREADIS, I. An anticipative crowd management system preventing clogging in exits during pedestrian evacuation processes. *IEEE Systems Journal 5*, 1 (2011), 129–141.

48. GERVAIS, E., LIU, H., NUSSBAUM, D., ROH, Y., SACK, J., AND YI, J. Intelligent map agents—A ubiquitous personalized GIS. *ISPRS Journal of Photogrammetry and Remote Sensing 62*, 5 (2007), 347–365.

49. GILBERT, N., AND TERNA, P. How to build and use agent-based models in social science. *Mind and Society: Cognitive Studies in Economics and Social Sciences 1*, 1 (2000), 57–72.

50. GOVINDAN, R., HELLERSTEIN, J. M., HONG, W., MADDEN, S., FRANKLIN, M., AND SHENKER, S. The sensor network as a database. Tech. Rep. 02-771, University of Southern California, Information Sciences Institute, September 2002.

51. GROSSGLAUSER, M., AND TSE, D. Mobility increases the capacity of ad hoc wireless networks. *IEEE/ACM Transactions on Networking 10*, 4 (2002), 477–486.

52. HAAS, Z., HALPERN, J., AND LI, L. Gossip-based ad hoc routing. In *Proc. 21st Annual Joint Conference of the IEEE Computer and Communications Societies (INFOCOM)* (2002), vol. 3, pp. 1707–1716.

53. HEIN, J. *Discrete Mathematics*. Jones and Bartlett, Sudbury, MA, 1996.

54. HIGHTOWER, J., AND BORIELLO, G. Location systems for ubiquitous computing. *IEEE Computer 34*, 8 (2001), 57–66.

55. HILL, J., SZEWCZYK, R., WOO, A., HOLLAR, S., CULLER, D. E., AND PISTER, K. S. J. System architecture directions for networked sensors. In *Proc. 9th International Conference on Architectural Support for Programming Languages and Operating Systems* (2000), pp. 93–104.

56. JENNINGS, N. Agent-based computing: Promise and perils. In *Proc. 16th International Joint Conference on Artificial Intelligence (IJCAI-99)* (1999), pp. 1429–1436.

57. JEUNG, H., YIU, M., ZHOU, X., JENSEN, C., AND SHEN, H. Discovery of convoys in trajectory databases. *Proceedings of the VLDB Endowment 1*, 1 (2008), 1068–1080.

58. JIANG, J., AND WORBOYS, M. Event-based topology for dynamic planar areal objects. *International Journal of Geographical Information Science 23*, 1 (2009), 33–60.

59. JONASSON, J., AND SCHRAMM, O. On the cover time of planar graphs. *Electronic Communications in Probability 5* (2000), 85–90.

60. KARL, H., AND WILLIG, A. *Protocols and Architectures for Wireless Sensor Networks*. Wiley, Chichester, England, 2005.

61. KARP, B., AND KUNG, H. GPSR: Greedy perimeter stateless routing for wireless networks. In *Proc. 6th Annual International Conference on Mobile Computing and N* (Boston, MA, 2000), ACM.

62. KEEFE, R., AND SMITH, P. *Vagueness: A Reader*. MIT Press, Cambridge, MA, 1996.

63. KEMBER, S., CHEVERST, K., CLARKE, K., DEWSBURY, G., HEMMINGS, T., HUGHES, J., MARTIN, D., ROUNCEFIELD, M., AND VILLER, S. Keep taking the medication: Assistive technologies for medication regimes in care settings. In *Universal Access and Assistive Technology*, S. Keates, P. Langdon, P. Clarkson, and P. Robinson, Eds. Springer, Berlin, 2004, pp. 285–294.

64. KINN, G. J. GIS operators requiring spatial context and their implications for remote sensing. In *Proc. International Society for Optical Engineering* (SPIE) (1993), vol. 1819, pp. 126–132.

65. KRANAKIS, E., SINGH, H., AND URRUTIA, J. Compass routing on geometric networks. In *Proc. 11th Canadian Conference on Computational Geometry* (1999), pp. 51–54.

66. KRISHNAMACHARI, B., ESTRIN, D., AND WICKER, S. Modeling data-centric routing in wireless sensor networks. In *Proc. IEEE INFOCOM* (2002).

67. KRYGIER, J. Sound and geographic visualization. In *Visualization in Modern Cartography*, A. MacEachren and D. Taylor, Eds. Pergamon, Oxford, 1994, pp. 1–12.

68. LATHAN, C. The role of telemedicine, or telecare, in rehabilitation and home care. In *Proc. EC/NSF Workshop on Universal Accessibility of Ubiquitous Computing: Providing for the Elderly* (Portugal, 2001), EC/NSF.

69. LAUBE, P., DUCKHAM, M., AND PALANISWAMI, M. Deferred decentralized movement pattern mining for geosensor networks. *International Journal of Geographical Information Science 25*, 2 (2011), 273–292.

70. LAUBE, P., DUCKHAM, M., AND WOLLE, T. Decentralized movement pattern detection amongst mobile geosensor nodes. In *GIScience*, T. Cova, K. Beard,

M. Goodchild, and A. Frank, Eds., no. 5266 in Lecture Notes in Computer Science. Springer, Berlin, 2008, pp. 199–218.

71. LI, N., HOU, J., AND SHA, L. Design and analysis of an MST-based topology control algorithm. *IEEE Transactions on Wireless Communications 4*, 3 (2005), 1195–1206.

72. LI, X.-Y., STOJMENOVIC, I., AND WANG, Y. Partial Delaunay triangulation and degree limited localized Bluetooth scatternet formation. *IEEE Transactions on Parallel and Distributed Systems 15* (2004), 350–361.

73. LI, X.-Y., WANG, Y., AND SONG, W.-Z. Applications of k-local MST for topology control and broadcasting in wireless ad hoc networks. *IEEE Transactions on Parallel and Distributed Systems 15* (2004), 1057–1069.

74. LIPSCHUTZ, S., AND LIPSON, M. *Discrete Mathematics*. Schaum's Outline Series. McGraw-Hill, New York, NY, 1997.

75. LISKOV, B. Practical uses of synchronized clocks in distributed systems. *Distributed Computing 6* (1993), 211–219.

76. LONGLEY, P. A., GOODCHILD, M. F., MAGUIRE, D. J., AND RHIND, D. W. *Geographic Information Systems and Science*. John Wiley & Sons, New York, 2001.

77. LYNCH, N. *Distributed Algorithms*. Morgan Kaufmann, San Mateo, CA, 1996.

78. MADDEN, S., FRANKLIN, M., HELLERSTEIN, J., AND HONG, W. TAG: A tiny aggregation service for ad-hoc sensor networks. In *5th Symposium on Operating System Design and Implementation (OSDI)* (2002).

79. MANDELBROT, B. *Fractals, Form, Chance and Dimension*. Freeman, San Francisco, 1977.

80. MANDELBROT, B. *The Fractal Geometry of Nature*. Freeman and Co., New York, 1982.

81. MILNER, R. *A Calculus of Communicating Systems*, vol. 92 of *Lecture Notes in Computer Science*. Springer, 1980.

82. MILNER, R. *Communicating and Mobile Systems: The π-calculus*. Cambridge University Press, 1999.

83. MILNER, R. Pure bigraphs: Structure and dynamics. *Information and Computation 204* (2006), 60–122.

84. NASA. NASA World Wind Java demo applications and applets. http://worldwind.arc.nasa.gov/java/demos/, 2011.

85. NITTEL, S., AND LEUNG, K. Parallelizing clustering of geoscientific data sets using data streams. In *Proc. 16th International Conference on Scientific and Statistical Database Management* (2004), pp. 73–84.

86. NITTEL, S., STEFANIDIS, A., CRUZ, I., EGENHOFER, M., GOLDIN, D., HOWARD, A., LABRINIDIS, A., MADDEN, S., VOISARD, A., AND WORBOYS, M. Report from the First Workshop on Geo Sensor Networks. *ACM SIGMOD Record 33*, 1 (2004), 141–144.

87. O'NEILL, R., KRUMMEL, J., GARDNER, R., SUGIHARA, G., JACKSON, B., DEANGELIS, D., MILNE, B., TURNER, M., ZYGMUNT, B., CHRISTENSEN, S., DALE, V., AND GRAHAM, R. Indices of landscape pattern. *Landscape Ecology 1*, 3 (1988), 153–162.

88. OPEN GEOSPATIAL CONSORTIUM. *OpenGIS Sensor Model Language (SensorML) Implementation Specification*, 2007.

89. PLAISANT, C. The challenge of information visualization evaluation. In *Proc. Conference on Advanced Visual Interfaces (AVI)* (2004), pp. 109–116.

90. RABAEY, J., AMMER, M., DA SILVA, J., PATEL, D., AND ROUNDY, S. Picoradio supports ad-hoc ultra-low power wireless networking. *IEEE Computer 33*, 7 (2003), 42–48.

91. RAVI, N., DANDEKAR, N., MYSORE, P., AND LITTMAN, M. Activity recognition from accelerometer data. In *IAAI'05: Proc. 17th conference on Innovative Applications of Artificial Intelligence* (Pittsburgh, PA, 2005), AAAI Press, pp. 1541–1546.

92. RICHARDSON, L. The problem of contiguity: An appendix of statistics of deadly quarrels. *General Systems Yearbook 6* (1961), 139–187.

93. RIGAUX, P., SCHOLL, M., AND VOISARD, A. *Spatial Databases with Application to GIS*. Morgan Kaufmann, 2002.

94. ROBERT, A., AND ROY, A. On the fractal interpretation of the mainstream length-drainage area relationship. *Water Resources Research 26*, 5 (1990), 839–842.

95. RÖMER, K., BLUM, P., AND MEIER, L. Time synchronization and calibration in wireless sensor networks. In *Handbook of sensor networks: Algorithms and architectures*, I. Stojmenović, Ed. Wiley, Hoboken, NJ, 2005, ch. 7, pp. 199–237.

96. RUSSELL, E. Field experiments: How they are made and what they are. *Journal of the Ministry of Agriculture 32* (1926), 989–1001.

97. SADEQ, M. *In network detection of topological change of regions with a wireless sensor network*. PhD thesis, University of Melbourne, 2009.

98. SADEQ, M., AND DUCKHAM, M. Decentralized area computation for spatial regions. In *Proc. SIGSPATIAL ACMGIS* (New York, 2009), ACM, pp. 432–435.

99. SANTORO, N. *Design and Analysis of Distributed Algorithms*. Wiley, New Jersey, 2007.

100. SARKAR, R., ZHU, X., GAO, J., GUIBAS, L., AND MITCHELL, J. Iso-contour queries and gradient descent with guaranteed delivery in sensor networks. In *Proc. 27th IEEE Conference on Computer Communications (INFOCOM)* (2008), pp. 960–967.

101. SHI, M., AND WINTER, S. Detecting change in snapshot sequences. In *GIScience*, S. Fabrikant, T. Reichenbacher, M. van Kreveld, and C. Schlieder, Eds., no. 6296 in Lecture Notes in Computer Science. Springer, Berlin, 2010, pp. 219–233.

102. SHNEIDERMAN, B. *Designing the User Interface*, 3rd ed. Addison-Wesley, Reading, MA, 1997.

103. SIVRIKAYA, F., AND YENER, B. Time synchronization in sensor networks: A survey. *IEEE Network 18*, 4 (2004), 45–50.

104. SKRABA, P., FANG, Q., NGUYEN, A., AND GUIBAS, L. Sweeps over wireless sensor networks. In *Proc. 5th Int'l Conference on Information Processing in Sensor Networks (IPSN)* (2006), pp. 143–151.

105. SMITH, B., AND VARZI, A. Fiat and bona fide boundaries. *Philosophy and Phenomenological Research 60*, 2 (2000), 401–420.

106. STOJMENOVIĆ, I., Ed. *Handbook of Sensor Networks: Algorithms and Architectures*. Wiley, Hoboken, NJ, 2005.

107. STONEHAM, G., CHAUDHRI, V., HA, A., AND STRAPPAZZON, L. Auctions for conservation contracts: An empirical examination of Victoria's BushTender trial. *Australian Journal of Agricultural and Resource Economics 47*, 4 (2003), 447–500.

108. SUNDARARAMAN, B., BUY, U., AND KSHEMKALYANI, A. Clock synchronization for wireless sensor networks: A survey. *Ad hoc networks 3* (2005), 281–323.

109. SZEWCZYK, R., OSTERWEIL, E., POLASTRE, J., HAMILTON, M., MAINWARING, A., AND ESTRIN, D. Habitat monitoring with sensor networks. *Communications of the ACM 47*, 6 (2004), 34–40.

110. WEISER, M. Some computer science issues in ubiquitous computing. *Communications of the ACM 36*, 7 (1993), 74–84.

111. WILLENSKY, U. Netlogo. Center for Connected Learning and Computer-Based Modeling, Northwestern University, Evanston, IL, 1999. http://ccl.northwestern.edu/netlogo/.

112. WINTER, S., AND NITTEL, S. Ad-hoc shared-ride trip planning by mobile geosensor networks. *International Journal of Geographical Information Science 20*, 8 (2006), 899–916.

113. WOOD, J., KIRSCHENBAUER, S., DÖLLNER, J., LOPES, A., AND BODUM, L. Using 3D in visualization. In *Exploring geovisualization*, J. Dykes, A. M. MacEachren, and M.-J. Kraak, Eds. Elsevier, 2005, pp. 295–312.

114. WOOD, Z., AND GALTON, A. A taxonomy of collective phenomena. *Applied Ontology 4*, 3–4 (2009), 267–292.

115. WOOLDRIDGE, M. Agent-based software engineering. *IEE Proceedings in Software Engineering 144* (1997), 26–37.

116. WORBOYS, M. Event-oriented aproaches to geographic phenomena. *International Journal of Geographic Information Science 19*, 1 (2005), 1–28.

117. WORBOYS, M., AND BOFAKOS, P. A canonical model for a class of areal spatial objects. In *Proc. Third International Symposium on Advances in Spatial Databases (SSD'93)* (Berlin, 1993), Springer, pp. 36–52.

118. WORBOYS, M., AND CLEMENTINI, E. Integration of imperfect spatial information. *Journal of Visual Languages and Computing 12*, 1 (2001), 61–80.

119. WORBOYS, M., AND DUCKHAM, M. *GIS: A Computing Perspective*, 2nd ed. CRC Press, Boca Raton, FL, 2004.

120. YAO, A. On constructing minimum spanning trees in k-dimensional spaces and related problems. *SIAM Journal on Computing 11*, 4 (1982), 721–736.

121. ZHAO, F., AND GUIBAS, L. *Wireless Sensor Networks—An Information Processing Approach*. Morgan Kaufmann Publishers, San Francisco, CA, 2004.

References

Appendix A: Discrete Mathematics Primer

THIS appendix provides a brief primer on the discrete mathematics structures and syntax used in this book. For more information the reader is referred to any basic text on discrete mathematics (e.g., [53, 74]).

Sets

A set is simply a "collection of things." Each "thing" in a set is referred to as a *member* or an *element* of a set. Conventionally, an uppercase letter is used to represent a set. For example, the set

$$A = \{a, b, c\}$$

defines A to be the set of three elements a, b, and c. The order of elements in a set is not significant, so $\{b, c, a\} = \{c, a, b\} = \{c, b, a\} = \cdots = \{a, b, c\} = A$.

An important special set is the set that contains no elements, called the *empty set*, written $\{\ \}$ or, more usually, $\varnothing$. There are also some other common special sets used in this book:

- $\mathbb{R}$, the set of real numbers (i.e., numbers on the number line), and $\mathbb{R}^+$, the set of positive real numbers;
- $\mathbb{N}$, the set of natural numbers (positive integers along with 0, i.e., 0, 1, 2, 3, 4, ...); and
- $\mathbb{B}$, the Boolean set (i.e., $\{0, 1\}$).

If we wish to assert that a is an element of set A, we may write $a \in A$. A set that contains all the elements of another set, and perhaps some additional elements, is termed a *superset*. Thus A is a superset of the set $C = \{a, c\}$. Conversely, C is a *subset* of A, written $C \subseteq A$. A set X is said to be a *proper subset* of Y, written $X \subset Y$, if X is a subset of Y but $X \neq Y$. For example, $\mathbb{N} \subset \mathbb{R}$. The empty set is a subset of every set, e.g., $\varnothing \subseteq A$.

The elements of a set can be specified by enumeration (as above, listing all the elements of that set). Alternatively, it is often easier to specify sets by

M. Duckham, *Decentralized Spatial Computing*, DOI 10.1007/978-3-642-30853-6,
© Springer-Verlag Berlin Heidelberg 2013

intension (particularly for sets with a large or infinite number of elements, such as $\mathbb{R}$). For example, the set $B = \{2, 3, 4, 5\}$ can be specified as $B = \{x \in \mathbb{N} | 1 < x < 6\}$ (read "the set B containing natural numbers x *such that* 1 is smaller than x is smaller than 6).

The *power set* of a set X is the set of all subsets of that set (including the empty set), written $\mathcal{P}(X)$ or sometimes 2^X. Thus, $\mathcal{P}(A) = \{\varnothing, \{a\}, \{b\}, \{c\}, \{a, b\}, \{a, c\}, \{b, c\}, \{a, b, c\}\}$. Note that the power set of a set is still a set: the elements of a set may themselves be sets (e.g., $\{a, c\} \in \mathcal{P}(A)$).

The *cardinality* of a set X is the number of elements in that set, written $|X|$. For example, $|A| = 3$, $|B| = 4$, $|C| = 2$.

The intersection of two sets X and Y, written $X \cap Y$ is the set of elements in X and in Y. The union of X and Y, written $X \cup Y$ is the set of elements in X or in Y. For example, $A \cap \{a, b, d\} = \{a, b\}$, $A \cup \{a, b, d\} = \{a, b, c, d\}$, $A \cup B = \{a, b, c, 2, 3, 4, 5\}$, and $A \cap B = \varnothing$. The set difference of two sets, $X - Y$, is the set of elements in X but not in Y. For example, $A - C = \{b\}$.

Products and relations

The elements of a set are not ordered. When it is necessary to specify the order of elements, a different structure called a *tuple* is needed. In general an n-tuple may have any number n of elements. However, we most frequently require *pairs*: tuples with two elements. A pair (or tuple) is written with enclosing parenthesis, for example, (a, c). For clarity (e.g., where elements in a tuple are themselves pairs or tuples) in this book, tuples may occasionally be enclosed with angled braces, $\langle a, c \rangle$. Unlike with sets, because a pair is ordered, $(a, c) \neq (c, a)$.

The (Cartesian) *product* of two sets X and Y, written $X \times Y$, is the set of all distinct pairs (x, y) with the first element $x \in X$ and the second element $y \in Y$. For example, the product $A \times C = \{(a, a), (a, c), (b, a), (b, c), (c, a), (c, c)\}$. Note that the product of two sets is, again, a set—albeit a set containing pairs as its elements. As another example, the set $\mathbb{R} \times \mathbb{R}$, often written $\mathbb{R}^2$, is the familiar points in the Cartesian plane (i.e., planar coordinates). Thus, for example, $(15.2, -1.9) \in \mathbb{R}^2$.

The product of two sets can be extended to an n-ary product (product of n sets) in a straightforward way (for example, $(b, 3, c) \in A \times B \times C$). Where there is no chance of confusion, the parenthesis and elements are often omitted from n-tuples if it helps with conciseness and clarity. For example, we may also write $ba \in A \times C$ or $b3c \in A \times B \times C$.

A relation between two sets is simply a subset of the product of those sets. Thus, the relation $L = A \times C$ is a relation from A to C. The relation $L' = \{(a, c), (b, a), (b, c), (c, a)\}$ is another relation from A to C.

Many important relations are from a set X to itself, termed a binary relation *on* X. For example, the relation $N = \{(1, 2), (2, 1), (2, 5), (5, 2)\}$ is a (binary) relation on $\mathbb{N}$ (i.e., $N \subseteq \mathbb{N}^2$). A relation R on X is said to be *reflexive* if for every $x \in X$, $(x, x) \in R$. A relation R on X is said to be *symmetric* if whenever $(x, y) \in R$ it is also the case that $(y, x) \in R$. Finally, a relation R on X is said to be *transitive* if whenever (x, y) and $(y, z) \in R$, $(x, z) \in R$. We also use in many places the set of (n-bit) binary numbers, e.g., $B^3 = \{000, 001, 010, 011, 100, 101, 110, 111\}$, the set of 3-bit binary numbers.

For example, relation N is neither symmetric nor transitive, but is reflexive. Conversely, the relation $\{(1, 2), (2, 5), (1, 5), (1, 1), (2, 2), (5, 5)\}$ is symmetric and transitive, but not reflexive.

Graphs

A *graph* is a pair $G = (V, E)$, where:

1. V is some set of elements, called *vertices* or *nodes*; and
2. E is a binary relation on V, $E \subseteq V \times V$.

If E is symmetric, then the graph G is termed *undirected* or *bidirected*; otherwise G is a *directed* graph. For an undirected graph $G = (V, E)$ we may sometimes use the shorthand $\{a, b\} \in E$ instead of writing both $(a, b) \in E$ and $(b, a) \in E$ (since in an undirected graph E is symmetric, so if $(a, b) \in E$ then necessarily $(b, a) \in E$, so the ordering does not matter).

Graphs (and relations) have a natural graphical representation, with vertices depicted as dots and edges depicted as lines connecting two vertices. For example, Fig. A.1 below shows a diagram depicting an undirected graph $G = (V, E)$ with vertices $V = \{a, b, c, d, e, f, g, h, i, j\}$ and edges $E = \{\{a, b\}, \{a, g\}, \{a, j\}, \{b, c\}, \{b, i\}, \{c, e\}, \{c, f\}, \{d, i\}, \{e, h\}, \{e, j\}, \{f, g\}, \{f, i\}, \{g, h\}, \{h, i\}\}$.

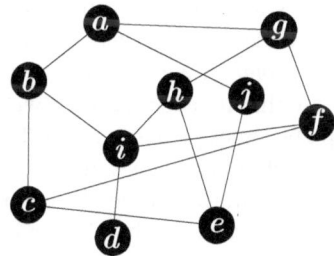

Fig. A.1. Example graphical representation of an undirected graph with vertices $a \ldots j$

Directed graphs may be similarly depicted, with the addition of arrows to indicate the direction(s) of edges.

Two nodes that are connected by an edge are said to be *adjacent* (e.g., a is adjacent to b in G, but a is not adjacent to c). The *degree* of a node is its number of adjacent neighbors. The degree of node d is 1 and of node h is 3, written $deg(d) = 1$ and $deg(h) = 3$. A *path* is a sequence of adjacent nodes. For example, a, b, c, e, h is a path in G; and a, b, c, d, e, f is not a path (e and f are not adjacent, and d is not adjacent to c or e). Two nodes are said to be *connected* if there exists a path between them. Two nodes are said to be n-connected if there exist n distinct paths between them (distinct in the sense that the paths share no vertices except the start and end nodes). For example, nodes a and e are 3-connected (by paths a, j, e; a, g, h, e; and a, b, c, e). A graph is said to be connected (or 1-connected) if every pair of nodes in the graph is connected. More generally, a graph is said to be n-connected if every pair of nodes in the graph is n-connected. The graph G is connected (or, equivalently, 1-connected), but not 2-connected (since node d is only 1-connected to the other nodes in the graph). If we were to remove edge $\{d, i\}$ from G, the graph would become disconnected (there would exist no path between node d and other nodes in the graph).

A *tree* is a (connected) graph where there exists exactly one path between any pair of nodes. A *rooted* tree is a tree with one node designated as the *root*. In a rooted tree, the *depth* of a node is the length of the (unique) path from that node to the root.

The *diameter* of a graph is the length (in terms of the number of edges) of the longest shortest path in the network. Thus, the diameter of the graph in Fig. A.1 is 4: the shortest path between d and j includes four edges (the path d, i, b, a, j). Related to the diameter, the *eccentricity* of a node is the length of the longest shortest path starting or ending with that node (i.e., a shortest path from a node that is longer than any other shortest path from that node). Thus, the eccentricity of nodes d and j is 4 in Fig. A.1; the eccentricity of nodes i and a is 3; the eccentricity of node b is 2. The set of nodes with eccentricity equal to the diameter of the graph is termed the *periphery* of the graph ($\{d, j\}$ in Fig. A.1).

A *planar* graph is a graph that can be drawn in the plane with edges only intersecting at nodes (i.e., no edges crossing). For example, the graph in Fig. A.1 is termed *non-planar*. We can prove that it is impossible to redraw this graph in such a way that no edges cross. If we remove the edge $\{f, g\}$ from G above, the graph becomes planar as it now becomes possible to redraw the graph in the plane such that no edges intersect (see Fig. A.2).

A graph that *can* be drawn in the plane with no crossing edges is called planar; a graph that *is* drawn in the plane with no crossing edges is called *plane*. Thus all plane graphs (such as the one in Fig. A.2) are also planar; and all non-planar graphs (such as the one in Fig. A.1) can never be depicted as a plane graph. However, it is possible for a planar graph to be depicted (or, more formally, "embedded in the plane") in a way that is not plane. For example, removing the edge $\{f, g\}$ from Fig. A.1 (but not repositioning any of the nodes) results in a graph that is planar (since it can be drawn with no crossing edges) but not plane (since it has been drawn with crossing edges, such as $\{f, i\}$ and $\{e, j\}$).

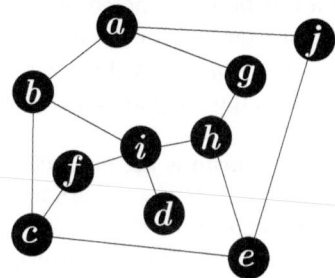

Fig. A.2. Example undirected planar graph (cf. Fig. A.1 without edge $\{f, g\}$)

Functions

A *function* is a special case of a relation between two sets, X and Y, where every element in X is related to a unique element in Y. The first set, X, is the set of all permitted inputs to the function, termed the *domain*. The second set, Y, is the set of all possible outputs of the function, termed the *codomain*.

Although functions are a special case of relations, they have their own special formal syntax to highlight their particular constraints. A function f with domain X and codomain Y is specified as $f : X \rightarrow Y$, read "f is a function from X to Y." We can indicate that the function f relates a particular input $x \in X$ to a particular result $y \in Y$ by writing $f(x) = y$. To highlight the correspondence with relations, if f were defined as relation (rather than as a function) we might instead write $(x, y) \in f$.

It is also often convenient to write the mapping rules for a function, which specify which elements in the domain map to which elements in the codomain. For example, consider a function $square : \mathbb{N} \rightarrow \mathbb{N}$ which generates the square of a natural number. The mapping rule for $square$ can be specified by writing $square(n) \mapsto n^2$ (read "the function $square$ maps n to n^2"). In a specific application of the function we may write $square(5) = 25$ or $square(52) = 2704$.

A function where every element in the domain maps to a distinct (different) element in the codomain is termed an *injection*. For example, a function that maps sensor nodes to their precise coordinate locations can be regarded as an injection. Every node must map to a different coordinate location (on the assumption that no two nodes can occupy exactly the same location).

A function where every element in the codomain has some element from the domain that maps to it is termed a *surjection*. For example, a function which maps address points in a state to the set of zip codes for that state is a surjection (every zip code will have at least one address point that maps to it) but not an injection (many address points will have the same zip code).

A function that is both an injection and a surjection is termed a *bijection*.

The *image* of a function f, denoted by f_*, is the set of elements that a function maps to. For example, for the function $f : X \to Y$, the image of the function $f_* = \{y \in Y | y = f(x)$ for some $x \in X\}$. Thus, the image of a function is necessarily a (possibly improper) subset of the codomain for that function (e.g., $f_* \subseteq Y$).

A function whose domain is the product of two sets is termed a *binary function*. For example, $f : X \times Y \to Z$ is a binary function. The special case of a binary function where $X = Y = Z$ (e.g., $f : X \times X \to X$) is termed a *binary operation*. A binary operation f is *associative* if it is always the case that $f(a, f(b, c)) = f(f(a, b), c)$. A binary operation f is *commutative* if it is always the case that $f(a, b) = f(b, a)$. An associative binary operation f along with its set X is termed a *semigroup*.

Logic

Logical expressions in mathematics and computer science can be assigned a *truth value*. In the most basic (two-valued) logic, expressions have a truth value of either true or false, but not both. For example, the truth value of the expression "$5 < 10$" is true; the truth value of the expression "$1 = 2$" is false.

The *conjunction* of two logical expressions $p \wedge q$ is true if *both p and q* are true; it is false otherwise. The *disjunction* of two logical expressions $p \vee q$ is true if *either p or q* are true; it is false if both p and q are false. For example, the truth value of "$\{a, b\} = \{b, a\} \wedge (a, b) = (b, a)$" is false; the truth value of "$5 < 10 \vee 1 = 2$" is true.

By convention, in a Boolean set $\mathbb{B}$, 1 is associated with true and 0 is associated with false. For example, $1 \vee 0 = 1$ and $1 \wedge 0 = 0$. Given two equal-length binary numbers, a *bitwise* logical operation performs its operation on each corresponding pair of bits. For example, given binary numbers in $\mathbb{B}^3$, $110 \wedge 011 = 010$ and $110 \vee 011 = 111$.

Appendix B: Relational Database and SQL Primer

R ELATIONAL databases are used for structured storage, manipulation, and retrieval of data. They are used for storage of the local data available to nodes in several more advanced algorithms. Consequently, these algorithms assume some familiarity with basic relational database concepts and SQL (structured query language), the standard language for managing data in a relational database. This primer contains a brief summary of the main concepts and SQL statements used in this book. For more information, the reader is referred to any introductory text on relational databases (e.g., [119]).

Relations

A relational database is a collection of tables (often also called *relations*). Each table has one or more labeled columns, called *attributes*, and zero or more rows, called *tuples* or *records*. The order of records in a table is not significant, just as the order of elements in a set is not significant. Each attribute has a data type (e.g., natural number, real number, time). Individual data items in a record must be drawn from the data type of their associated attribute.

For example, the table $m\langle oid : \mathbb{N}, enter : T, exit : T, in : \mathbb{N}, out : \mathbb{N}\rangle$ describes a table called m with five attributes (see §6.3.1). The attributes *oid*, *in*, and *out* have data type $\mathbb{N}$ (natural numbers, e.g., node identifiers), while attributes *enter* and *exit* have data type T (the set of timestamps, $t_1, t_2, \ldots$). Example data in table m might be:

oid	enter	exit	in	out
2	t_1	t_2	9	14
1	t_2	t_3	6	9
1	t_3	NULL	9	NULL
3	t_2	t_4	14	8

The following sections describe the SQL commands for manipulating tables used in this book.

M. Duckham, *Decentralized Spatial Computing*, DOI 10.1007/978-3-642-30853-6,
© Springer-Verlag Berlin Heidelberg 2013

CREATE TABLE command

Tables can be created with the create table command:

CREATE TABLE *table* (*attribute1 type1, attribute2 type2, …*)

For example:

CREATE TABLE *m* (*oid* N, *enter T*, *exit T*, *in* N, *out* N)

Basic SELECT command

The SELECT command selects a subset of named attributes (termed a *projection*) and a subset of records (termed a *restriction*) that match some specified restriction conditions. The general form of the basic SELECT statement is:

SELECT *attribute1, attribute2, …* FROM *table* WHERE *conditions*

For example, the SQL query

SELECT *oid, enter, in* FROM *m* WHERE *out*=9

would generate the resulting table:

oid	enter	in
1	t_2	6

More complex restriction conditions can be specified using a range of operators, including the logical AND, OR and NOT and the set-based IN. The set-based IN operator is compatible with the results of other SELECT statements. For example,

SELECT *oid* FROM *m* WHERE *in* IN (SELECT *out* FROM *m*) AND NOT (*out*=NULL)

would result in the table:

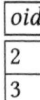

oid
2
3

When *two* tables are specified in the SELECT statement, every record in the first table is paired with every other record in the second table (as in the set-based Cartesian product). Where confusion might arise, attributes and tables can always be renamed with the AS clause, which specifies an alias for the attribute/table. The dot operator can also be used to disambiguate which attribute comes from which table. For example, the SQL query

SELECT m_1.*oid* AS *node*, m_1.*in*, m_2.*out* FROM *m* AS m_1, *m* AS m_2 WHERE m_1.*oid*=m_1.*oid* AND m_2.*in*=m_1.*out* generates the table:

node	$m_1.in$	$m_2.out$
1	t_2	NULL

Advanced SELECT command

SQL also makes available a number of functions for summary statistics over attributes, such as COUNT, MAX, MIN, and AVG. For example:

SELECT COUNT ($*$) AS *num* FROM m

results in the table:

num
4

The asterisk ($*$) is used as a "wildcard" to mean "all attributes" in a SELECT statement. In several algorithms, it is necessary to extract data from a table into a local data structure for further processing. To achieve this, we can use the SELECT INTO statement, which can assign the data from a table containing a singleton data item (i.e., one column and one row) into a variable. For example, assuming a local variable d has already been defined, the SQL statement

SELECT MAX (*enter*) INTO d FROM m

will result in the data item t_3 being stored in the local variable d.

More complex summary statistics tables can be constructed using the GROUP BY clause in a SELECT statement, which is able to group together records based on a named attribute. For example,

SELECT *enter*, COUNT (*oid*) AS *records* FROM m GROUP BY *enter*

generates the table:

enter	records
t_1	1
t_2	2
t_3	1

Data from tables in SQL can be represented as *views*. Views store the results of a query for subsequent access and manipulation as if these query results were another table. The SQL syntax for creating a view is:

CREATE VIEW *view* AS *select statement*

The table resulting from the specified *select statement* can then be accessed using the named *view* just like any other table.

CREATE VIEW *recordcount* AS SELECT *enter*, COUNT (*oid*) AS *records* FROM m GROUP BY *enter*

The contents of the view are automatically updated to reflect any changes in the underlying table(s) on which the view is based.

Inserting, deleting, and updating records

New records can be inserted into a table using the INSERT INTO command, with the general form:

INSERT INTO *table* VALUES (*att1value, att2value, ...*)

For example,

INSERT INTO m VALUES (2, t_2, NULL, 14, NULL)

will result in the statement

SELECT * FROM m

generating the table:

oid	enter	exit	in	out
2	t_1	t_2	9	14
1	t_2	t_3	6	9
1	t_3	NULL	9	NULL
3	t_2	t_4	14	8
2	t_2	NULL	14	NULL

Records can be deleted using the DELETE command:

DELETE FROM *table* WHERE *conditions*

The restriction conditions can be as for any SELECT statement. For example, the statement

DELETE FROM m WHERE *enter* $< t_2$

will result in the table m containing the records:

oid	enter	exit	in	out
1	t_2	t_3	6	9
1	t_3	NULL	9	NULL
3	t_2	t_4	14	8
2	t_2	NULL	14	NULL

Finally, existing records in a table can be updated using the UPDATE statement:

UPDATE *table* SET *attribute1=value1, attribute1=value1, . . .* WHERE *conditions*

For example:

UPDATE m SET *exit=t_5, out=7* WHERE *oid=1* AND *enter=t_3*

will further alter the data in m to:

oid	enter	exit	in	out
1	t_2	t_3	6	9
1	t_3	t_5	9	7
3	t_2	t_4	14	8
2	t_2	NULL	14	NULL

Index

M. Duckham, *Decentralized Spatial Computing*, DOI 10.1007/978-3-642-30853-6,
© Springer-Verlag Berlin Heidelberg 2013